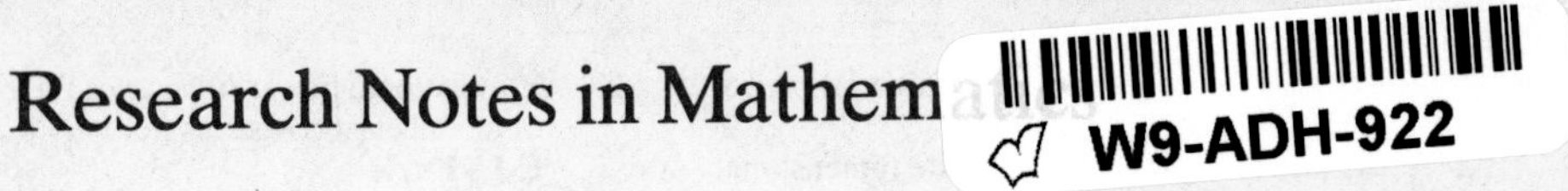

Research Notes in Mathem

Main Editors
A. Jeffrey, University of Newcastle-upon-Tyne
R. G. Douglas, State University of New York at Stony Brook

Editorial Board
F. F. Bonsall, University of Edinburgh
H. Brezis, Université de Paris
R. J. Elliott, University of Hull
G. Fichera, Università di Roma
R. P. Gilbert, University of Delaware
K. Kirchgässner, Universität Stuttgart
B. Lawson, State University of New York at Stony Brook
R. E. Meyer, University of Wisconsin-Madison
J. Nitsche, Universität Freiburg
L. E. Payne, Cornell University
G. F. Roach, University of Strathclyde
I. N. Stewart, University of Warwick
S. J. Taylor, University of Virginia

Submission of proposals for consideration
Suggestions for publication, in the form of outlines and representative samples, are invited by the editorial board for assessment. Intending authors should contact either the main editor or another member of the editorial board, citing the relevant AMS subject classifications. Refereeing is by members of the board and other mathematical authorities in the topic concerned, located throughout the world.

Preparation of accepted manuscripts
On acceptance of a proposal, the publisher will supply full instructions for the preparation of manuscripts in a form suitable for direct photo-lithographic reproduction. Specially printed grid sheets are provided and a contribution is offered by the publisher towards the cost of typing.

Illustrations should be prepared by the authors, ready for direct reproduction without further improvement. The use of hand-drawn symbols should be avoided wherever possible, in order to maintain maximum clarity of the text.

The publisher will be pleased to give any guidance necessary during the preparation of a typescript, and will be happy to answer any queries.

Important note
In order to avoid later retyping, intending authors are strongly urged not to begin final preparation of a typescript before receiving the publisher's guidelines and special paper. In this way it is hoped to preserve the uniform appearance of the series.

Titles in this series

1. Improperly posed boundary value problems
A Carasso and A P Stone
2. Lie algebras generated by finite dimensional ideals
I N Stewart
3. Bifurcation problems in nonlinear elasticity
R W Dickey
4. Partial differential equations in the complex domain
D L Colton
5. Quasilinear hyperbolic systems and waves
A Jeffrey
6. Solution of boundary value problems by the method of integral operators
D L Colton
7. Taylor expansions and catastrophes
T Poston and I N Stewart
8. Function theoretic methods in differential equations
R P Gilbert and R J Weinacht
9. Differential topology with a view to applications
D R J Chillingworth
10. Characteristic classes of foliations
H V Pittie
11. Stochastic integration and generalized martingales
A U Kussmaul
12. Zeta-functions: An introduction to algebraic geometry
A D Thomas
13. Explicit *a priori* inequalities with applications to boundary value problems
V G Sigillito
14. Nonlinear diffusion
W E Fitzgibbon III and H F Walker
15. Unsolved problems concerning lattice points
J Hammer
16. Edge-colourings of graphs
S Fiorini and R J Wilson
17. Nonlinear analysis and mechanics: Heriot-Watt Symposium Volume I
R J Knops
18. Actions of fine abelian groups
C Kosniowski
19. Closed graph theorems and webbed spaces
M De Wilde
20. Singular perturbation techniques applied to integro-differential equations
H Grabmüller
21. Retarded functional differential equations: A global point of view
S E A Mohammed
22. Multiparameter spectral theory in Hilbert space
B D Sleeman
24. Mathematical modelling techniques
R Aris
25. Singular points of smooth mappings
C G Gibson
26. Nonlinear evolution equations solvable by the spectral transform
F Calogero
27. Nonlinear analysis and mechanics: Heriot-Watt Symposium Volume II
R J Knops
28. Constructive functional analysis
D S Bridges
29. Elongational flows: Aspects of the behaviour of model elasticoviscous fluids
C J S Petrie
30. Nonlinear analysis and mechanics: Heriot-Watt Symposium Volume III
R J Knops
31. Fractional calculus and integral transforms of generalized functions
A C McBride
32. Complex manifold techniques in theoretical physics
D E Lerner and P D Sommers
33. Hilbert's third problem: scissors congruence
C-H Sah
34. Graph theory and combinatorics
R J Wilson
35. The Tricomi equation with applications to the theory of plane transonic flow
A R Manwell
36. Abstract differential equations
S D Zaidman
37. Advances in twistor theory
L P Hughston and R S Ward
38. Operator theory and functional analysis
I Erdelyi
39. Nonlinear analysis and mechanics: Heriot-Watt Symposium Volume IV
R J Knops
40. Singular systems of differential equations
S L Campbell
41. N-dimensional crystallography
R L E Schwarzenberger
42. Nonlinear partial differential equations in physical problems
D Graffi
43. Shifts and periodicity for right invertible operators
D Przeworska-Rolewicz
44. Rings with chain conditions
A W Chatters and C R Hajarnavis
45. Moduli, deformations and classifications of compact complex manifolds
D Sundararaman
46. Nonlinear problems of analysis in geometry and mechanics
M Atteia, D Bancel and I Gumowski
47. Algorithmic methods in optimal control
W A Gruver and E Sachs
48. Abstract Cauchy problems and functional differential equations
F Kappel and W Schappacher
49. Sequence spaces
W H Ruckle
50. Recent contributions to nonlinear partial differential equations
H Berestycki and H Brezis
51. Subnormal operators
J B Conway
52. Wave propagation in viscoelastic media
F Mainardi
53. Nonlinear partial differential equations and their applications: Collège de France Seminar. Volume I
H Brezis and J L Lions
54. Geometry of Coxeter groups
H Hiller
55. Cusps of Gauss mappings
T Banchoff, T Gaffney and C McCrory

56 An approach to algebraic K-theory
A J Berrick
57 Convex analysis and optimization
J-P Aubin and R B Vintner
58 Convex analysis with applications in the differentiation of convex functions
J R Giles
59 Weak and variational methods for moving boundary problems
C M Elliott and J R Ockendon
60 Nonlinear partial differential equations and their applications: Collège de France Seminar. Volume II
H Brezis and J L Lions
61 Singular systems of differential equations II
S L Campbell
62 Rates of convergence in the central limit theorem
Peter Hall
63 Solution of differential equations by means of one-parameter groups
J M Hill
64 Hankel operators on Hilbert space
S C Power
65 Schrödinger-type operators with continuous spectra
M S P Eastham and H Kalf
66 Recent applications of generalized inverses
S L Campbell
67 Riesz and Fredholm theory in Banach algebra
B A Barnes, G J Murphy, M R F Smyth and T T West
68 Evolution equations and their applications
F Kappel and W Schappacher
69 Generalized solutions of Hamilton-Jacobi equations
P L Lions
70 Nonlinear partial differential equations and their applications: Collège de France Seminar. Volume III
H Brezis and J L Lions
71 Spectral theory and wave operators for the Schrödinger equation
A M Berthier
72 Approximation of Hilbert space operators I
D A Herrero
73 Vector valued Nevanlinna Theory
H J W Ziegler
74 Instability, nonexistence and weighted energy methods in fluid dynamics and related theories
B Straughan
75 Local bifurcation and symmetry
A Vanderbauwhede
76 Clifford analysis
F Brackx, R Delanghe and F Sommen
77 Nonlinear equivalence, reduction of PDEs to ODEs and fast convergent numerical methods
E E Rosinger
78 Free boundary problems, theory and applications. Volume I
A Fasano and M Primicerio
79 Free boundary problems, theory and applications. Volume II
A Fasano and M Primicerio
80 Symplectic geometry
A Crumeyrolle and J Grifone
81 An algorithmic analysis of a communication model with retransmission of flawed messages
D M Lucantoni
82 Geometric games and their applications
W H Ruckle
83 Additive groups of rings
S Feigelstock
84 Nonlinear partial differential equations and their applications: Collège de France Seminar. Volume IV
H Brezis and J L Lions
85 Multiplicative functionals on topological algebras
T Husain
86 Hamilton-Jacobi equations in Hilbert spaces
V Barbu and G Da Prato
87 Harmonic maps with symmetry, harmonic morphisms and deformations of metrics
P Baird
88 Similarity solutions of nonlinear partial differential equations
L Dresner
89 Contributions to nonlinear partial differential equations
C Bardos, A Damlamian, J I Díaz and J Hernández
90 Banach and Hilbert spaces of vector-valued functions
J Burbea and P Masani
91 Control and observation of neutral systems
D Salamon
92 Banach bundles, Banach modules and automorphisms of C*-algebras
M J Dupré and R M Gillette
93 Nonlinear partial differential equations and their applications: Collège de France Seminar. Volume V
H Brezis and J L Lions
94 Computer algebra in applied mathematics: an introduction to MACSYMA
R H Rand
95 Advances in nonlinear waves. Volume I
L Debnath
96 FC-groups
M J Tomkinson
97 Topics in relaxation and ellipsoidal methods
M Akgül
98 Analogue of the group algebra for topological semigroups
H Dzinotyiweyi
99 Stochastic functional differential equations
S E A Mohammed
100 Optimal control of variational inequalities
V Barbu
101 Partial differential equations and dynamical systems
W E Fitzgibbon III
102 Approximation of Hilbert space operators. Volume II
C Apostol, L A Fialkow, D A Herrero and D Voiculescu
103 Nondiscrete induction and iterative processes
V Ptak and F-A Potra
104 Analytic functions – growth aspects
O P Juneja and G P Kapoor
105 Theory of Tikhonov regularization for Fredholm equations of the first kind
C W Groetsch

106 Nonlinear partial differential equations and free boundaries. Volume I
J I Díaz
107 Tight and taut immersions of manifolds
T E Cecil and P J Ryan
108 A layering method for viscous, incompressible L_p flows occupying R^n
A Douglis and E B Fabes
109 Nonlinear partial differential equations and their applications: Collège de France Seminar. Volume VI
H Brezis and J L Lions
110 Finite generalized quadrangles
S E Payne and J A Thas
111 Advances in nonlinear waves. Volume II
L Debnath
112 Topics in several complex variables
E Ramírez de Arellano and D Sundararaman
113 Differential equations, flow invariance and applications
N H Pavel
114 Geometrical combinatorics
F C Holroyd and R J Wilson
115 Generators of strongly continuous semigroups
J A van Casteren
116 Growth of algebras and Gelfand–Kirillov dimension
G R Krause and T H Lenagan
117 Theory of bases and cones
P K Kamthan and M Gupta
118 Linear groups and permutations
A R Camina and E A Whelan
119 General Wiener–Hopf factorization methods
F-O Speck
120 Free boundary problems: applications and theory, Volume III
A Bossavit, A Damlamian and M Fremond
121 Free boundary problems: applications and theory, Volume IV
A Bossavit, A Damlamian and M Fremond
122 Nonlinear partial differential equations and their applications: Collège de France Seminar. Volume VII
H Brezis and J. L. Lions
123 Geometric methods in operator algebras
H Araki and E G Effros
124 Infinite dimensional analysis–stochastic processes
S Albeverio

Analytic functions – growth aspects

O P Juneja & G P Kapoor
Indian Institute of Technology, Kanpur

Analytic functions – growth aspects

Pitman Advanced Publishing Program
BOSTON · LONDON · MELBOURNE

PITMAN PUBLISHING INC
1020 Plain Street, Marshfield, Massachusetts 02050

PITMAN PUBLISHING LIMITED
128 Long Acre, London WC2E 9AN

Associated Companies
Pitman Publishing Pty Ltd, Melbourne
Pitman Publishing New Zealand Ltd, Wellington
Copp Clark Pitman, Toronto

© O P Juneja and G P Kapoor 1985

First published 1985

AMS Subject Classifications: (main) 30-02, 32-02, 41-02
(subsidiary) 30B50, 32A22, 41A25

ISSN 0743-0337

Library of Congress Cataloging in Publication Data

Juneja, O. P.
Analytic functions—growth aspects.

(Research notes in mathematics; 104)
"Pitman advanced publishing program."
Bibliography: p.
Includes index.
1. Analytic functions. I. Kapoor, G. P. II. Title.
III. Series.
QA331.J86 1985 515.9 84-14810
ISBN 0-273-08630-8

British Library Cataloguing in Publication Data

Juneja, O. P.
Analytic functions: growth aspects—(Research notes in mathematics, ISSN 0743-0337; 104)
1. Calculus 2. Analytic functions
I. Title II. Kapoor, G. P. III. Series
515'.223 QA331

ISBN 0-273-08630-8

All rights reserved. No part of this publication may be reproduced, stored in a retrieval system, or transmitted, in any form or by any means, electronic, mechanical, photocopying, recording and/or otherwise, without the prior written permission of the publishers. This book may not be lent, resold, hired out or otherwise disposed of by way of trade in any form of binding or cover other than that in which it is published, without the prior consent of the publishers.

Reproduced and printed by photolithography
in Great Britain by Biddles Ltd, Guildford

Math-Sci
sep

To

R A J *and* S U M A N

Contents

Preface

Work on the classical theory of measurement of growth of entire functions, which started toward the end of the last century, has been the subject of numerous standard texts, research monographs and conferences over the past sixty years. Some of the well-known publications in this field include the books by Valiron, Boas, Cartwright and Levin, recent monographs by Holland and Evgrafov and the proceedings of the international conferences on entire functions held at La Jolla (1966) and Athens, Ohio (1970). However, efforts to develop systematically an analogous theory for functions analytic in a finite disc are of more recent origin, having been extensively pursued only during the last three decades or so. Although this subject is developing fast, there are still many aspects of the theory of entire functions whose analogues for functions analytic in a finite disc have not yet been touched upon.

The aim of the present monograph is twofold: first, to present those aspects of the theory whose development has reached a definitive stage; second, to make the reader aware of some applications in certain areas of this subject. Much of the material included here is scattered in research papers and has not so far appeared in book form. In fact, this seems to be the first monograph dealing with the growth aspects of analytic functions.

Chapter 1 starts with factorization of functions analytic in the unit disc and, after introducing various growth parameters, deals with their relation to maximum term, central index and Wiman-Valiron theory. Chapter 2 is devoted to coefficient characterizations of different growth parameters and related results, and Chapter 3 contains results that throw light on the behaviour of the minimum modulus of the functions analytic in the unit disc. Recent applications of growth results to approximation and interpolation problems are discussed in Chapter 4.

The study of growth of entire functions of several complex variables, though of comparatively recent origin, has received wide attention during the past twenty years. The monographs written by Fuks and Ronkin give a beautiful account of the progress made in this field. Surprisingly, nothing seems to have been done so far to study the growth of functions of several complex variables that are analytic in a finite domain. The last chapter (Chpater 5) of the present monograph is an attempt in this direction wherein, for the first time, a foundation is laid for this study by introducing various growth scales for analytic functions of several complex variables. The results in this chapter are new and are mostly unpublished.

In the choice of topics discussed the authors have been guided solely by their individual interest and many interesting and related topics such as asymptotic values, value distribution theory etc., could not be touched upon for want of space; indeed, to extend the discussion in these directions would need another monograph about the same size.

An effort has been made to make the book self-contained; the reader with a background of basic courses in real and complex analysis should be able to follow the results of all the chapters. Thus, for the sake of clarity, in some places the temptation to prove results in their most general form had to be resisted. However, the exposition given here should allow the experienced reader to percieve what is needed to prove similar results in more general situations.

Each chapter is divided into sections which (apart from the introductory section) are followed by exercises and problems. Open problems are marked with an asterisk. Usually, references are cited in exercises only where the particular result has not already appeared in some form in a textbook or monograph. The references listed at the ends of chapters are by no means exhaustive - only those that are directly relevant and available are included. The complete list of references for the entire work is given at the end of the book.

The subject matter of the first chapter is used throughout the book, except in the last chapter. Chapters 1,2 and 4 should be read in this order, but the reader who is more interested in the minimum modulus theorems may prefer to go directly from Chapter 1 to Chapter 3. Chapter 5 may be read after completing the first two chapters, although its contents are essentially independent of earlier chapters.

Finally, it is a pleasure to express our affectionate gratitude to our families for their understanding and cooperation during the preparation of this work.

Kanpur
February, 1984

O P Juneja
G P Kapoor

1 Zeros, maximum modulus and maximum term of analytic functions

1.1. INTRODUCTION

Let f be analytic in the disc $\mathcal{U}_R = \{z \in \mathbb{C}: |z| < R\}$ where $0 < R < \infty$ and $\mathbb{C}$ denotes the complex plane. Without loss of generality we may assume that $R = 1$, since otherwise, throughout, $f(Rz)$ could be considered in place of $f(z)$. Also we shall use the symbol $\mathcal{U}$ for $\mathcal{U}_1$.

We start with the factorization, in terms of Blaschke product, of functions analytic in $\mathcal{U}$ and having bounded characteristic. The factorization of analytic functions with unbounded characteristic but having finite Nevalinna order in $\mathcal{U}$ is considered in Section 1.3. The concepts of maximum term and central index for functions analytic in $\mathcal{U}$ are introduced in Section 1.4 and their fundamental properties are studied. Some basic relations involving the maximum modulus, the maximum term and the central index also find a place in this section. For the measurement of the growth of maximum modulus of functions analytic in $\mathcal{U}$, the growth parameters, i.e.,the order, the lower order, the type and the lower type, are introduced in Section 1.5 and their relations with maximum term and central index are found. Growth measurement is further refined by introducing, in Section 1.6, the concept of proximate order. It is established that for every function analytic in $\mathcal{U}$ and having nonzero finite order there exists a proximate order. Some fundamental relations involving the maximum term, the central index and the proximate order are also derived in this section. Finally, in Section 1.7, it is shown that there exist analytic functions in $\mathcal{U}$ having prescribed asymptotic growth. The existence of functions analytic in $\mathcal{U}$ whose maximum modulus increases arbitrarily fast or arbitrarily slowly on some sequences approaching 1 is also established.

1.2 FACTORIZATION OF ANALYTIC FUNCTIONS WITH BOUNDED CHARACTERISTIC

We start by proving a theorem, usually known as the Blaschke theorem, which is basic for obtaining factorization of functions analytic in $\mathcal{U}$.

Theorem 1.2.1. *Let* $\{a_n\}_{n=1}^{\infty}$ *be a sequence of non-zero points in* U. *Then the infinite product*

$$\prod_{n=1}^{\infty} \frac{\overline{a}_n}{|a_n|} \cdot \frac{a_n - z}{1 - z\,\overline{a}_n} \tag{1.2.1}$$

converges uniformly on compact subsets of U *if and only if*

$$\sum_{n=1}^{\infty} (1-|a_n|) < \infty . \tag{1.2.2}$$

Proof. We can assume, without loss of generality, that $0 < |a_1| \leqq |a_2| \leqq \ldots < 1$. Since $w = (a-z)/(1-\overline{a}z)$, $|a| < 1$, maps the open unit disc U in the z-plane onto the open unit disc in the w-plane, it follows that if

$$B_n(z) = \prod_{k=1}^{n} \frac{\overline{a}_k}{|a_k|} \frac{a_k - z}{1 - \overline{a}_k z} \tag{1.2.3}$$

then $|B_n(z)| < 1$ for $z \in U$ and $n = 1,2,\ldots$. Each B_n is analytic in U.

Now suppose (1.2.2) is satisfied. If u_k is defined by $u_k(z) = (\overline{a}_k/|a_k|) \cdot ((a_k-z)/(1-\overline{a}_k z))$, $z \in U$, then we have

$$1 - u_k(z) = 1 - \frac{|a_k|}{a_k} \frac{a_k - z}{1 - \overline{a}_k z}$$

$$= \frac{(1-|a_k|)\,(a_k+|a_k|z)}{a_k(1-\overline{a}_k z)}$$

so that, for $|z| \leqq r < 1$,

$$|1-u_k(z)| \leqq \frac{2(1-|a_k|)}{1-r} .$$

Thus, in view of (1.2.2), $\Sigma_{k=1}^{\infty} |1-u_k(z)|$ converges uniformly in $|z| \leqq r < 1$. This implies that the infinite product $\Pi_{k=1}^{\infty} u_k(z)$ converges absolutely and uniformly in $|z| \leqq r < 1$ and hence on any compact subset of U.

Conversely, if (1.2.1) converges uniformly on compact subsets of U, then, in particular, the infinite product converges at $z = 0$. Thus, the product $\Pi_{k=1}^{\infty} |a_k|$ converges. But this is equivalent to saying that the series $\Sigma_{k=1}^{\infty} (1-|a_k|)$ converges, which is (1.2.2).□

Since $B_n(z) = \Pi_{k=1}^{n} (\bar{a}_k/|a_k|) \cdot ((a_k-z)/(1-\bar{a}_k z))$, $n = 1,2,\ldots$ are analytic in U and, on compact subsets of U, $\lim_{n\to\infty} B_n(z) = \Pi_{k=1}^{\infty} (\bar{a}_k/|a_k|) \cdot ((a_k-z)/(1-\bar{a}_k z)) = B(z)$ (say) uniformly, it follows, by Weierstrass theorem (cf. Appendix A.4) that B is analytic in U. Further, since B_n vanishes at $a_1, a_2, \ldots, a_n$ and nowhere else, it follows, by Hurwitz theorem (cf. Appendix A.4), that B vanishes at the points $\{a_k\}_{k=1}^{\infty}$ and nowhere else. The infinite product $B(z)$ is called the *Blaschke product* formed by the zeros $a_1, a_2, \ldots a_k, \ldots$. A Blaschke product may also be defined having a zero of order m, say, at the origin. In that case, (1.2.1) must be multiplied by a factor z^m.

Our next result interrelates the growth of an analytic function with the distribution of its zeros in U. Throughout, in the sequel, we assume that $f(0) \neq 0$.

Theorem 1.2.2. *Let* f *be analytic in* U *and have zeros at the points* $\{a_k\}_{k=1}^{\infty}$ *in* U. *Further, let*

$$L(r,f) = \frac{1}{2\pi} \int_0^{2\pi} \log |f(re^{i\theta})| \, d\theta.$$

Then

(i) $L(r,f)$ *is an increasing function of* r *for* $0 < r < 1$;

(ii) $\lim_{r\to 1} L(r,f) < \infty$ *if and only if the series* $\Sigma_{k=1}^{\infty} (1-|a_k|)$ *is convergent.*

Proof. We assume that $0 < |a_1| \leq |a_2| \leq \ldots \leq |a_k| \leq \ldots < 1$. Let $0 < r_1 < r_2 < 1$. If f has no zeros in the annulus $r_1 < |z| \leq r_2$ and $|a_N| \leq r_1 < |a_{N+1}|$, then, by Jensen's theorem (cf. Appendix A.9)

$$L(r_1,f) = \log |f(0)| + \log \frac{r_1^N}{|a_1| \ldots |a_N|}$$

$$\leq \log |f(0)| + \log \frac{r_2^N}{|a_1| \ldots |a_N|}$$

$$= \frac{1}{2\pi} \int_0^{2\pi} \log |f(r_2 e^{i\theta})| \, d\theta = L(r_2,f) .$$

If f has zeros $a_{N+1},\ldots,a_M$ in the annulus $r_1 < |z| \leq r_2$, then

$$L(r_1,f) \leqq \log |f(0)| + \log \frac{r_2^N}{|a_1|\ldots|a_N|}$$

$$\leqq \log |f(0)| + \log \frac{r_2^N}{|a_1|\ldots|a_N|} \cdot \frac{r_2^{M-N}}{|a_{N+1}|\ldots|a_M|}$$

$$= \frac{1}{2\pi} \int_0^{2\pi} \log |f(r_2 e^{i\theta})| \, d\theta = L(r_2,f).$$

This proves that $L(r,f)$ is an increasing function of r for $0 < r < 1$. Thus, $\lim_{r\to 1} L(r,f)$ exists as an extended real number. Let $\lim_{r\to 1} L(r,f) \leqq M < \infty$. Fix a positive integer n and let r be such that $|a_n| < r < 1$; then

$$\log |f(0)| < \log (|f(0)| \frac{r^n}{|a_1|\ldots|a_n|}) = L(r,f) \leqq M < \infty .$$

This gives, on making $r \to 1$, that

$$0 < \log (\prod_{k=1}^{n} \frac{1}{|a_k|}) \leqq M < \infty .$$

Since this holds for every n, the infinite product $\Pi_{k=1}^{\infty} |a_k|$ converges which, in turn, implies that $\Sigma_{k=1}^{\infty} (1-|a_k|)$ converges.

Conversely, if $\Sigma_{k=1}^{\infty} (1-|a_k|)$ converges, then $\Pi_{k=1}^{\infty} |a_k|$ converges and, by Jensen's theorem, one has

$$L(r,f) = \log(|f(0)| \frac{r^n}{|a_1|\ldots|a_n|}) \leqq \log (|f(0)| \frac{1}{\Pi_{k=1}^{\infty}|a_k|}) < \infty \quad .\square$$

From Theorems 1.2.1 and 1.2.2 it follows that $\lim_{r\to 1} L(r,f) < \infty$ if and only if the Blaschke product $B(z)$ formed by zeros of f in $\mathcal{U}$ converges. This observation leads to an interesting factorization of functions analytic in $\mathcal{U}$.

Theorem 1.2.3. *Let* $f \not\equiv 0$ *be analytic in* U. *Then* $\lim_{r \to 1} L(r,f) < \infty$ *if and only if* f *is of the form* $f(z) = B(z)e^{g(z)}$ *where* $B(z)$ *is the Blaschke product formed by the zeros of* f *and* g *is an analytic function in* U.

Proof. Let f be given by $f(z) = B(z)\, e^{g(z)}$. It is clear that zeros of f are only at the zeros of B, i.e., at $a_1, a_2, \ldots$. Since $B(z)$ converges, by Theorem 1.2.1, $\Sigma_{k=1}^{\infty} (1-|a_k|) < \infty$. This, by Theorem 1.2.2, gives that $\lim_{r \to 1} L(r,f) < \infty$.

Conversely, suppose $\lim_{r \to 1} L(r,f) < \infty$; by the previous theorem, $\Sigma_{k=1}^{\infty} (1-|a_k|) < \infty$ and so the Blaschke product formed by the zeros of f is convergent. Hence the function h given by $h(z) = f(z)/B(z)$ is analytic in U and has no zeros there. Thus there exists an analytic branch of $\log h$. Denoting this by g we have the desired factorization. □

The next theorem gives an integral representation for functions analytic in U which is of great practical use.

Theorem 1.2.4. *Let* f *be analytic in* U. *Then a necessary and sufficient condition that* f *be represented in the form*

$$f(z) = i\lambda + \int_0^{2\pi} \frac{e^{it} + z}{e^{it} - z}\, d\Psi(t) \tag{1.2.4}$$

where $\lambda = \operatorname{Im} f(0)$ *and* Ψ *is a real-valued function of bounded variation on* $[0,2\pi]$, *is that*

$$\sup_{r<1} \int_0^{2\pi} |\operatorname{Re} f(re^{i\theta})| \; d\theta < \infty . \tag{1.2.5}$$

Proof. Suppose f is analytic in U having representation (1.2.4) where Ψ is a real-valued function of bounded variation on $[0,2\pi]$. Set for $t \in [0,2\pi]$, $\Psi(t) = \Psi_1(t) - \Psi_2(t)$ where Ψ_1 and Ψ_2 are increasing functions of t. Then, for $z = re^{i\theta}$, $r < 1$,

$$\operatorname{Re} f(re^{i\theta}) = \int_0^{2\pi} \frac{1-r^2}{1-2r\cos(\theta-t)+r^2}\, d\Psi_1(t)$$

$$- \int_0^{2\pi} \frac{1-r^2}{1-2r\cos(\theta-t)+r^2}\, d\Psi_2(t)$$

$$= u_1(re^{i\theta}) - u_2(re^{i\theta}), \text{ say.}$$

By Poisson's integral formula (cf. Appendix A.8) both u_1 and u_2 are harmonic functions in $\mathcal{U}$ and, since Ψ_1 and Ψ_2 are increasing, it follows that u_1 and u_2 are always positive. Hence

$$\int_0^{2\pi} |\operatorname{Re} f(re^{i\theta})|\, d\theta \leq \int_0^{2\pi} u_1(re^{i\theta})\, d\theta + \int_0^{2\pi} u_2(re^{i\theta})\, d\theta$$

$$= 2\pi\, [u_1(0) + u_2(0)].$$

Thus

$$\sup_{r<1} \int_0^{2\pi} |\operatorname{Re} f(re^{i\theta})|\, d\theta < \infty, \quad \text{which is (1.2.5).}$$

Now, let f be analytic in $\mathcal{U}$ and suppose (1.2.5) is satisfied. Let, for $z \in \mathcal{U}$,

$$f(z) = \Sigma_{n=0}^{\infty} (c_n + i\, d_n)\, z^n \tag{1.2.6}$$

so that

$$\operatorname{Re} f(re^{i\theta}) = \Sigma_{n=0}^{\infty} (c_n r^n \cos n\theta - d_n r^n \sin n\theta). \tag{1.2.7}$$

Set, for $0 < r < 1$,

$$\Psi_r(t) = \frac{1}{2\pi} \int_0^t \operatorname{Re} f(re^{i\theta})\, d\theta. \tag{1.2.8}$$

Then $\Psi_r(0) = 0$, $\Psi_r(2\pi) = c_o$. Further, by (1.2.7), one has

$$\int_0^{2\pi} 2\cos nt\, d\Psi_r(t) = r^n c_n, \quad n = 1,2,\ldots$$

and

$$\int_0^{2\pi} -2 \sin nt \, d\Psi_r(t) = r^n d_n, \quad n = 1,2,\ldots .$$

Now, for $0 = t_o < t_1 < \ldots < t_n = 2\pi$,

$$\Sigma_{k=1}^n |\Psi_r(t_k) - \Psi_r(t_{k-1})| \leq \frac{1}{2\pi} \int_0^{2\pi} |\text{Re } f(re^{i\theta})| \, d\theta \leq A < \infty.$$

Hence the functions $\{\Psi_r : 0 < r < 1\}$ are of uniformly bounded variation on $[0,2\pi]$. By Helley's selection principle (cf. Appendix A.2), there exists a sequence $\{r_k\}$ tending to 1 and a function Ψ of bounded variation on $[0,2\pi]$ such that $\Psi_{r_k}(t) \to \Psi(t)$ as $k \to \infty$ uniformly on $[0,2\pi]$, and

$$\begin{aligned}\int_0^{2\pi} 2 \cos nt \, d\Psi(t) &= \lim_{k\to\infty} \int_0^{2\pi} 2 \cos nt \, d\Psi_{r_k}(t) \\ &= \lim_{k\to\infty} r_k^n c_n \\ &= c_n, \quad n = 1,2,\ldots .\end{aligned}$$

Similarly,

$$\int_0^{2\pi} -2 \sin nt \, d\Psi(t) = d_n, \quad n = 1,2,\ldots .$$

Substituting for c_n and d_n in (1.2.6), one gets

$$\begin{aligned}f(z) &= i\, d_o + \int_0^{2\pi} [1+2 \Sigma_{n=1}^\infty z^n e^{int}] \, d\Psi(t) \\ &= i\, d_o + \int_0^{2\pi} \frac{e^{it} + z}{e^{it} - z} \, d\Psi(t)\end{aligned}$$

which is (1.2.4). □

Remark. The function $\Psi_r(t)$, defined by (1.2.8), is monotonically increasing if $\text{Re } f(re^{i\theta}) > 0$, therefore, in this case, the function $\Psi(t) = \lim_{k\to\infty} \Psi_{r_k}(t)$ in the proof of the theorem is also monotonically increasing. Thus, it follows that $\text{Re } f(re^{i\theta}) > 0$ if and only if the function $\Psi(t)$ occurring in the representation (1.2.4) is monotonically increasing.

We now introduce the class of functions with *bounded characteristics*, denoted by N after R. Nevanlinna who made a deep study of the properties of this class. A function f, analytic in $\mathcal{U}$, is said to be of class N if its *characteristic function*

$$T(r,f) = \frac{1}{2\pi} \int_0^{2\pi} \log^+ |f(re^{i\theta})| \, d\theta \tag{1.2.9}$$

is bounded for $0 \leq r < 1$; in other words, there exists a constant A such that

$$\sup_{0 \leq r < 1} T(r,f) \leqq A < \infty . \tag{1.2.10}$$

Since $\log^+ |f|$ is subharmonic whenever f is analytic and the mean value of a subharmonic function over $|z| = r$ forms an increasing function of r, it follows that T(r,f) is an increasing function of r. Thus (1.2.10) is equivalent to

$$\lim_{r \to 1} T(r,f) \leqq A < \infty . \tag{1.2.11}$$

The function T(r,f) has some interesting properties. If f, analytic in $\mathcal{U}$, has no zeros there, then, since

$$\log a = \log^+ a - \log^+ \frac{1}{a} ,$$

Jensen's formula (cf. Appendix A.9) immediately leads to the useful relation

$$T(r,f) = T(r, \frac{1}{f}) + \log |f(0)|. \tag{1.2.12}$$

We now obtain an intrinsic characterization of functions of class N in terms of factorization.

Theorem 1.2.5. *If* f *is analytic in* $\mathcal{U}$, *then* f *is of class* N *if and only if*

$$f(z) = B(z) \, e^{h(z)-k(z)} , \quad z \in \mathcal{U} \tag{1.2.13}$$

where B *is the Blaschke product formed with the zeros of* f *and* h *and* k *are analytic in* $\mathcal{U}$ *having positive real parts there.*

Proof. If f is of the form (1.2.13), then for $r < 1$,

$$\log |f(re^{i\theta})| = \log |B(re^{i\theta})| + \text{Re } h(re^{i\theta}) - \text{Re } k(re^{i\theta});$$

since $|B(re^{i\theta})| < 1$ and h and k have positive real parts in U, this leads to

$$\int_0^{2\pi} \log^+ |f(re^{i\theta})| \, d\theta \leqq \int_0^{2\pi} \mathrm{Re}\, \{h(re^{i\theta})\} \, d\theta + \int_0^{2\pi} \mathrm{Re}\, \{k(re^{i\theta})\} \, d\theta$$

$$= 2\pi \, \{h(0)+k(0)\}.$$

Thus

$$\sup_{r<1} T(r,f) \leqq h(0) + k(0) < \infty$$

i.e., $f \in N$.

Conversely, let $\sup_{r<1} T(r,f) < \infty$, then $\lim_{r\to 1} L(r,f) \leqq \sup_{r<1} T(r,f) < \infty$ and so, by Theorem 1.2.3, f has the form $f = B\, e^g$ where B is the Blaschke product formed by the zeros of f and g is analytic in U. We show that

$$\sup_{r<1} \int_0^{2\pi} |\mathrm{Re}\, g(re^{i\theta})| \, d\theta < \infty. \tag{1.2.14}$$

Since, by Theorem 1.2.1, the sequence B_n, defined by (1.2.3), converges to B uniformly on compact subsets of U, it follows that, given $\epsilon > 0$ and $0 < r < 1$, there exists an integer n_o such that $|B_n(z)| < (1+\epsilon)\, |B(z)|$ for $n \geq n_o$ and $|z| \leqq r$. Let $N \geqq n_o$ be so large that $f(z)/B_N(z)$ has no zeros inside $|z| \leq r$. Then

$$|e^{g(re^{i\theta})}| < (1+\epsilon)^{-1} \left|\frac{f(re^{i\theta})}{B_N(re^{i\theta})}\right|$$

so that

$$\mathrm{Re}\, g(re^{i\theta}) \leqq \log \left|\frac{f(re^{i\theta})}{B_N(re^{i\theta})}\right| - \log\,(1+\epsilon) \leqq \log^+ \left|\frac{f(re^{i\theta})}{B_N(re^{i\theta})}\right| \quad . \tag{1.2.15}$$

But

$$|e^{-g(re^{i\theta})}| = \left|\frac{B(re^{i\theta})}{f(re^{i\theta})}\right| \leqq \left|\frac{B_N(re^{i\theta})}{f(re^{i\theta})}\right|$$

gives

$$-\text{Re } g(re^{i\theta}) \leqq \log \left|\frac{B_N(re^{i\theta})}{f(re^{i\theta})}\right| \leqq \log^+ \left|\frac{B_N(re^{i\theta})}{f(re^{i\theta})}\right| . \tag{1.2.16}$$

(1.2.15) and (1.2.16) lead to

$$|\text{Re } g(re^{i\theta})| \leqq \max \left(\log^+ \left|\frac{f(re^{i\theta})}{B_N(re^{i\theta})}\right|, \log^+ \left|\frac{B_N(re^{i\theta})}{f(re^{i\theta})}\right|\right).$$

Thus, by (1.2.12),

$$\frac{1}{2\pi} \int_0^{2\pi} |\text{Re } g(re^{i\theta})| \, d\theta \leqq \max \left(T(r, \frac{f}{B_N}), T(r, \frac{B_N}{f})\right)$$

$$= \max \left(T(r, \frac{f}{B_N}), T(r, \frac{f}{B_N}) + \log |f(0)|\right),$$

$$\leqq \lim_{r\to 1} T(r, \frac{f}{B_N}) + A < \infty,$$

i.e.

$$\sup_{r<1} \int_0^{2\pi} |\text{Re } g(re^{i\theta})| \, d\theta < \infty .$$

Thus, g satisfies (1.2.14). In view of Theorem 1.2.4, g can be expressed as

$$g(z) = i\lambda + \int_0^{2\pi} \frac{e^{it} + z}{e^{it} - z} \, d\Psi(t)$$

where Ψ is a real-valued function of bounded variation on $[0,2\pi]$. Set, for $t \in [0,2\pi]$, $\Psi(t) = \phi_1(t) - \phi_2(t)$ where ϕ_1 and ϕ_2 are increasing. Thus,

$$g(z) = i\lambda + \int_0^{2\pi} \frac{e^{it} + z}{e^{it} - z} \, d\phi_1(t) - \int_0^{2\pi} \frac{e^{it} + z}{e^{it} - z} \, d\phi_2(t)$$

$$= h(z) - k(z)$$

where h and k are analytic in $\mathcal{U}$ and have positive real parts in $\mathcal{U}$. □

Finally, we have

Theorem 1.2.6. *If* f *is analytic in* U, *then* f *has bounded characteristic if and only if* $f = p/q$ *where each of* p *and* q *is bounded and analytic in* U.

Proof. If f has bounded characteristic then, by Theorem 1.2.5, f is of the form $f = B\,e^{h-k}$ where B is the Blaschke product and h and k are analytic in U with positive real parts there. Putting $p = B\,e^{-k}$ and $q = e^{-h}$, we have the desired representation.

Conversely, if $f = p/q$, we may assume $q \neq 0$ in U. Then, using (1.2.12) for the function q(z), it follows that

$$\lim_{r\to 1} T(r,f) = \lim_{r\to 1}\left[\frac{1}{2\pi}\int_0^{2\pi} \log^+ |p(re^{i\theta})/q(re^{i\theta})|\,d\theta\right]$$

$$\leqq \lim_{r\to 1}\left[\frac{1}{2\pi}\int_0^{2\pi} \log^+|p(re^{i\theta})|\,d\theta + \frac{1}{2\pi}\int_0^{2\pi}\log^+|q(re^{i\theta})|\,d\theta\right] + A$$

$$\leqq \sup_{r<1} |p(re^{i\theta})| + \sup_{r<1} |q(re^{i\theta})| + A$$

$$< \infty ,$$

where A is a constant. □

EXERCISES 1.2

1. Let B be the Blaschke product formed with zeros $\{a_n\}$ satisfying $0 < |a_n| < 1$ and $\Sigma_{n=1}^{\infty} (1-|a_n|) < \infty$. Then, prove that $\lim_{r\to 1} B(re^{i\theta}) = B(e^{i\theta})$ exists and $|B(e^{i\theta})| = 1$ almost everywhere in $0 \leq \theta \leq 2\pi$.
2. If B is a Blaschke product having only real zeros in U, then prove that the zeros of B'(z) are also real and are separated by the zeros of B(z).
3. Find a sequence $\{a_n\}$ in $|z| < 1$ such that $\Sigma_{n=1}^{\infty} (1-|a_n|) < \infty$ and every number $e^{i\theta}$ is a limit point of $\{a_n\}$.
4. Suppose $\{a_n\}$ is a sequence such that $|a_n| < 1$, $n = 1,2,\ldots$, $0 < |a_1| \leq |a_2| \leq \ldots$ and $\Sigma_{n=1}^{\infty} (1-|a_n|) = \infty$. If f and g are analytic and bounded in U and $f(a_n) = g(a_n)$, $n = 1,2,\ldots$, then prove that $f \equiv g$.

5. Show that there exists a sequence $\{a_n\}$ with $0 < a_n < 1$ such that $a_n \to 1$ so rapidly that the Blaschke product B with zeros at the points $\{a_n\}$ satisfies

$$\lim_{r\to 1} |B(r)| = 1.$$

6. If B is a Blaschke product, then prove that

$$\limsup_{r\to 1} (1-r) \log |B(r)| = 0.$$

(Heins [1], Shapiro and Shields [1])

7. Show that a function f, analytic in $\mathcal{U}$, is a Blaschke product with a multiple of a constant of modulus 1 if and only if

$$\lim_{r\to 1} \int_0^{2\pi} |\log |f(re^{i\theta})|| \, d\theta = 0.$$

(Shapiro and Shields [1])

8. Suppose f is analytic, $f(0) \neq 0$ and $L(r,f)$, (cf. Theorem 1.2.2), is bounded in $0 < r < 1$. If $n(r)$ is the number of zeros of f in $|z| \leq r$, then prove that

$$\lim_{r\to 1} n(r) \log r = 0.$$

9. Let B be Blaschke product with zeros $\{a_n\}$, $0 < |a_n| < 1$. For any positive integer p, define

$$m_p(r,B) = \{\frac{1}{2\pi} \int_0^{2\pi} |\log |f(re^{i\theta})||^p \, d\theta\}^{1/p}.$$

If $m_2(r,B) \geq (1-r)^{-\beta}$ on some sequence of values of r approaching 1, then show that $\beta \leq 1/2$.

(Sons [4])

*10. Does a result analogous to Exercise 9 hold for any positive integer p?

11. Let B and $m_p(r,B)$ be the same as in Exercise 9. If $n(r)$, as in Exercise 8, satisfies

$$n(r) \leq C(1-r)^{-\alpha}, \qquad 0 < r < 1$$

for some constant C and some constant α in $0 < \alpha < 1$, then prove that

$$m_p^p(r,B) < K(\alpha,p)\, C^p (1-r)^{1-\alpha p}, \quad 0 < r < 1. \tag{1.2.17}$$

(Linden [7])

Let

$$M_p(r,f) = \{\frac{1}{2\pi}\int_0^{2\pi} |f(re^{i\theta})|^p \, d\theta\}^{1/p}, \quad 0 < p < \infty$$

$$M_\infty(r,f) = \max_{|z|=r} |f(z)|, \quad 0 < r < 1.$$

A function f, analytic in U is said to belong to the class H^p, $0 < p \leq \infty$, if $M_p(r,f)$ is bounded in $0 < r < 1$. A function f, analytic in U, is said to be in class N^+ if

$$\lim_{r\to 1}\int_0^{2\pi} \log^+ |f(re^{i\theta})| \, d\theta = \int_0^{2\pi} \log^+ |f(e^{i\theta})| \, d\theta < \infty.$$

12. For $0 < p < q \leq \infty$, show that

$$H^q \subset H^p \subset N^+ \subset N$$

and the inclusions are proper. (Duren [1])

13. Let $f \not\equiv 0$ be in H^p, $0 < p \leq \infty$, then prove that $f(z) = B(z)h(z)$, where $B(z)$ is the Blaschke product formed with zeros of f and $h(z)$ is a nonvanishing function in H^p (compare with Theorem 1.2.5).

14. Find a result analogous to that in Exercise 12 for the class N^+. (Duren [1])

*15. Prove that a function f, analytic in $|z| < 1$, is in H^p, $0 < p < \infty$, if and only if, for $z = x + iy$,

$$\iint_{|z|<1} (1-|z|)f_p^*(z) \, dx \, dy < \infty$$

where $f_p^*(z) = \frac{p}{2} |f|^{\frac{p}{2}-1} |f'|$. (Yamashita [1])

The concept of functions with bounded characteristic has been extended to the class of functions meromorphic in $|z| < 1$, i.e. to functions whose only singularities in $|z| < 1$ are poles. For a meromorphic function f in U let $p(t)$ denote the number of poles of f in $|z| \leq t$. Define

$$T^*(r,f) = \frac{1}{2\pi}\int_0^{2\pi} \log^+ |f(re^{i\theta})| \, d\theta + N(r,f)$$

where $f(0) \neq \infty$ and $N(r,f) = \int_0^r (p(t)/t)\, dt$. A function f meromorphic in U is said to have the *bounded characteristic* if $T^*(r,f)$ is bounded for all r in $0 < r < 1$. Denote the Nevanlinna class of all such functions by M. Let M^p denote the Hardy class of meromorphic functions, i.e. $f \in M^p$ whenever,

$$\iint_{|z|<1} (1-|z|)(f_p^{\#}(z))^2\, dx\, dy < \infty, \qquad 0 < p < \infty,$$

where

$$f_p^{\#}(z) = \frac{p}{2}\, |f|^{\frac{p}{2}-1}\, \frac{|f'|}{1 + |f|^p}.$$

From Exercise 11, $\bigcup_{p>0} H^p \subset N$ and the inclusion is proper. In sharp contrast to this result, for meromorphic functions we have

16. $M^p = M$ for each p, $\qquad 0 < p < \infty$.

(Yamashita [2])

1.3 FACTORIZATION OF ANALYTIC FUNCTIONS HAVING FINITE NEVANLINNA ORDER

A function f, analytic in U, is said to be of finite *Nevanlinna order* if there exists a number A such that Nevanlinna characteristic $T(r,f)$ of f defined by (1.2.9) satisfies

$$T(r,f) < (1-r)^{-A} \tag{1.3.1}$$

for all r in $0 < r_0(A) < r < 1$. If no A satisfying (1.3.1) can be found, then f is said to be of infinite Nevanlinna order. The greatest lower bound of all numbers A satisfying (1.3.1) is called Nevanlinna order of f. Thus, the Nevanlinna order $\sigma(f)$ of f is given by

$$\sigma(f) = \limsup_{r \to 1} \frac{\log T(r,f)}{-\log (1-r)}. \tag{1.3.2}$$

The function $\exp (1-z)^{-\sigma}$, $1 \le \sigma < \infty$, has Nevanlinna order $\sigma-1$.

If f_1 and f_2 are analytic in $|z| < 1$ each having Nevanlinna order not greater than $\sigma(f)$, then, in view of the inequalities

$$\log^+ |f_1 f_2| \le \log^+ |f_1| + \log^+ |f_2|$$

and

$$\log^+ |f_1+f_2| \leq \log^+ |f_1| + \log^+ |f_2| + \log 2,$$

it follows that each of the functions $f_1 f_2$ and f_1+f_2 has Nevanlinna order not exceeding $\sigma(f)$.

Theorem 1.3.1. *Let* f, *analytic in* U, *be of finite Nevanlinna order* $\sigma(f)$. *Let* $\{a_n\}_{n=1}^{\infty}$ *be the sequence of zeros of* f *in* U, *counted according to the multiplicity. Then, for any* $\epsilon > 0$

$$\Sigma_{n=1}^{\infty} (1-|a_n|)^{\sigma(f)+1+\epsilon} < \infty.$$

Proof. Let $n(t)$ denote the number of zeros of f in $|z| \leq t$, $0 \leq t < 1$. For $0 < r < 1$, define

$$N(r) = \int_0^r \frac{n(t) - n(0)}{t} dt + n(0) \log r.$$

Then, for any $\lambda > 0$, the three integrals $\int_0^1 (1-t)^{\lambda+1} dn(t)$, $\int_0^1 (1-t)^{\lambda} n(t)\, dt$ and $\int_0^1 (1-t)^{\lambda-1} N(t)\, dt$ converge or diverge simultaneously. Further, since $\sigma(f) < \infty$, for any $\epsilon > 0$,

$$\int_0^1 T(t)\ (1-t)^{\sigma(f)-1+\epsilon} < \infty.$$

Thus, by Jensen's theorem (cf. Appendix A.9)

$$\int_0^1 N(t)(1-t)^{\sigma(f)-1+\epsilon} < \infty$$

and consequently

$$\Sigma_{n=1}^{\infty} (1-|a_n|)^{\sigma(f)+1+\epsilon} = \int_0^1 (1-t)^{\sigma(f)+1+\epsilon} dn(t) < \infty. \quad \square$$

The *exponent of convergence* of the zeros $\{a_n\}_{n=1}^{\infty}$ of a function f, analytic in $|z| < 1$, is the smallest number μ, $0 \leq \mu < \infty$, such that

$$\Sigma_{n=1}^{\infty} (1-|a_n|)^{\mu+1} < \infty. \tag{1.3.3}$$

If no nonnegative number μ satisfying (1.3.3) exists, then the exponent of convergence of f is defined to be ∞.

An analytic function in U having zeros at $a_n = 1 - (\frac{1}{n\pi})^{\alpha}$, $0 < \alpha \leq 1$, $n = 1,2,\ldots$, has an exponent of convergence $(1/\alpha) - 1$. The exponent of convergence of zeros of an analytic function having zeros at $1 - (1/\log n)$, $n \geq 2$, is ∞. It is clear from Definition 1.3.1 and Theorem 1.3.1 that

$$0 \leq \mu \leq \sigma(f)$$

so that, for a function of finite Nevanlinna order, the exponent of convergence of its zeros is also finite.

The following theorem is useful in determining the exponent of convergence of a given analytic function.

Theorem 1.3.2. *The exponent of convergence* μ, $0 \leq \mu \leq \infty$, *of the zeros* $\{a_n\}_{n=1}^{\infty}$ *of a function* f, *analytic in* U, *is given by*

$$1+\mu = \limsup_{r \to 1} \frac{\log n(r)}{-\log (1-r)} = \limsup_{n \to \infty} \frac{\log n}{-\log (1-|a_n|)} . \tag{1.3.4}$$

Proof. Without loss of generality we may assume that $f(0) \neq 0$. The equality of lim sup in both the expressions of the right-hand side of (1.3.4) is obvious. Set

$$1 + \mu_o = \limsup_{n \to \infty} \frac{\log n}{-\log (1-|a_n|)} .$$

First, let $\mu_o < \infty$. Let $\epsilon > 0$ be given. Then, for $\epsilon' = \epsilon/2$, there exists an integer $n_o = n_o(\epsilon')$ such that

$$\Sigma_{n=n_o}^{\infty} (1-|a_n|)^{\mu_o+1+\epsilon} < \Sigma_{n=n_o}^{\infty} n^{-\frac{\mu_o+1+\epsilon}{\mu_o+1+\epsilon'}} < \infty$$

so that

$$\mu \leq \mu_o . \tag{1.3.5}$$

Inequality (1.3.5) obviously holds when $\mu_o = \infty$.

To prove the reverse inequality, for $\lambda > 0$, let $\Sigma(1-|a_n|)^{\lambda+1}$ be convergent. Then, the integral $\int_0^r (1-t)^{\lambda+1}\, dn(t)$ is bounded in $0 < r < 1$.

Now,

$$\int_0^r (1-t)^{\lambda+1}\, dn(t) = (1-r)^{\lambda+1}\, n(r) + (\lambda+1) \int_0^r (1-t)^{\lambda}\, n(t)\, dt.$$

Since both the terms on the right-hand side of this expression are positive and the second term is a nondecreasing function of r, the integral $\int_0^1 (1-t)^{\lambda}\, n(t)\, dt < \infty$. Thus, we have $(1-r)^{\lambda+1}\, n(r) \to 0$ as $r \to 1$ and it follows that

$$1 + \mu_o = \limsup_{r \to 1} \frac{\log n(r)}{-\log(1-r)} \le \lambda+1.$$

Now, first let $\mu < \infty$. Then, the above inequality holds for any $\lambda > \mu$ and we have

$$\mu_o \le \mu . \tag{1.3.6}$$

The inequality (1.3.6) obviously holds when $\mu = \infty$. On combining (1.3.5) and (1.3.6), we get (1.3.4). □

In view of the relation $\mu \le \sigma(f)$, it follows from Theorem 1.3.2 that if f, analytic in U, is of finite Navanlinna order, then $n(r) = O((1-r)^{-\sigma(f)-1-\epsilon})$ as $r \to 1$, for every $\epsilon > 0$. A stronger result than this is the following:

Theorem 1.3.3. *For a function* f, *analytic in* U *and having an exponent of convergence of its zeros* μ,

$$\liminf_{r \to 1} \frac{(1-r)\, n(r)}{T(r)} \le \liminf_{r \to 1} \frac{\log n(r)}{-\log(1-r)} \le \mu-1$$

where n(r) *is the number of zeros of* f *in* $|z| \le r < 1$.

Proof. Let

$$\lambda = \liminf_{r \to 1} \frac{\log n(r)}{-\log(1-r)} .$$

It is enough to show that

$$\liminf_{r \to 1} \frac{(1-r)\, n(r)}{N(r)} \le \lambda-1 \tag{1.3.7}$$

where, $N(r) = \int_0^r \frac{n(t)}{t}\, dt$; for, by Jensen's theorem $N(r) \leq T(r)$.

Suppose (1.3.7) does not hold. Then, there exists an $\epsilon > 0$ such that

$$\frac{(1-r)\, n(r)}{N(r)} > \lambda - 1 + 2\epsilon$$

for r in $0 < r_o(\epsilon) < r < 1$. Thus, for $r_o < R$, $S < 1$,

$$(\lambda-1+2\epsilon) \int_R^S (1-x)^{\lambda-2+\epsilon} N(x)\, dx < \int_R^S (1-x)^{\lambda-1+\epsilon} n(x)\, dx$$

$$= \int_R^S (1-x)^{\lambda-1+\epsilon} N'(x)\, x\, dx$$

$$\leq (1-S)^{\lambda-1+\epsilon} N(S)$$

$$+ (\lambda+\epsilon-1) \int_R^S (1-x)^{\lambda-2+\epsilon} N(x)\, dx$$

where $N'(x)$ is the first derivative of $N(x)$ wherever it exists. The above inequalities lead to

$$\epsilon \int_R^S (1-x)^{\lambda-2+\epsilon} N(x)\, dx \leq N(S)\, (1-S)^{\lambda-1+\epsilon} \tag{1.3.8}$$

$$< \frac{n(S)\, (1-S)^{\lambda+\epsilon}}{\lambda-1+2\epsilon} .$$

By the definition of λ, we get $n(S) < (1-S)^{-\lambda-(\epsilon/2)}$ for a sequence $\{S_n\}$ in U satisfying $S_n \to 1$ as $n \to \infty$. Thus, the right-hand side of inequality (1.3.8) tends to zero for the sequence $\{S_n\}$. However, the left-hand side of (1.3.8) is positive and R is independent of S. Thus, (1.3.8) is not true. This proves that inequality (1.3.7) must hold. □

The *genus* of the zeros $\{a_n\}_{n=1}^{\infty}$ of a function f, analytic in U, is the smallest positive integer p such that

$$\Sigma_{n=1}^{\infty} (1-|a_n|)^{p+1} < \infty .$$

The exponent of convergence μ and the genus p of the zeros of a function analytic in U satisfy the following relations:

(i) if μ is not an integer, then $p = [\mu] + 1$; where $[\mu]$ is the greatest integer not exceeding μ;

(ii) if μ is an integer and $\Sigma_{n=1}^{\infty} (1-|a_n|)^{\mu+1} < \infty$, then $p = \mu$;

(iii) if μ is an integer and $\Sigma_{n=1}^{\infty} (1-|a_n|)^{\mu+1} = \infty$, then $p = \mu+1$.

Thus, the following relation among the exponent of convergence μ, the genus p and the Navanlinna order $\sigma(f)$ of an analytic function is always satisfied

$$p-1 \leq \mu \leq \sigma(f).$$

For a function f, analytic in U, whose zeros $\{a_n\}_{n=1}^{\infty}$ have genus $p > 1$, the *canonical product* P formed with its nonzero zeros is defined as

$$P(z) = \Pi_{n=1}^{\infty} \left(1 - \frac{1-|a_n|^2}{1 - \bar{a}_n z}\right) \exp\left(\frac{1-|a_n|^2}{1-\bar{a}_n z} + \ldots + \frac{1}{p}\left(\frac{1-|a_n|^2}{1-\bar{a}_n z}\right)^p\right). \qquad (1.3.9)$$

The function P is analytic in U and has zeros only at the points $\{a_n\}_{n=1}^{\infty}$. The following upper bound of $\log^+ |P(z)|$ is needed in factorization of analytical functions.

Theorem 1.3.4. *Let* f, *analytic in* U, *have finite Nevanlinna order. Then*

$$\log^+ |P(z)| \leq 2^{p+1} \Sigma_{n=1}^{\infty} \left|\frac{1-|a_n|^2}{1 - \bar{a}_n z}\right|^{\mu+1+\varepsilon}$$

where $\varepsilon = 0$ *if* $p = \mu$ *and* $0 < \varepsilon \leq p-\mu$ *if* $p \neq \mu$.

Proof. It is clear from (1.3.9) that

$$\log^+ |P(z)| \leq \Sigma_1 \operatorname{Re}^+(E(z,a_n)) + \Sigma_2 |E(z,a_n)|, \qquad (1.3.10)$$

where Σ_1 is summation over those n for which $(1-|a_n|^2)/|1-\bar{a}_n z| \geq \frac{1}{2}$ and Σ_2 is summed over n satisfying $(1-|a_n|^2)/|1-\bar{a}_n z| < \frac{1}{2}$, and where $\operatorname{Re}^+ \zeta = \max(\operatorname{Re} \zeta, 0)$ and

$$E(z,a_n) = \log\left(1 - \frac{1-|a_n|^2}{1-\bar{a}_n z}\right) + \frac{1-|a_n|^2}{1-\bar{a}_n z} + \ldots + \frac{1}{p}\left(\frac{1-|a_n|^2}{1-\bar{a}_n z}\right)^p . \tag{1.3.11}$$

If $(1-|a_n|^2)/(|1-\bar{a}_n z|) < \frac{1}{2}$, then

$$|E(z,a_n)| \le \frac{1}{p+1} \left|\frac{1-|a_n|^2}{1-\bar{a}_n z}\right|^{p+1} + \frac{1}{p+2} \left|\frac{1-|a_n|^2}{1-\bar{a}_n z}\right|^{p+2} + \ldots \tag{1.3.12}$$

$$< \left|\frac{1-|a_n|^2}{1-\bar{a}_n z}\right|^{p+1} \left(1 + \frac{1-|a_n|^2}{|1-\bar{a}_n z|} + \left|\frac{1-|a_n|^2}{1-\bar{a}_n z}\right|^2 + \ldots\right)$$

$$< 2 \left|\frac{1-|a_n|^2}{1-\bar{a}_n z}\right|^{p+1} .$$

If $p = \mu$, then

$$|E(z,a_n)| \le 2 \left|\frac{1-|a_n|^2}{1-\bar{a}_n z}\right|^{\mu+1} . \tag{1.3.13}$$

If $p \neq \mu$, then for any ε satisfying $0 < \varepsilon \le 1 - (\mu-[\mu])$, we get $p \ge \mu+\varepsilon$ so that (1.3.12) gives

$$|E(z,a_n)| \le 2 \left|\frac{1-|a_n|^2}{1-\bar{a}_n z}\right|^{\mu+1+\varepsilon} \left|\frac{1-|a_n|^2}{1-\bar{a}_n z}\right|^{p-(\mu+\varepsilon)}$$

$$\le 2 \left|\frac{1-|a_n|^2}{1-\bar{a}_n z}\right|^{\mu+1+\varepsilon} . \left(\frac{1}{2}\right)^{p-(\mu+\varepsilon)}$$

$$\le 2 \left|\frac{1-|a_n|^2}{1-\bar{a}_n z}\right|^{\mu+1+\varepsilon} .$$

It follows now that

$$\Sigma_2 |E(z,a_n)| \le 2\, \Sigma_2 \left|\frac{1-|a_n|^2}{1-\bar{a}_n z}\right|^{\mu+1+\varepsilon} \tag{1.3.14}$$

where $\varepsilon = 0$ if $p = \mu$ and $0 < \varepsilon \le p-\mu$, if $p \neq \mu$.

To estimate Σ_1, it is to be observed that, if $(1-|a_n|^2)/|1-\bar{a}_n z| \geq \frac{1}{2}$, then

$$\log \left|1 - \frac{1-|a_n|^2}{1-\bar{a}_n z}\right| = \log \left|\frac{\bar{a}_n(a_n-z)}{1-\bar{a}_n z}\right| \leq 0 .$$

Thus,

$$\mathrm{Re}^+(E(z,a_n)) \leq \frac{1-|a_n|^2}{|1-\bar{a}_n z|} + \dots + \frac{1}{p}\left(\frac{1-|a_n|^2}{|1-\bar{a}_n z|}\right)^p \tag{1.3.15}$$

$$\leq \left|\frac{1-|a_n|^2}{1-\bar{a}_n z}\right|^p \left(1 + \left|\frac{1-\bar{a}_n z}{1-|a_n|^2}\right| + \dots + \left|\frac{1-\bar{a}_n z}{1-|a_n|^2}\right|^{p-1}\right)$$

$$\leq \left|\frac{1-|a_n|^2}{1-\bar{a}_n z}\right|^p \quad (1+2+\dots+2^{p-1})$$

$$< 2^p \left|\frac{1-|a_n|^2}{1-\bar{a}_n z}\right|^p .$$

If $p = \mu$, then since $|1-\bar{a}_n z|/(1-|a_n|^2) \leq 2$ in Σ_1, by (1.3.15),

$$\mathrm{Re}^+(E(z,a_n)) < 2^{p+1} \left|\frac{1-|a_n|^2}{1-\bar{a}_n z}\right|^{\mu+1} .$$

If $p \neq \mu$ and ε is such that $0 < \varepsilon \leq p-\mu$, then $0 < \mu+1+\varepsilon-p \leq 1$. This gives, by (1.3.15), that

$$\mathrm{Re}^+(E(z,a_n)) < 2^p \left|\frac{1-|a_n|^2}{1-\bar{a}_n z}\right|^{\mu+1+\varepsilon} \left|\frac{1-\bar{a}_n z}{1-|a_n|^2}\right|^{\mu+1+\varepsilon-p}$$

$$\leq 2^{\mu+1+\varepsilon} \left|\frac{1-|a_n|^2}{1-\bar{a}_n z}\right|^{\mu+1+\varepsilon}$$

$$\leq 2^{p+1} \left|\frac{1-|a_n|^2}{1-\bar{a}_n z}\right|^{\mu+1+\varepsilon} .$$

This gives,

$$\Sigma_1 \operatorname{Re}^+(E(z,a_n)) \leq 2^{p+1} \Sigma_1 \left|\frac{1-|a_n|^2}{1-\bar{a}_n z}\right|^{\mu+1+\varepsilon} \tag{1.3.16}$$

where $\varepsilon = 0$ if $p = \mu$ and $0 < \varepsilon \leq p-\mu$ if $p \neq \mu$.
Inequalities (1.3.14) and (1.3.16) together with (1.3.10) give the required estimate of the theorem. □

Theorem 1.3.5. *The Navanlinna order* $\sigma^*(P)$ *of the canonical product* P, *given by* (1.3.9), *is equal to the exponent of convergence of the zeros of* P.

Proof. Since $\mu \leq \sigma^*(P)$, we need only to show that $\sigma^*(P) \leq \mu$. The estimate in Theorem 1.3.9, for $0 < r < 1$, gives

$$T(r,P) = \frac{1}{2\pi} \int_0^{2\pi} \log^+ |P(re^{i\theta})| \, d\theta$$

$$\leq \frac{2^p}{\pi} \Sigma_{n=1}^{\infty} \int_0^{2\pi} \left|\frac{1-|a_n|^2}{1-\bar{a}_n re^{i\theta}}\right|^{\mu+1+\varepsilon} d\theta \tag{1.3.17}$$

where $\varepsilon = 0$ if $p = \mu$ and $\varepsilon > 0$ is arbitrarily small if $p \neq \mu$. Let

$$I = \int_{-\pi}^{\pi} \frac{d\theta}{|1-re^{i\theta}|^{\mu+1+\varepsilon}}$$

$$= \int_{|\theta|\leq 1-r} \frac{d\theta}{|1-re^{i\theta}|^{\mu+1+\varepsilon}} + \int_{1-r\leq|\theta|\leq\pi} \frac{d\theta}{|1-re^{i\theta}|^{\mu+1+\varepsilon}}$$

$$= I_1 + I_2 .$$

Since $\mu+\varepsilon > 0$, we have

$$I_1 \leq \frac{1}{(1-r)^{\mu+1+\varepsilon}} \int_{|\theta|\leq 1-r} d\theta = \frac{2}{(1-r)^{\mu+\varepsilon}}$$

and since $|1-re^{i\theta}|^2 = 1-2r \cos \theta+r^2 \geq (1-r)^2+a^2\theta^2$,

$$I_2 = 2 \int_{1-r}^{\pi} \frac{d\theta}{|1-re^{i\theta}|^{\mu+1+\varepsilon}} \leq 2 \int_{1-r}^{\pi} \frac{d\theta}{((1-r)^2+a^2\theta^2)^{(\mu+1+\varepsilon)/2}}$$

$$\leq \frac{2}{(1-r)^{\mu+\varepsilon}} \int_1^{\infty} \frac{dt}{(1+K^2t^2)^{(\mu+1+\varepsilon)/2}}$$

$$= O(\frac{1}{(1-r)^{\mu+\varepsilon}}), \text{ as } r \to 1,$$

where K is a constant. These two estimates give

$$I = O((1-r)^{-(\mu+\varepsilon)}) \qquad \text{as } r \to 1. \tag{1.3.18}$$

Now (1.3.17) and (1.3.18) give

$$T(r,P) = O((1-r)^{-(\mu+\varepsilon)}) \text{ as } r \to 1. \tag{1.3.19}$$

Thus, in view of (1.3.19), $\sigma^*(P) \leq \mu$.□

Theorem 1.3.6. *Let* f *be analytic in* U *and have Nevanlinna order* $\sigma(f) < \infty$. *Then,*

$$f(z) = z^m P(z) g(z)$$

where P, *the canonical product formed with nonzero zeros of* f, *is given by* (1.3.9), *and* $g(z) \neq 0$ *is analytic in* U *and has Nevanlinna order not greater than* $\sigma(f)$.

Proof. Let F be meromorphic in U, i.e., the only singularities of F in U are poles. Assume $F(0) \neq 0, \infty$ and set,

$$T^*(r,F) = \frac{1}{2\pi} \int_0^{2\pi} \log^+ |F(re^{i\theta})| \, d\theta + \int_0^r \frac{p(t)}{t} dt \tag{1.3.20}$$

where p(t) denotes the number of poles of F in $|z| \leq t < 1$. By Poisson Jensen's formula (cf. Appendix A.10) it follows that

$$T^*(r,F) = \frac{1}{2\pi} \int_0^{2\pi} \log^+ \frac{1}{|F(re^{i\theta})|} d\theta + \int_0^r \frac{n(t)}{t} dt + K(r) \tag{1.3.21}$$

where $n(t)$ is the number of zeros of F in $|z| \leq t < 1$ and $K(r)$ is bounded in $0 < r < 1$. Comparing (1.3.20) and (1.3.21), for the function F we have

$$T^*(r,F) = T^*(r, \frac{1}{F}) + K(r).$$

Now, define $g(z) = f(z)/z^m P(z)$. Then g is analytic and nonzero in $\mathcal{U}$. We prove that the Nevanlinna order of g is not greater than $\sigma(f)$. We have

$$T(r,g) = T^*(r,g) \leq T^*(r,f) + T^*(r, \frac{1}{z^m P})$$

$$= T(r,f) + T^*(r,z^m P) - K(r)$$

$$= T(r,f) + T(r,z^m P) - K(r).$$

Thus, by Theorem 1.3.5 and the definition of Nevanlinna order, $T(r,g) = O((1-r)^{-(\sigma(f)+\varepsilon)}) + O((1-r)^{-(\mu+\varepsilon)})$, as $r \to 1$, where μ is the exponent of convergence of zeros of f. Since $\mu \leq \sigma(f)$, it follows that g is of Nevanlinna order not exceeding $\sigma(f)$. □

EXERCISES 1.3

1. Let f be analytic in $\mathcal{U}$ and have Nevanlinna order σ. Let σ' be the Nevanlinna order of its derivative f'. Then prove that $\sigma' = \sigma$.

   For a function f, analytic in $\mathcal{U}$ and having nonzero finite Nevanlinna order σ, define

$$\tau = \limsup_{r \to 1} \frac{T(r,f)}{(1-r)^{-\sigma}} .$$

   Then τ is called the *Nevanlinna type* of f.

2. If f_1 and f_2 are analytic in $\mathcal{U}$ and have Nevanlinna types τ_1 and τ_2 respectively, then find relations of Nevanlinna types of f_1+f_2 and f_1f_2 with τ_1 and τ_2.

*3. Can the result of Exercise 1.2.2 be extended to the functions analytic in $\mathcal{U}$ and having genus 1?

4. Let f be analytic in $\mathcal{U}$ and $k > 0$. If the integral $\int_0^1 (1-t)^{k-1} \log M(t)\, dt$ is convergent, then show that the series

$\sum_{n=1}^{\infty} (1-r_n)^{k+1}$ is also convergent, where $M(t)$ is the maximum of $|f|$ on $|z| = t$ and $\{r_n\}$ is the sequence of modului of zeros of f arranged in the increasing order. Show that the converse need not hold.

Let $n(t)$ be the number of zeros of a function f, analytic in $\mathcal{U}$, in $|z| \leq t < 1$ and $N(r) = \int_0^r (n(t)/t)\, dt$. For $0 < \sigma < \infty$, define

$$\lim_{r\to 1} \begin{matrix}\sup\\ \inf\end{matrix} \frac{n(r)}{r(1-r)^{-\sigma-1}} = \begin{matrix}L\\ \ell\end{matrix}$$

and

$$\lim_{r\to 1} \begin{matrix}\sup\\ \inf\end{matrix} \frac{N(r)}{(1-r)^{-\sigma}} = \begin{matrix}\xi\\ \eta\end{matrix} .$$

5. Prove that $0 < \ell \leq L < \infty$ if and only if $0 < \eta \leq \xi < \infty$.

6. Prove that $0 < \ell = L < \infty$ if and only if $0 < \eta = \xi < \infty$.

7. (i) $L + \ell \leq \frac{(\sigma+1)^{\sigma+1}}{\sigma^{\sigma}} \xi$

(ii) $\ell \leq \sigma \xi$.

The inequality cannot simultaneously hold in (i) and (ii).

8. Let P be a canonical product given by (1.3.9) and having positive Nevanlinna order σ. Let $\exp(g(z))$ be analytic in $\mathcal{U}$ and have Nevanlinna order less than σ. Define f by

$f(z) = P(z) \exp(g(z))$.

Then prove that the Nevanlinna order σ^* of f is given by

$$\sigma^* = \sigma = \limsup_{r \to 1} \frac{\log N(r)}{-\log (1-r)} .$$

Further, if the genus p of zeros of f is not equal to σ, show that there exists a positive constant $C = C(\sigma,p)$ such that

$$\limsup_{r \to 1} \frac{N(r)}{T(r)} \geq C.$$

(Sons [3])

9. Let f be analytic in U and $m_2(r,f)$ be defined as in Exercise 1.2.9. Then show that there exists a nonnegative integer k and a function g, analytic in U, such that

$$f(z) = z^k \exp(g(z))\, P(z)$$

where P(z) is either a Blaschke product as defined in Section 1.2 or the canonical product, given by (1.3.9), formed with nonzero zeros of f or $P(z) \equiv 1$. (Sons [4])

*10. Does the result of Exercise 1.3.9 hold if $m_2(r,f)$ is replaced by $m_q(r,f)$ (cf. Exercise 1.2.9), for any integer q satisfying $1 \leq q < \infty$?

1.4 MAXIMUM TERM AND CENTRAL INDEX

Let $f(\not\equiv 0)$, defined by

$$f(z) = \sum_{n=0}^{\infty} a_n z^n \tag{1.4.1}$$

be analytic in U. Then, for any $z \neq 0$, with $|z| = r < 1$, the series in (1.4.1) is absolutely convergent so that $|a_n| r^n \to 0$ as $n \to \infty$. Let a_K be the first nonzero coefficient in (1.4.1). If n_o is a nonnegative integer such that $|a_n| r^n < |a_K| r^K$ for all $n > n_o$, then there exists at least one nonnegative integer k such that

$$|a_k| r^k = \max\,(|a_K| r^K, |a_{K+1}| r^{K+1}, \ldots, |a_{n_o}| r^{n_o}). \tag{1.4.2}$$

The term $|a_k| r^k$ is called a maximum term of f for $|z| = r$. If, for $|z| = r$, (1.4.2) is satisfied for the nonnegative integers $k_1, \ldots, k_p$ and $\nu \equiv \nu(r) = \max\,(k_1, \ldots, k_p)$, then $\nu(r)$ is called the central index of f for $|z| = r$.* Thus, we have the following definitions of the *maximum term* $\mu(r)$ and the *central index* $\nu(r)$ of f:

$$\mu(r) = \max\,\{|a_n| r^n : n = 0,1,2,\ldots\}$$

and

$$\nu(r) = \max\,\{n : \mu(r) = |a_n| r^n\}.$$

---

*If r = 0 and a_K is the first nonzero coefficient of the series in (1.4.1), then we agree to define $\nu(0) = K$ and $\mu(0) = |a_o|$ if K = 0 and $\mu(0) = 0$ if K = 1,2,... .

The functions $\mu(r)$ and $\nu(r)$ are nondecreasing. For, if $0 < r_1 \le r_2 < 1$ and $\mu(r_1) = |a_{N_1}| r_1^{N_1}$; $\mu(r_2) = |a_{N_2}| r_2^{N_2}$, then

$$|a_{N_2}| r_2^{N_2} \ge |a_{N_1}| r_2^{N_1}$$

and

$$|a_{N_1}| r_1^{N_1} \ge |a_{N_2}| r_1^{N_2}.$$

It follows from the above inequalities that

$$\mu(r_2) = |a_{N_2}| r_2^{N_2} \ge |a_{N_1}| r_2^{N_1} \ge |a_{N_1}| r_1^{N_1} = \mu(r_1)$$

and

$$\left(\frac{r_2}{r_1}\right)^{N_2 - N_1} \ge 1$$

so that $\nu(r_2) = N_2 \ge N_1 = \nu(r_1)$.

We assume that the coefficients a_n in Taylor series (1.4.1) of the function f satisfy

$$\sup_n |a_n| = \infty. \tag{1.4.3}$$

It follows from (1.4.3) that $\mu(r) \to \infty$ and $\nu(r) \to \infty$ as $r \to 1^-$. For, if there exists a nonzero finite constant M such that $\mu(r) \le M$ for all r in $0 < r < 1$, than $|a_n| \le M\, r^{-n}$ for all $n = 0,1,2,\ldots$, and all r in $0 < r < 1$. Now, making $r \to 1^-$ in this inequality, $|a_n| \le M$ for all $n = 0,1,2,\ldots$. This gives $\sup_n |a_n| \le M$ and we arrive at a contradiction of (1.4.3). Similarly, if there exists a nonzero finite constant M such that $\nu(r) \le M$ for all r in $0 < r < 1$, then let S be an integer such that

$$|a_S| r^S = \max \{|a_K| r^K, \ldots, |a_{[M]}| r^{[M]}\}$$, where $[M]$ is integral part of M,

a_K being the first nonzero coefficient in (1.4.1). Thus, for all $n = 0,1,2,\ldots$ and all r in $0 < r < 1$, we have $|a_n| r^n \le |a_S| r^S$. Consequently

$|a_n| \le |a_S| r^{S-n}$. Making $r \to 1$ in this inequality, we get $|a_n| \le |a_S|$ for all $n = 0,1,2,\ldots$, giving thereby $\sup_n |a_n| \le |a_S|$, a contradiction to (1.4.3).

Conversely, it is easily seen that if $\mu(r) \to \infty$ or $\nu(r) \to \infty$, then (1.4.3) is satisfied.

We also assume from here onwards that

$$f(0) = 1.$$

There is no loss of generality in doing so, for, if a_K is the first nonzero coefficient in (1.4.1), then $g(z) = f(z)/a_K z^K$ is analytic in $|z| < 1$, satisfies $g(0) = 1$ and, if $\mu_g(r)$ and $\nu_g(r)$ are the maximum term and central index of g for $|z| = r \neq 0$, then $\mu_g(r) = \mu(r)/a_K r^K$ and $\nu_g(r) = \nu(r) - K$.

The central index $\nu(r)$ of f is an integer-valued step function of r. The elements $\{n_k\}_{k=0}^{\infty}$ in the range set of $\nu(r)$ are called the *principal indices* of f. We may assume without loss of generality that $0 \le n_o < \ldots < n_{i-1} < n_i < \ldots$ so that $n_k \to \infty$ as $k \to \infty$. We define

$$R(n_o) = 0 \quad \text{and} \quad R(n_k) = \sup \{r : \nu(r) = n_{k-1}\},\ k = 1,2,\ldots . \tag{1.4.4}$$

Then, $\nu(r)$ has only left-hand discontinuities at the points $R(n_k)$, since,

$$\lim_{r \to R^-(n_k)} \nu(r) = n_{k-1} < n_k = \nu(R_{(n_k)}) = \lim_{r \to R^+(n_k)} \nu(r) .$$

The points $R_{(n_k)}$ are called the *jump points* of $\nu(r)$. It is clear that, for $k = 0,1,2,\ldots,$

$$\nu(r) = n_k \quad \text{if and only if} \quad R(n_k) \le r < R(n_{k+1}) , \tag{1.4.5}$$

so that $\{R(n_k)\}_{k=0}^{\infty}$ is a strictly increasing sequence. Since (1.4.3) is satisfied, $n_k \to \infty$ as $k \to \infty$ and it follows that $R(n_k) \to 1$ as $k \to \infty$.

Hadamard's Polygon. Let $\{n_k\}_{k=0}^{\infty}$ be the principal indices of a function $f(z) = \Sigma_{n=0}^{\infty} a_n z^n$, analytic in $\mathcal{U}$ and satisfying (1.4.3). Set $g_n = \log |a_n|$ and construct a polygon $\pi(f)$, called Hadamard's polygon of f, having the points $A_{n_k} = (n_k, g_{n_k})$ as its vertices. The slope w_k of a side D_k of $\pi(f)$ having $A_{n_{k-1}}$ and A_{n_k} as the end points is given by

$$w_k = \log (1/R^*(n_k))$$

where

$$R^*(n_k) = |a_{n_{k-1}}/a_{n_k}|^{1/(n_k-n_{k-1})} .$$

For a given z with $|z| = r$, $0 \leq r < 1$, if n_k is the central index of f, then

$$|a_{n_k}|r^{n_k} \geq |a_{n_{k-1}}|r^{n_{k-1}} \text{ and } |a_{n_k}|r^{n_k} > |a_{n_{k+1}}|r^{n_{k+1}}$$

so that

$$R^*(n_k) \leq r < R^*(n_{k+1}) \qquad (1.4.6)$$

and it follows that $\{R^*(n_k)\}_{k=0}^{\infty}$ is a strictly increasing sequence. Since (1.4.3) is satisfied, we have $R^*(n_k) \to 1$ as $k \to \infty$. Thus the slopes w_k of the sides D_k, $k = 1,2,\ldots$, decrease to zero as $k \to \infty$. Consequently, the polygon $\pi(f)$ is concave downwards.

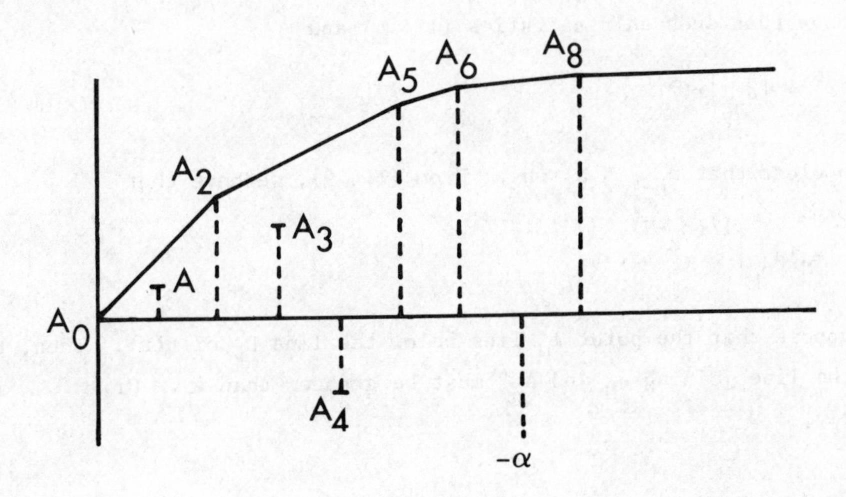

Figure 1

Using (1.4.6), we have that if $\nu(r) = n_{k-1}$, then $R^*(n_{k-1}) \leq r < R^*(n_k)$, so that

$$\nu(R^*(n_k)) = n_k. \qquad (1.4.7)$$

This, together with (1.4.6), gives that $\nu(r) = n_k$ if and only if

$R^*(n_k) \leq r < R^*(n_{k+1})$. In view of (1.4.7), we therefore have

$$R^*(n_k) = R(n_k),$$

where $R(n_k)$ is defined by (1.4.4).

Now, for any r in $R^*(n_k) \leq r < R^*(n_{k+1})$, since $\nu(r) = n_k$, we have for all $m = 0,1,2,\ldots$

$$\log |a_m| - m \log \frac{1}{r} \leq \log |a_{n_k}| - n_k \log \frac{1}{r} \qquad (1.4.8)$$

with the convention that $\log |a_m| = -\infty$ if $a_m = 0$. Set $g_m = \log |a_m|$, $m = 0,1,2,\ldots$. Then it follows from inequality (1.4.8) that the points $A_m = (m,g_m)$ lie on or below the line passing through A_{n_k} and having the slope $\log (1/r)$. In particular, choosing $r = R^*(n_k)$ in this inequality, we have that the points A_m lie on or below the line passing through A_{n_k} and having the slope w_k. Consequently, all the points A_m lie on or below $\pi(f)$.

Let, for $|z| = r$, $0 \leq r < 1$, $|a_N|r^N$ be one of the maximum terms of f and $\nu(r) = n_k$. Then, the point $A_N = (N,g_N)$ lies on the side D_k of $\pi(f)$. To see this, we have that such an r satisfies (1.4.6) and

$$|a_N|r^N = |a_{n_k}|r^{n_k}. \qquad (1.4.9)$$

It is also clear that $n_{k-1} < N < n_k$. From (1.4.9), we have that

$$r = |a_N/a_{n_k}|^{1/(n_k-N)}.$$

Now, suppose that the point A_N lies below the line D_k of $\pi(f)$. Then, the slope of the line joining A_N and A_{n_k} must be greater than w_k. Or, equivalently,

$$\frac{g_{n_k} - g_N}{n_k - N} > \frac{g_{n_k} - g_{n_{k-1}}}{n_k - n_{k-1}}$$

which gives

$$r = |a_N/a_{n_k}|^{1/(n_k-N)} < |a_{n_{k-1}}/a_{n_k}|^{1/(n_k-n_{k-1})} = R^*(n_k) ,$$

a contradiction to (1.4.6). This proves that the point $A_N = (N, g_N)$ lies on the side D_k of $\pi(f)$. It is obvious now that if a point $A_m = (m, g_m)$ does not correspond to any of the maximum terms of f, then it lies below $\pi(f)$.

We have the following interrelations among the maximum modulus $M(r)$, the maximum term $\mu(r)$ and the central index $\nu(r)$ of an analytic function:

Theorem 1.4.1. *Let* $f(z) = \Sigma_{n=0}^{\infty} a_n z^n$ *be analytic in* U *and* $f(0) = 1$. *Suppose* a_n *satisfy* (1.4.3). *Then, for* $0 < r_o < r < 1$,

$$\log \mu(r) = \log \mu(r_o) + \int_{r_o}^{r} \frac{\nu(t)}{t}\, dt \tag{1.4.10}$$

and

$$\mu(r) \leq M(r) < \mu(r) \quad [1+2\nu(r + \frac{1-r}{\nu(r)})] \frac{1}{1-r} . \tag{1.4.11}$$

where $M(r) = \max_{|z|=r} |f(z)|$.

Proof. Let G_n be the ordinate of the points on the Hadamard Polygon $\pi(f)$ whose abscissa is n, n=0,1,2,... . Then, $g_n \leq G_n$ for all n. Since $f(0) = 1$ we have $G_o = 0$. Let

$$W(z) = \Sigma_{n=0}^{\infty} e^{G_n} z^n, \quad |z| < 1.$$

Since $\log (1/R_n)$, the slope of the line joining the points $A^*_{n-1} = (n-1, G_{n-1})$ and $A^*_n = (n, G_n)$, is a nonincreasing function of n and tends to zero as $n \to \infty$, it follows that W is analytic in U. The number $R_n = e^{G_{n-1} - G_n}$, n=1,2,... is called the *rectified ratio* of $|a_{n-1}|$ and $|a_n|$. Further, if $n_{k-1} < n \leq n_k$ then $R_n = R_{n_k}$.

Now, for any r in $0 \leq r < 1$,

$$\mu(r) = e^{G_{\nu(r)}} r^{\nu(r)}$$

$$= \frac{r^{\nu(r)}}{R_1 \cdots R_{\nu(r)}}$$

since $G_o = 1$. Thus,

$$\log \mu(r) = \nu(r) \log r - \Sigma_{n=1}^{\nu(r)} \log R_n$$

$$= \nu(r)\left[\log r - \log R_{\nu(r)}\right]$$

$$+ (\nu(r)-1)\left[\log R_{\nu(r)} - \log R_{\nu(r)-1}\right] + \ldots + \left[\log R_2 - \log R_1\right]$$

$$= \int_{R_{\nu(r)}}^{r} \frac{\nu(t)}{t}\, dt + \ldots + \int_{R_1}^{R_2} \frac{\nu(t)}{t}\, dt$$

$$= \int_0^r \frac{\nu(t)}{t}\, dt$$

since $\nu(r) = 0$ for r in $0 \leq r < R_1$. If r_o is such that $0 \leq r_o < r < 1$, then it follows from the above that

$$\log \mu(r) = \log \mu(r_o) + \int_{r_o}^{r} \frac{\nu(t)}{t}\, dt.$$

This proves (1.4.10).

Next, since

$$M(r) = \max_{|z|=r} |f(z)| \leq W(r),$$

we have, for any nonnegative integer S, to be chosen later,

$$M(r) \leq W(r) = \Sigma_{n=0}^{\infty} e^{G_n} r^n$$

$$\leq (1+S)\, \mu(r) + \Sigma_{p=S+1}^{\infty} e^{G_p} r^p. \qquad (1.4.12)$$

The upper estimate of the kth term in the above series is obtained as

$$e^{G_{S+k}} r^{S+k} = e^{G_{S+k} - G_{S+k-1}} \ldots . e^{G_{S+1} - G_S} e^{G_S} r^{S+k}$$

$$\leq \frac{\mu(r)\, r^k}{R_{S+1} \cdot \ldots \cdot R_{S+k}} .$$

Since R_n is a nondecreasing function of n, this gives

$$\Sigma_{p=S+1}^{\infty} e^{G_p} r^p \leq \mu(r) \Sigma_{p=0}^{\infty} \left(\frac{r}{R_{S+1}}\right)^p = \frac{\mu(r)}{1-(r/R_{S+1})} . \tag{1.4.13}$$

Now, for any r' such that $r < r' < 1$, if $S = \nu(r')$, then $r' < R_{S+1}$. This gives

$$\frac{1}{1-(r/R_{S+1})} < \frac{1}{1-(r/r')} < \frac{1}{r'-r}$$

and the inequality in (1.4.13) becomes

$$\Sigma_{p=S+1}^{\infty} e^{G_p} r^p < \frac{\mu(r)}{r'-r} .$$

Now, let r_o be such that r', defined by

$$r' = r + \frac{1-r}{\nu(r)} ,$$

satisfies $r' < 1$ for all r in $0 < r_o < r < 1$. Such an r_o exists since $\nu(r) \to \infty$ as $r \to 1^-$. The above inequality now gives that, for all r satisfying $0 < r_o < r < 1$,

$$\Sigma_{p=S+1}^{\infty} e^{G_p} r^p < \frac{\mu(r)\ \nu(r)}{1-r} . \tag{1.4.14}$$

Combining (1.4.12) and (1.4.14), for all r satisfying $0 < r_o < r < 1$,

$$M(r) < \mu(r) \left[1 + \nu(r') + \frac{\nu(r)}{1-r}\right]$$

$$< \mu(r) \left[1+2\ \nu(r+\frac{1-r}{\nu(r)})\right] \frac{1}{1-r}$$

and we get the right-hand inequality in (1.4.11). By Cauchy's inequality, $\mu(r) \leq M(r)$ for all r in $0 \leq r < 1$. This gives the left-hand inequality in (1.4.11). □

Theorem 1.4.1 gives a general relationship that exists among the maximum modulus, the maximum term and the central index of an analytic function f for all r sufficiently close to 1. Now, we obtain a sharp Wiman-Valiron-type inequality involving the maximum modulus and the maximum term of f that holds for all r outside an exceptional set.

Let S be a Lebesgue measurable subset of $-1 \leq x < 0$. The *logarithmic measure* $\ell(S)$ and the *upper logarithmic density* $\overline{d\ell}(S)$ of S are defined as

$$\ell(S) = \int_S \frac{dx}{-x} \tag{1.4.15}$$

and

$$\overline{d\ell}(S) = \limsup_{x \to 0} \frac{\ell(S,x)}{-\log(-x)}$$

where $\ell(S,x) = \ell(S \cap (-1,x))$. If limit exists on the right-hand side of (1.4.16), then $d\ell(S)$ is called the *logarithmic density* of S.

If E is a Lebesgue measurable subset of $(0,1)$ and E^* is its image under the mapping $x = \log r$, then the *logarithmic measure* $m(E)$ of E is defined as

$$m(E) = \ell(E^* \cap (-1,0)).$$

The upper logarithmic density $\overline{dm(E)}$ and the logarithmic density $dm(E)$ of E are defined as the upper logarithmic and the logarithmic density of $E^* \cap (-1,0)$ respectively.

We have

Theorem 1.4.2. *Let* $\Psi_1(y)$ *and* $\Psi_2(y)$ *be positive and increasing in* $0 < y < \infty$ *such that*

$$\int^{\infty} \frac{dy}{\Psi_j(y)} < \infty, \quad j=1,2 \tag{1.4.17}$$

and

$$\frac{\Psi_j(y)}{y}, \ j=1,2, \textit{ is greater than or equal to } 1, \textit{ is increasing and tends to infinity as } y \to \infty. \tag{1.4.18}$$

Further, let, for $A > 1$,

$$\Psi_2(y_1 y_2) \leq A(y_2 \Psi_2(y_1) + y_1 \Psi_2(y_2)). \tag{1.4.19}$$

Suppose $\Psi(y) = \Psi_2(\Psi_1(y))$. *Let* $f(z) = \Sigma_{n=0}^{\infty} a_n z^n$ *be analytic in* U *and* (1.4.3) *be satisfied. Then,*

$$\frac{M(r)}{\sqrt{\Psi(\log M(r))}} \le \frac{C\sqrt{A}}{1-r}\,\mu(r) \tag{1.4.20}$$

outside an exceptional set of (0,1) *having zero logarithmic density, where* C *is an absolute constant.*

The following lemmas are needed in the proof of Theorem 1.4.2.

Lemma 1.4.1 *Let* $f(x) = \Sigma_{n=0}^{\infty} a_n x^n$, $a_n \ge 0$ *for all* n, *be convergent for* x *in* $-1 < x < 1$. *Let* $F(x) = f(e^x)$ *and* $g(x) = \log F(x)$ *be such that* $g''(x) \ge 1$. *Then, for every* $x < 0$,

$$F(x) \le \mu(e^x)\,(\sqrt{Kg''(x)} + \frac{\sqrt{K}}{2}) \tag{1.4.21}$$

where $\mu(x) = \max_{n \ge 0} a_n x^n$ *and* K *may be chosen as* 3^3.

Proof. Let Σ' denote the summation over all n such that $|n-g'(x)| \ge C\sqrt{g''(x)}$, C being any constant greater than 1. Then, for $x < 0$,

$$C^2 g''(x)\ \Sigma' \frac{a_n e^{nx}}{F(x)} \le \Sigma'(n-g'(x))^2\ \frac{a_n e^{nx}}{F(x)}$$

$$\le \Sigma(n-g'(x))^2 \frac{a_n e^{nx}}{F(x)}$$

$$= \Sigma n^2\ \frac{a_n e^{nx}}{F(x)} - 2g'(x)\ \Sigma n\ \frac{a_n e^{nx}}{F(x)}$$

$$+ (g'(x))^2\ \Sigma\ \frac{a_n e^{nx}}{F(x)}$$

$$= \frac{F''(x)}{F(x)} - (g'(x))^2$$

$$= \frac{F''(x)}{F(x)} - (\frac{F'(x)}{F(x)})^2 = g''(x).$$

Thus,

$$\Sigma'\ \frac{a_n e^{nx}}{F(x)} \le C^{-2}\ .$$

Now, let Σ" denote the summation over all n such that $|n-g'(x)| < C\sqrt{g''(x)}$. Then, for $x < 0$,

$$\Sigma'' \frac{a_n e^{nx}}{F(x)} = 1 - \Sigma' \frac{a_n e^{nx}}{F(x)} \geq 1-C^{-2}.$$

Consequently, for all x satisfying $x < 0$,

$$F(x) \leq \frac{1}{1-C^{-2}} \Sigma'' a_n e^{nx}$$

$$\leq \left(\frac{2C\sqrt{g''(x)}+1}{1-C^{-2}}\right) \mu(e^x)$$

$$< \left(\sqrt{Kg''(x)} + \frac{\sqrt{K}}{2}\right) \mu(e^x),$$

where $K = (2C/(1-C^{-2}))^2$. The value of K in the lemma corresponds to $C=\sqrt{3}$. □

Lemma 1.4.2. *Suppose* $g(x) \geq 1$ *is a twice differentiable function for* $-1 \leq x < 0$. *Let* (i) $\lim_{x\to 0} g(x) = \infty$, (ii) $g'(x)$ *and* $g''(x)$ *be greater than* 1 *in* $-1 \leq x < 0$. *Then,*

$$g'(x) < \frac{1}{-x} \Psi_1(g(x)) \tag{1.4.22}$$

outside an exceptional set E_1 *of* $(-1,0)$ *having finite logarithmic measure, and, for any constant* $A > 1$,

$$g''(x) \leq \frac{2A}{x^2} \Psi(g(x)) \tag{1.4.23}$$

outside an exceptional set E *of* $(-1,0)$ *having zero logarithmic density. The functions* Ψ_1 *and* Ψ *in* (1.4.22) *and* (1.4.23) *satisfy the hypotheses of Theorem* 1.4.2.

Proof. Let

$$E_1 = \{x: g'(x) \geq -\frac{1}{x} \Psi_1(g(x))\}$$

and

$$E_2 = \{x: g''(x) \geq -\frac{1}{x} \Psi_2(g'(x))\}$$

where Ψ_1 and Ψ_2 are as in Theorem 1.4.2. We prove that $\ell(E_j) < \infty$ for $j=1,2$.

$$\ell(E_1) = \int_{E_1} \frac{dx}{-x} \le \int_{E_1} \frac{g'(x)dx}{\Psi_1(g(x))} \le \int_{-1}^{0} \frac{g'(x)dx}{\Psi_1(g(x))}$$

$$\le \int_{g(-1)}^{\infty} \frac{dy}{\Psi_1(y)} \le \int_{1}^{\infty} \frac{dy}{\Psi_1(y)} < \infty .$$

Similarly, using the hypotheses on g' and g", it follows that $\ell(E_2) < \infty$.

Let

$$\Gamma = \{x : g(x) \ge \frac{1}{-x}\}$$

and

$$E = \{x : g''(x) \ge \frac{2A}{x^2} \Psi(g(x))\} .$$

We first need to prove

$$E \cap \Gamma \subset E_1 \cup E_2 . \tag{1.4.24}$$

If $x \notin E_1 \cup E_2$ and $x \in \Gamma$, then

$$g'(x) < \frac{1}{-x} \Psi_1(g(x)),\ g''(x) < \frac{1}{-x} \Psi_2(g'(x))$$

and

$$g(x) \ge -\frac{1}{x}$$

Using the condition (1.4.19), we have

$$g''(x) < -\frac{1}{x} \Psi_2(g'(x)) < -\frac{1}{x} \Psi_2(-\frac{1}{x} \Psi_1(g(x)))$$

$$\le \frac{A}{x} \{-\frac{1}{x} \Psi_2(\Psi_1(g(x)) + \Psi_2(-\frac{1}{x}) \Psi_1(g(x))\}$$

$$= \frac{A}{x^2} \Psi_1(g(x)) \{\frac{\Psi_2(\Psi_1(g(x)))}{\Psi_1(g(x))} + \frac{\Psi_2(-\frac{1}{x})}{-\frac{1}{x}}\}$$

$$< \frac{2A}{x^2} \Psi_2(\Psi_1(g(x)) = \frac{2A}{x^2} \Psi(g(x))$$

since $(\Psi_2(y)/y)$ increases and $\Psi_1(g(x)) \ge g(x) \ge -(1/x)$. Thus, $x \notin E$ and (1.2.24) is proved.

Now, we observe that, in view of (1.4.24) and the fact that $\ell(E_j) < \infty$, $j = 1,2$,

$$\ell(E,x) = \ell(E-\Gamma,x) + \ell(E\cap\Gamma\ ,x)$$

$$\leq \ell(E-\Gamma,x) + \ell(E_1 \cup E_2,x)$$

$$\leq \ell(E-\Gamma,x) + B$$

where B is a constant. Thus, in order to prove that E is of zero logarithmic density, it is sufficient to prove that $E-\Gamma$ is of zero logarithmic density.

In order to prove that $E-\Gamma$ is of zero logarithmic density, for $\delta > 0$, we define

$$E_\delta^1 = \{x:g'(x) \geq \frac{\delta^2}{-x} g(x)\}$$

$$E_\delta^2 = \{x:g'(x) < \frac{\delta}{-x} g(x) \text{ and } g''(x) \geq \frac{\delta}{-x} g'(x)\}$$

and

$$E_\delta^3 = \{x:g''(x) \geq \frac{\delta^2}{x^2} g(x)\}.$$

Then, $E_\delta^1 \cap E_\delta^2 = \phi$ and $E_\delta^3 \subset E_\delta^1 \cup E_\delta^2$. Further,

$$\ell(E_\delta^1-\Gamma,x) = \int_{(E_\delta^1-\Gamma)\cap(-1,x)} \frac{dt}{-t} \leq \frac{1}{\delta} \int_{-1}^{x_1} \frac{g'(t)}{g(t)}\, dt$$

where x_1 is the least upper bound of the set $(E_\delta^1-\Gamma) \cap (-1,x)$.

Hence,

$$\ell(E_\delta^1-\Gamma,x) \leq \frac{1}{\delta} \{\log g(x_1)-\log g(-1)\}$$

$$\leq \frac{1}{\delta} \log \frac{1}{-x_1} \leq \frac{1}{\delta} \log \frac{1}{-x} .$$

Similarly,

$$\ell(E_\delta^2-\Gamma,x) \leq \frac{1}{\delta} \{\log g'(x_2)-\log g'(-1)\}$$

where x_2 is the least upper bound of the set $(E_\delta^2-\Gamma)\cap(-1,x)$. Since E_δ^1 and E_δ^2 are disjoint, it follows that, for $x > x_3(\delta)$,

$$\ell(E_\delta^2-\Gamma,x) \le \frac{1}{\delta}\{\log (\frac{\delta}{-x_2} g(x_2))-\log g'(-1)\}$$

$$< \frac{1}{\delta}\{2 \log \frac{1}{-x} + \log \frac{\delta}{g'(-1)}\}$$

$$< \frac{3}{\delta} \log \frac{1}{-x} .$$

Thus, for $x > x_3(\delta)$,

$$\ell(E_\delta^3-\Gamma,x) \le \ell(E_\delta^1-\Gamma,x) + \ell(E_\delta^2-\Gamma,x) \tag{1.4.25}$$

$$< \frac{4}{\delta} \log \frac{1}{-x} .$$

We now show that there exists a number ξ such that

$$E-\Gamma \subset [-1,\xi] \cup (E_\delta^3-\Gamma) . \tag{1.4.26}$$

Since $(y/\Psi(y)) \to 0$ as $y \to \infty$ and $g(x) \to \infty$ as $x \to 0$, we choose ξ such that

$$\frac{g(\xi)}{\Psi(g(\xi))} \le \frac{1}{\delta^2} .$$

Then, for $x \ge \xi$, $x \notin E_\delta^3 \cup \Gamma$ we have

$$g''(x) < \frac{\delta^2}{x^2} g(x) \le \frac{1}{x^2} \frac{\Psi(g(\xi))}{g(\xi)} g(x)$$

$$< \frac{1}{x^2} \Psi(g(x))$$

since $(\Psi(g(x)))/g(x)$ increases. The inclusion (1.4.26) follows from the above inequality.

Now, in view of (1.4.26) and (1.4.25),

$$\ell(E-\Gamma,x) < \log \frac{1}{-\xi} + \ell(E_\delta^3-\Gamma,x)$$

$$< \frac{5}{\delta} \log \frac{1}{-x}$$

for $x > x_4(\delta)$. Since $\delta > 0$ is arbitrary, we get that $E-\Gamma$ is of zero logarithmic density. □

Proof of Theorem 1.4.2. If we replace a_n by $|a_n|$, we do not change $\mu(r)$ and do not decrease $M(r)$. Hence, it is sufficient to prove the theorem for nonnegative coefficients. Let

$$g(x) = \log f(e^x) = \log M(e^x).$$

We assume without loss of generality that $g(x) \geq 1$ and $\lim_{x \to 0} g(x) = \infty$. Since $g(x)$ satisfies the conditions of Lemma 1.4.2, we get

$$g''(x) \leq \frac{2A}{x^2} \Psi(g(x)) \tag{1.4.27}$$

outside a set E of zero logarithmic density. Since $g(x)$ satisfies the conditions of Lemma 1.4.1 also, we get that, if $x \notin E$ and x is sufficiently close to 1, then

$$f(e^x) \leq \frac{K_1 \sqrt{2A}}{-x} \mu(e^x) \sqrt{\Psi(g(x))}.$$

This proves (1.4.20). □

EXERCISES 1.4

1. For $f(z) = \Sigma_{n=1}^{\infty} (z^n/n)$, show that the maximum term $\mu(r) = r$ and the central index $\nu(r) = 1$ for all r in $0 \leq r < 1$.
2. Let f be analytic in $\mathcal{U}$ and $f(0) = 0$. If $0 < r_1 \leq r_2 < 1^-$, then prove that the maximum term $\mu(r)$ of f satisfies $\mu(r_2)/\mu(r_1) \geq r_2/r_1$.
3. For $0 < \alpha < 1$, prove that the central index $\nu(r)$ of the function

   $$f(z) = 1 + \Sigma_{n=1}^{\infty} e^{\alpha^{-1} n^{\alpha}} z^n,$$

   analytic in $|z| < 1$, assumes all the values 0,1,2,... . Further, show that the maximum term $\mu(r)$ of f satisfies

   $$\mu(r) \sim \exp\left(\frac{1-\alpha}{\alpha}\left(-\frac{1}{\log r}\right)^{\alpha/(1-\alpha)}\right), \text{ as } r \to 1.$$

4. With f as in Exercise 3, prove that, as $r \to 1$,

   $$\Sigma_{n=0}^{\infty} e^{\alpha^{-1} n^{\alpha}} r^n \sim \frac{\sqrt{2\alpha\pi}}{1-\alpha} \left(\frac{\alpha}{1-\alpha}\right)^{(1-\alpha)/\alpha} (\log \mu(r))^{\frac{1}{2} + \frac{1-\alpha}{\alpha}} \mu(r).$$

   (Polya and Szego [1])

5. Prove that every term of the series in (1.4.1) is a maximum term for some r in $0 \leq r < 1$ if and only if $\Psi(n) = |a_{n-1}/a_n|$ forms a strictly increasing sequence of n.

6. Let f be analytic in U and a_K be the first nonzero coefficient in the Taylor series (1.4.1). Then, for $0 < r_o < r < 1$, prove that

$$\log \mu(r) = \log \mu(r_o) + \int_{r_o}^{r} \frac{\nu(t)-K}{t}\, dt .$$

7. If $0 \leq r < t < 1$, then prove that

$$\mu(r) \leq M(r) \leq \mu(r) \left(\nu(t) + \frac{t}{t-r}\right).$$

(Bogda and Shankar [1])

8. Let f be analytic in U and $\{R_n\}$ be the sequence of jump points of the central index of f arranged in increasing order of magnitude. Then show that, for any $k > 0$, the series $\Sigma_1^{\infty}(1-R_n)^{k+1}$ and the integral $\int_0^1 (1-t)^{k-1} \log \mu(t)dt$ converge or diverge together.

9. If f is analytic in U and (1.4.3) is satisfied, then, for $0 < \alpha < 1$, prove that the maximum term $\mu(r)$ of f satisfies

$$\lim_{r\to 1} \frac{\mu(\alpha r)}{\mu(r)} = 0.$$

Does the above result also hold for the maximum modulus of f on $|z| = r$?

For a function f, analytic in U and satisfying (1.4.3), let

$$\mu_k(r) = \mu(r,f^{(k)}) \text{ and } \nu_k(r) = \nu(r,f^{(k)})$$

where $f^{(k)}$, $k = 1,2,\ldots$, denotes the kth derivative of f. We calculate $\nu_k(r)$, $k = 0,1,2,\ldots$, from the first term of the series of f. Let $\mu_o(r) \equiv \mu(r)$ and $\nu_o(r) \equiv \nu(r)$. It is easily seen that

$$\nu_k(r) \geq \nu(r),\ k = 1,2,\ldots .$$

Define

$$\phi(r,k) = \nu_k(r) - \nu(r),\ k = 1,2,\ldots .$$

10. Show that for $k = 1,2,\ldots$

$$\nu_k(r) \leq r\,\frac{\mu_{k+1}(r)}{\mu_k(r)} \leq \nu_{k+1}(r)$$

and

$$\nu(r) \le r\left\{\frac{\mu_{k+1}(r)}{\mu(r)}\right\}^{1/(k+1)} \le \nu_{k+1}(r).$$

(Kapoor and Gopal [3])

The maximum term $\mu(r)$ of a function f, analytic in $|z| < 1$, is a differentiable function of r for almost all r in (0,1). Let $\mu^{(k)}(r)$ denote the kth derivative of $\mu(r)$ at the points of existence.

11. Prove that

$$r\,\frac{\mu^{(k)}(r)}{\mu^{(k-1)}(r)} + k-1 = \nu(r)$$

and

$$\mu^{(1)}(r) \le \mu(r,f^{(1)}).$$

(Kapoor and Gopal [3])

12. Let $f(z) = \exp\,(1-z)^{-\rho}$, $0 < \rho < \infty$, $a(r) = d\log M(r)/d\log r$ and $b(r) = da(r)/d\log r$. Then, as $r \to 1$,

$$\mu(r) \sim \frac{f(r)}{\sqrt{2\pi b(r)}} = \frac{1}{\sqrt{2\pi\rho(\rho+1)}}\,(1-r)^{1+\frac{\rho}{2}}\,f(r).$$

(Hayman [1])

The asymptotic relation in Exercise 12 shows that the order of estimate in (1.4.20) is the best possible in the sense that $(1-r)^{-1}$ on the right-hand side of inequality (1.4.20) cannot be replaced by $(1-r)^{-\alpha}$ for any α satisfying $0 < \alpha < 1$.

1.5 GROWTH PARAMETERS.

The growth of a function f, analytic in U, is quite effectively and conveniently measured by comparing the growth of its maximum modulus function $M(r) = \max_{|z|=r} |f(z)|$ with the growth of $\exp\,(\beta(1-r)^{-\alpha})$, $0 < \alpha,\beta < \infty$ as $r \to 1$. Thus, if there exists a positive number α such that

$$M(r) < \exp\left(\frac{1}{1-r}\right)^{\alpha}$$

for all r in $0 \le r_o < r < 1$, then f is said to be of finite order. The greatest lower bound of the numbers α satisfying the above inequality is called the *order* of f. If no number satisfying this inequality can be found, then f is said to be of infinite order. Equivalently, the order $\rho, 0 \le \rho \le \infty$, of a function f, analytic in U, is defined as

$$\rho = \limsup_{r \to 1} \frac{\log^+ \log^+ M(r)}{-\log (1-r)} . \tag{1.5.1}$$

The *lower order* λ, $0 \leq \lambda \leq \rho \leq \infty$ of a function f, analytic in U, is similarly defined as

$$\lambda = \liminf_{r \to 1} \frac{\log^+ \log^+ M(r)}{-\log (1-r)} . \tag{1.5.2}$$

For the comparison of rates of growth of analytic functions having the same nonzero finite order, the concepts of type and lower type are introduced. Thus, a function f, analytic in U and having order ρ, $0 < \rho < \infty$, is said to be of *type* T and *lower type* t, $0 \leq t \leq T \leq \infty$, if

$$T = \limsup_{r \to 1} \frac{\log M(r)}{(1-r)^{-\rho}} \tag{1.5.3}$$

and

$$t = \liminf_{r \to 1} \frac{\log M(r)}{(1-r)^{-\rho}} . \tag{1.5.4}$$

The function f is said to be of *regular growth* if $0 < \lambda = \rho < \infty$ and of *perfectly regular growth* if $0 < t = T < \infty$. It is easily seen that if f is of perfectly regular growth then it is of regular growth.

The function $\exp (\beta(1-z)^{-\alpha})$, $0 < \alpha, \beta < \infty$, with an analytic branch of $(1-z)^{-\alpha}$ in U and taking real values for real z, is analytic in U, is of regular growth and has order α. In fact, this function is of perfectly regular growth and has type β. The function $(1-z)^{-1}$ has order zero and the function $\exp (\exp (1-z)^{-1})$ has order infinity.

The following theorems give relations between the growth parameters defined by (1.5.1) to (1.5.4) and the maximum term or the central index of a function, analytic in U. We assume in the sequel that the Taylor series coefficients of f satisfy (1.4.3) so that $\log^+$ in (1.5.1) and (1.5.2) can be replaced by log.

Theorem 1.5.1. *Let f be analytic in U and have order ρ and lower order λ. Then,*

$$\frac{\rho}{\lambda} = \lim_{r\to 1} \frac{\sup}{\inf} \frac{\log \log \mu(r)}{-\log (1-r)}. \tag{1.5.5.}$$

Proof. Let lim sup and lim inf on the right-hand side of (1.5.5) be denoted by ϕ_1 and ϕ_2 respectively. In view of the inequality $\mu(r) \leq M(r)$ for all r in $0 \leq r < 1$, $\phi_1 \leq \rho$ and $\phi_2 \leq \lambda$ are obvious. We prove the inequalities $\rho \leq \phi_1$ and $\lambda \leq \phi_2$.

Using (1.4.10), for any $b > 2$ there exists an $r_o(b)$ such that, for all r satisfying $0 \leq r_o(b) < r < 1$,

$$(1-r)\,\nu(r)/b \leq \nu(r) \log \frac{r+(1-r)/b}{r} \tag{1.5.6}$$

$$\leq \int_r^{r+(1-r)/b} \frac{\nu(t)}{t}\, dt \leq \log \mu(r+(1-r)/b).$$

It follows from inequality (1.5.6) that

$$\nu(r+(1-r)/b) \leq \frac{\log \mu(r + \frac{1-r}{b} + \frac{1-r}{b}(1 - \frac{1}{b}))}{((1-r) - (1-r)/b)/b} \tag{1.5.7}$$

$$\leq \frac{b^2}{(b-1)(1-r)} \log \mu(r+2\,\frac{1-r}{b}).$$

Now, from (1.4.11) and (1.5.7), for all r satisfying $0 \leq r_o < r < 1$,

$$\log M(r) < \log \mu(r) + \log \frac{1}{1-r} + \log 3 + \log \nu(r + \frac{1-r}{b}) \tag{1.5.8}$$

$$\leq \log \mu(r) + 2 \log \frac{1}{1-r} + \log 3 + \log \frac{b^2}{b-1}$$

$$+ \log \log \mu(r + 2\,\frac{1-r}{b})$$

$$\leq 5(\log \mu(r+2\,\frac{1-r}{b})) \log \frac{1}{1-r}.$$

It follows easily from (1.5.8) that $\rho \leq \phi_1$ and $\lambda \leq \phi_2$. This completes the proof of the theorem. □

Theorem 1.5.2. *Let* f *be analytic in* $\mathcal{U}$ *and have order* ρ *and lower order* λ. *Then,*

$$1 + \rho = \limsup_{r \to 1} \frac{\log \nu(r)}{-\log (1-r)} \tag{1.5.9}$$

and

$$\lambda \le \liminf_{r \to 1} \frac{\log \nu(r)}{-\log (1-r)} \le 1+\lambda. \tag{1.5.10}$$

Proof. From (1.5.6) and (1.5.5), we have

$$\lim_{r\to 1} \begin{matrix}\sup\\ \inf\end{matrix} \frac{\log \nu(r)}{-\log (1-r)} \le \begin{matrix}1+\rho\\ 1+\lambda\end{matrix} . \tag{1.5.11}$$

Now, from the definition of order, for any $\varepsilon > 0$, there exists $r_o(\varepsilon)$ such that for $0 \le r_o(\varepsilon) < r < 1$,

$$\log M(r) < (1-r)^{-(\rho+\varepsilon)} .$$

Using Cauchy inequality, for all n and all r in $0 \le r_o(\varepsilon) < r < 1$,

$$\log |a_n| < (1-r)^{-(\rho+\varepsilon)} + n(1-r).$$

Let $(1-r) = ((\rho+\varepsilon)/n)^{1/(\rho+\varepsilon+1)}$. Then, the above inequality gives that, for all $n > n_o(\varepsilon)$,

$$\log |a_n| < K n^{(\rho+\varepsilon)/(\rho+\varepsilon+1)}$$

where $K = (1+\rho+\varepsilon)(\rho+\varepsilon)^{-(\rho+\varepsilon)/(\rho+\varepsilon+1)}$. Suppose $\mu(r) = |a_{\nu(r)}| r^{\nu(r)}$. Then, it follows that for all r in $0 \le r_o(\varepsilon) < r < 1$,

$$\log \mu(r) \le \log |a_{\nu(r)}| \le K(\nu(r))^{(\rho+\varepsilon)/(\rho+\varepsilon+1)} .$$

Now, using Theorem 1.5.1, we get

$$1 + \rho \le \limsup_{r \to 1} \frac{\log \nu(r)}{-\log(1-r)} .$$

Combining (1.5.11) and the above inequality we get (1.5.9).

To prove the remaining inequality in (1.5.10), we observe that from (1.4.10), for $\frac{1}{2} < r_o < r < 1$,

$$\log \mu(r) < \log \mu(r_o) + \nu(r) \log 2$$

so that

$$\lambda \leq \liminf_{r \to 1} \frac{\log \nu(r)}{-\log (1-r)} .$$

This inequality together with (1.5.11) gives (1.5.10). □

The equality $1+\lambda = \liminf_{r \to 1} ((\log \nu(r))/(-\log (1-r)))$ in (1.5.10) need not always hold. For, consider

$$f_o(z) = \Sigma_{n=0}^{\infty} \exp (\lambda_n) z^{\lambda_n^2}$$

where $\lambda_o > 1$ and $\lambda_{n+1} = \lambda_n^2$. Then, f_o is analytic in U and the function

$$\Psi(n) = |a_{n-1}/a_n|^{1/(\lambda_n - \lambda_{n-1})} = \exp (-(\lambda_n + \lambda_{n-1})^{-1})$$

is increasing. Further, the maximum term $\mu(r)$ and the central index $\nu(r)$ of f_o are given by

$$\log \mu(r) = \lambda_n + \lambda_{n+1} \log r, \ \nu(r) = \lambda_{n+1}$$

for r satisfying $\Psi(n) \leq r < \Psi(n+1)$. Since $-\log (1-r) \sim -\log \log (1/r)$ as $r \to 1$, for r sufficiently close to 1,

$$\frac{1}{2} \leq \frac{\log \log \mu(r)}{-\log (1-r)} \leq 1$$

and the lower bound is attained at $r = \Psi(n)$. It follows therefore, in view of (1.5.5), that $\lambda = \frac{1}{2}$. However, for r sufficiently close to 1,

$$1 \leq \frac{\log \nu(r)}{-\log (1-r)} \leq 2$$

and the lower bound is attained at $r = \Psi(n)$, so that

$$\liminf_{r \to 1} \frac{\log \nu(r)}{-\log (1-r)} = 1.$$

Thus, strict inequality holds on the right-hand inequality in (1.5.10).

Theorem 1.5.3. *Let* f *be analytic in* U *and have nonzero finite order. If*

$$\lim_{r \to 1} \frac{\log M(r)}{-\log (1-r)} = \infty$$

then, as $r \to 1$

$$\log \mu(r) \sim \log M(r).$$

Proof. Since $\mu(r) \le M(r)$ we clearly have

$$\limsup_{r \to 1} \frac{\log \mu(r)}{\log M(r)} \le 1.$$

Thus, we have only to show that

$$\liminf_{r \to 1} \frac{\log \mu(r)}{\log M(r)} \ge 1.$$

To prove this inequality we use the first inequality in (1.5.8) and (1.5.9). By (1.5.9), for any $\varepsilon > 0$ and r sufficiently close to 1,

$$\log \nu(r) < -(1 + \rho + \varepsilon) \log (1-r).$$

Using the above inequality in (1.5.8), for r sufficiently close to 1,

$$\log M(r) < \log \mu(r) + (2+\rho+\varepsilon) \log \frac{1}{1-r} + \log 3(1 - \frac{1}{b})^{-(1+\rho+\varepsilon)} .$$

Thus, in view of

$$\lim_{r\to 1} \frac{\log M(r)}{-\log (1-r)} = \infty$$

we easily have from the above inequality that

$$\liminf_{r \to 1} \frac{\log \mu(r)}{\log M(r)} \ge 1. \square$$

Corollary 1. *Let* f *be analytic in* U *having order* ρ *and lower order* λ *such that* $0 < \lambda \le \rho < \infty$. *Then, as* $r \to 1$,

$$\log \mu(r) \sim \log M(r).$$

Corollary 2. *Let* f *be analytic in* U *having nonzero finite order* ρ, *type* T *and lower type* t. *If*

$$\limsup_{r \to 1} \frac{\log M(r)}{-\log (1-r)} = \infty$$

then,

$$\begin{matrix} T \\ t \end{matrix} = \lim_{r\to 1} \begin{matrix} \sup \\ \inf \end{matrix} \frac{\log \mu(r)}{(1-r)^{-\rho}} .$$

The following theorem gives a sharp Wiman-Valiron type of inequality for functions of nonzero finite order.

Theorem 1.5.4. *Let* f *be analytic in* U *and have nonzero finite order* ρ. *Then,*

$$\frac{M(r)}{\sqrt{\log M(r)}} \leq \frac{Cw}{1-r} \sqrt{\rho(\rho+1)} \quad \mu(r) \tag{1.5.12}$$

outside an exceptional set of upper logarithmic density not greater than 1/w, C *being a constant.*

We need the following lemma:

Lemma 1.5.1. *Let* $g(x) \geq 1$ *be an increasing differentiable function for* $-1 \leq x < 0$ *that satisfies*

$$g(x) \leq \frac{B}{(-x)^{\alpha}}$$

for $-1 \leq x_o < x$, B *being a constant. Then, outside an exceptional set* E *such that*

$$\ell(E,x) < (\frac{2}{w} + \varepsilon) \log (1/-x)$$

for $-1 < x_o \leq x_1(\varepsilon) < x$,

$$g'(x) < \frac{w\alpha}{-x} g(x) \tag{1.5.13}$$

and if, further, $g'(x)$ *is increasing, differentiable and* $g'(x) \geq 1$ *in* $-1 \leq x < 0$, *then outside* E

$$g''(x) < \frac{w^2\alpha(\alpha+1)}{x^2} \; g(x). \tag{1.5.14}$$

Proof. Let

$$E_1 = \{x:g'(x) \geq \frac{w\alpha}{-x} g(x)\}$$

Then

$$\begin{aligned}\ell(E_1,x) = \int_{E_1 \cap [-1,x]} \frac{dt}{-t} &\leq \frac{1}{w\alpha} \int_{E_1 \cap [-1,x]} \frac{g'(t)}{g(t)} dt \\ &\leq \frac{1}{w\alpha} \int_{-1}^{x} \frac{g'(t)}{g(t)} dt \\ &= \frac{1}{w\alpha} \{\log g(x) - \log g(-1)\} \\ &< \frac{1}{w\alpha} \log g(x) \\ &\leq \frac{1}{w} \left[\frac{\log B}{\alpha} + \log \frac{1}{-x}\right] \\ &< (\log \frac{1}{-x}) \; (\frac{1}{w} + \varepsilon)\end{aligned}$$

for $-1 \leq x_o(\varepsilon) < x$. This proves (1.5.13). For x not in E_1,

$$g'(x) < \frac{w\alpha B}{(-x)^{\alpha+1}}$$

and a repeated application gives (1.5.14). □

Proof of Theorem 1.5.4. As in the proof of Theorem 1.4.2, we can replace a_n by $|a_n|$ and prove the theorem for nonnegative coefficients.

Let

$$g(x) = \log f(e^x) = \log M(e^x)$$

and assume that $g(x) \geq 1$ and $\lim_{x \to 0} g(x) = \infty$. Since f is of finite order ρ, we have

$$g(x) < \left(\frac{2}{-x}\right)^{2\rho}$$

for $-1 \leq x_o < x$, since $-x/2 < 1-e^x$ for $-1 \leq x < 0$. Using Lemma 1.5.1, with $\alpha = 2\rho$, we have, for x not in E,

$$g''(x) < \frac{w^2 2\rho(2\rho+1)}{x^2} \; g(x).$$

The above inequality combined with Lemma 1.4.1 gives that outside a set E_1 of logarithmic density not greater than 1/w,

$$f(e^x) < \mu(e^x)\left[\frac{2w\sqrt{K\rho(\rho+1)}}{-x} \; \sqrt{g(x)} + \frac{\sqrt{K}}{2}\right].$$

The inequality (1.5.12) outside E follows from the above inequality. □

<u>EXERCISES 1.5</u>

1. Let ρ,λ,T and t denote the order, lower order, type and lower type respectively of an analytic function f in $\mathcal{U}$ and let ρ',λ',T' and t' denote the corresponding quantities for its derivative f'. Show that
$$\rho = \rho', \quad \lambda = \lambda', \quad T = T' \quad \text{and} \quad t = t'.$$
2. Let f_1 and f_2 be analytic in $\mathcal{U}$ having order ρ_1 and ρ_2 respectively. Show that if ρ and ρ* denote the orders of f_1+f_2 and f_1f_2 respectively, then
$$\rho \leq \max(\rho_1,\rho_2)$$
and
$$\rho^* \leq \max(\rho_1,\rho_2)$$
with equality if $\rho_1 \neq \rho_2$. What is the corresponding result for types?
3. Extend the result of Exercise 2 to the case of lower orders and lower types.

4. Let f be analytic in $\mathcal{U}$ and have order ρ and lower order λ. Then, for $k=1,2,\ldots$, prove that

$$1+\rho = \limsup_{r\to 1} \frac{\log \nu(r)}{-\log(1-r)} = \limsup_{r\to 1} \frac{\log (r\{\mu_k(r)/\mu(r)\}^{1/k})}{-\log(1-r)}$$

and

$$\lambda \leq \liminf_{r\to 1} \frac{\log \nu(r)}{-\log(1-r)} = \liminf_{r\to 1} \frac{\log (r\{\mu_k(r)/\mu(r)\}^{1/k}}{-\log(1-r)} \leq 1+\lambda$$

where $\mu_k(r)$ is the maximum term of the kth derivative of f for $|z| = r < 1$, (Kapoor and Gopal [3])

5. Let, for a function f, analytic in $\mathcal{U}$, having order ρ and lower order λ,

$$\begin{matrix}\alpha_k \\ \beta_k\end{matrix} = \lim_{r\to 1} \begin{matrix}\sup \\ \inf\end{matrix} \frac{\phi(r,k)}{(1-r)^{-1}}, \quad k = 1,2,\ldots$$

where $\phi(r,k)$ is as defined in Exercises 1.4. Then prove that

$$k(\rho+1) = k \limsup_{r\to 1} \frac{\log \nu(r)}{-\log(1-r)} \leq \alpha_k$$

and

$$\beta_k \leq k \liminf_{r\to 1} \frac{\log \nu(r)}{-\log(1-r)} \leq k(\lambda+1).$$

Further, show that

$$k(\rho+1) = \limsup_{r\to 1} \frac{1}{\log(1-r)^{-1}} \int_{r_o}^{r} \frac{\phi(x,k)}{x}\,dx$$

$$k(\lambda+1) = \liminf_{r\to 1} \frac{1}{\log(1-r)^{-1}} \int_{r_o}^{r} \frac{\phi(x,k)}{x}\,dx.$$

(Kapoor and Gopal [3])

6. For a function f analytic in $\mathcal{U}$ having order ρ and lower order λ, show that

$$\liminf_{r\to 1} \frac{(1-r)\,\nu(r)}{\log \mu(r)} \leq \lambda \leq \rho \leq \limsup_{r\to 1} \frac{(1-r)\,\nu(r)}{\log \mu(r)}.$$

7. For a function analytic in $\mathcal{U}$ having nonzero finite order ρ, define

$$\begin{matrix}\gamma \\ \delta\end{matrix} = \lim_{r\to 1} \begin{matrix}\sup \\ \inf\end{matrix} \frac{\nu(r)}{r(1-r)^{-\rho-1}}.$$

Then, prove that

$$\frac{\delta}{\gamma\rho} \leq \liminf_{r\to 1} \frac{\log \mu(r)}{(1-r)\,\nu(r)} \leq \limsup_{r\to 1} \frac{\log \mu(r)}{(1-r)\,\nu(r)} \leq \frac{\gamma}{\delta\rho}.$$

8. Show that if f is analytic in $\mathcal{U}$ having order ρ, $0 < \rho < \infty$, and type T, then

$$\frac{(\rho+1)^{\rho+1}}{\rho^{\rho}} T = \limsup_{r \to 1} \frac{\nu(r)}{(1-r)^{-\rho-1}} \left[1 + \frac{\log \mu(r)}{(1-r)\,\nu(r)}\right]^{\rho+1}.$$

(Srivastava and Juneja [1])

Does the corresponding result hold for the lower type also?

9. Suppose f is analytic in $\mathcal{U}$ and has nonzero finite order ρ, type T and lower type t. Let γ and δ be as defined in Exercise 7. Prove that

$$\delta \leq \rho t \leq \rho\delta \left[\left(\frac{\rho+1}{\rho}\right)\left(\frac{\rho T}{\delta}\right)^{1/(\rho+1)} - 1\right] \leq \rho T$$

$$\delta \leq \rho t \leq \rho\gamma \left[\left(\frac{\rho+1}{\rho}\right)\left(\frac{\rho T}{\gamma}\right)^{1/(\rho+1)} - 1\right] \leq \rho T$$

$$\rho T \leq \gamma \leq \gamma\left(\frac{\rho\gamma+\rho t}{\rho\gamma+\delta}\right)^{\rho+1} \leq \rho T \left(\frac{\rho+1}{\rho}\right)^{\rho+1}\left(1 + \frac{\delta}{\rho\gamma}\right)^{-\rho-1}.$$

With the help of the above inequalities show that

$$0 < \gamma = \delta < \infty \quad \text{if and only if } 0 < T = t < \infty.$$

(Srivastava and Juneja [1])

*10. Are the inequalities in Exercise 9 sharp?

11. Let $f(z) = \Sigma_{n=0}^{\infty} a_n z^n$ and $g(z) = \Sigma_{n=0}^{\infty} b_n z^n$ be analytic in $\mathcal{U}$. Suppose the Taylor series of their convolution $f * g$ defined by

$$(f * g)(z) = \Sigma_{n=0}^{\infty} a_n b_n z^n$$

has radius of convergence 1. Then show that the order ρ and lower order λ of $f * g$ satisfy

$$2(\rho+1) = \limsup_{r \to 1} \frac{\log (\mu(r,f' * g')/\mu(r,f * g))}{-\log (1-r)}$$

$$2(\lambda+1) \geq \liminf_{r \to 1} \frac{\log (\mu(r,f' * g'))/\mu(r,f * g))}{-\log (1-r)}$$

(Bogda and Shankar [2])

where f' and g' denote the derivatives of f and g respectively.

A function f analytic in $\mathcal{U}$ is said to be of *index* q, q=2,3,... if its q-*order* $\rho(q)$ defined by

$$\rho(q) = \limsup_{r \to 1} \frac{\log^{[q]} M(r)}{-\log (1-r)}$$

satisfies $\rho(q) < \infty$ and $\rho(q-1) = \infty$, where $\log^{[0]} M(r) = M(r)$ and $\log^{[q]} M(r) = \log (\log^{[q-1]} M(r))$, q=1,2,3,... . For the function f having index q, define its *lower* q-*order* $\lambda(q)$ by

$$\lambda(q) = \liminf_{r \to 1} \frac{\log^{[q]} M(r)}{-\log(1-r)} .$$

12. For a function f analytic in U and having index q, $\rho(q) > 0$, $q = 2,3,\ldots$, prove that

$$\frac{\rho(q)}{\lambda(q)} = \lim_{r\to 1} \frac{\sup}{\inf} \frac{\log^{[q]} \mu(r)}{-\log(1-r)}$$

and

$$\rho(q) + A(q) = \limsup_{r \to 1} \frac{\log^{[q-1]} \nu(r)}{-\log(1-r)}$$

where $A(q) = 1$ if $q = 2$ and $A(q) = 0$ if $q=3,4,\ldots$. Further, if $q \geq 3$, then

$$\lambda(q) = \liminf_{r \to 1} \frac{\log^{[q-1]} \nu(r)}{-\log(1-r)} .$$

(Kapoor and Gopal [1])

13. Let f be analytic in U and have the *rational order* $\rho_o (< \infty)$, defined as

$$\rho_o = \limsup_{r \to 1} \frac{\log^+ M(r)}{-\log(1-r)} .$$

Prove that

$$M(r) \leq \frac{C\omega\sqrt{\rho_o}}{1-r} \mu(r)$$

outside an exceptional set of upper logarithmic density not greater than $1/\omega$, where C is a constant.

(Kovari [1])

14. Let f be analytic in U having order ρ and lower order λ. Suppose the Nevanlinna order σ and lower Navanlinna order ζ of f are defined as

$$\frac{\sigma}{\zeta} = \lim_{r\to 1} \frac{\sup}{\inf} \frac{\log T(r,f)}{-\log(1-r)} .$$

Then prove that

$\sigma \leq \rho \leq \sigma + 1$

$\zeta \leq \lambda \leq \zeta + 1$

and that

(i) if $\rho = \sigma$ and $\sigma = \zeta$, then $\rho = \lambda$;

(ii) if $\lambda = \zeta$ and $\rho = \lambda$, then $\sigma = \zeta$.

(Sons [1]; see also Linden [3])

A function f, analytic in U and having q-order $\rho(q) > 0$, $q = 2,3,\ldots$, is said to be of *q-type* $T(q)$ and *lower q-type* $t(q)$ if

$$\begin{matrix} T(q) \\ t(q) \end{matrix} = \lim_{r\to 1} \begin{matrix} \sup \\ \inf \end{matrix} \frac{\log^{[q-1]} M(r)}{(1-r)^{-\rho(q)}} .$$

15. Show that

$$\begin{matrix} T(q) \\ t(q) \end{matrix} = \lim_{r\to 1} \begin{matrix} \sup \\ \inf \end{matrix} \frac{\log^{[q-1]} \mu(r)}{(1-r)^{-\rho(q)}} .$$

A function f, analytic in U, is said to be of *slow growth* if its order $\rho = 0$. To study precisely the rates of growth of such functions, the *logarithmic order* ρ_o and the *lower logarithmic order* λ_o of f are defined as

$$\begin{matrix} \rho_o \\ \lambda_o \end{matrix} = \lim_{r\to 1} \begin{matrix} \sup \\ \inf \end{matrix} \frac{\log^+ \log^+ M(r)}{\log \log \frac{1}{1-r}} .$$

If $0 < \rho_o < \infty$, the *logarithmic type* T_o and *lower logarithmic type* t_o of f are defined as

$$\begin{matrix} T_o \\ t_o \end{matrix} = \lim_{r\to 1} \begin{matrix} \sup \\ \inf \end{matrix} \frac{\log M(r)}{(\log \frac{1}{1-r})^{\rho_o}} .$$

16. Show that, if $1 \le \lambda_o \le \rho_o < \infty$, then

$$\begin{matrix} \rho_o \\ \lambda_o \end{matrix} = \lim_{r\to 1} \begin{matrix} \sup \\ \inf \end{matrix} \frac{\log^+ \log^+ \mu(r)}{\log \log (\frac{1}{1-r})} . \qquad \text{(Gopal [1])}$$

17. Show that, if $1 < \rho_o < \infty$, then

$$\begin{matrix} T_o \\ t_o \end{matrix} = \lim_{r\to 1} \begin{matrix} \sup \\ \inf \end{matrix} \frac{\log \mu(r)}{(\log \frac{1}{1-r})^{\rho_o}} . \qquad \text{(Gopal [1])}$$

Let L^o be the class of all functions h such that $h(x)$ is positive, continuous and strictly increasing on $[a,\infty)$, $h(x) \to \infty$ as $x \to \infty$ and

$$\lim_{x\to\infty} \frac{h(x(1+\delta(x)))}{h(x)} = 1$$

for every $\delta(x)$ such that $\delta(x) \to 0$ as $x \to \infty$.

Let Λ^o be the class of all functions h such that $h(x)$ is positive, continuous and strictly increasing on $[a,\infty)$ $\alpha(x) \to \infty$, and

$$\lim_{x\to\infty} \frac{\alpha(cx)}{\alpha(x)} = 1, \text{ for all } c, \quad 0 < c < \infty.$$

It is easily seen that $\Lambda^o \subset L^o$.

A function f, analytic in U, is said to be of (α,β)-*order* $\rho(\alpha,\beta)$ and *lower* (α,β)-*order* $\lambda(\alpha,\beta)$, if

(i) $\lim_{x\to\infty} \frac{\alpha(x/F(x,c))}{\alpha(x)} = 1$, for every c, $0 < c < \sigma$,

where $F(x,c) = \beta^{-1}(c(\alpha(x)))$; $\sigma = \infty$ if $\alpha \not\equiv \beta$ and $\sigma=1$ if $\alpha\equiv\beta\not\equiv\log x$

or

(ii) $\alpha(x) \equiv \beta(x) = \log x$

and

$$\begin{matrix}\rho(\alpha,\beta) \\ \lambda(\alpha,\beta)\end{matrix} = \lim_{r\to 1} \begin{matrix}\sup \\ \inf\end{matrix} \frac{\alpha(\log M(r))}{\beta(\frac{1}{1-r})} .$$

18. Show that the following pairs of functons satisfy conditions (i) and (ii):

(I) $\alpha(x) = \log^{[p+q]} x$, $\beta(x) = (\log^{[q]} x)^d$; $p \geq 1$, $q \geq 0$, $0 < d < \infty$.

(II) $\alpha(x) = \log x$, $\beta(x) = (\log x)^d$, $1 < d < \infty$.

(III) $\alpha(x) = \log^{[p]} x$, $\beta(x) = (\log^{[p]} x)^d$, $p \geq 2$, $1 \leq d < \infty$.

(IV) $\alpha(x) = (\log x)^d$ or $\alpha(x) = (\log^{[p]} x)^{d'}$, $\beta(x) = \exp(\log x)^d$, $0 < d < 1$, $0 < d' < \infty$, $p \geq 2$.

(V) $\alpha(x) = \log x$, $\beta(x) = \exp(\log^{[2]} x)^d$, $1 < d < \infty$.

19. Let β be positive, continuous and strictly increasing on $[a,\infty)$. Let $\beta(x) \to \infty$ as $x \to \infty$. Then $\beta \in L^o$, if and only if,

$$\lim_{c\to 1} \{\liminf_{x\to\infty} \frac{\beta(cx)}{\beta(x)}\} = 1$$

(Kapoor and Nautiyal [2])

20. Prove that, if $\alpha \not\equiv \beta$,

$$\begin{matrix}\rho(\alpha,\beta) \\ \lambda(\alpha,\beta)\end{matrix} = \lim_{r\to 1} \begin{matrix}\sup \\ \inf\end{matrix} \frac{\alpha(\log \mu(r))}{\beta(\frac{1}{1-r})} = \lim_{r\to 1} \begin{matrix}\sup \\ \inf\end{matrix} \frac{\alpha(\nu(r))}{\beta(\frac{1}{1-r})} .$$

(Nautiyal [1])

21. Prove that, if $\alpha(x) \equiv \beta(x) \not\equiv \log x$ and $\rho(\alpha,\alpha)\{\lambda(\alpha,\alpha)\} \geq 1$, then

$$\rho(\alpha,\alpha)\{\lambda(\alpha,\alpha)\} = \max\left(1, \limsup_{r\to 1} \{\inf\} \frac{\alpha(\log \mu(r))}{\alpha(\frac{1}{1-r})}\right)$$

$$= \max\left(1, \limsup_{r\to 1} \{\inf\} \frac{\alpha(\nu(r))}{\alpha(\frac{1}{1-r})}\right) .$$

(Nautiyal [1])

Let Λ denote the class of differentiable functions in Λ^o. The *generalized order* $\rho(h)$ and *lower generalized order* $\lambda(h)$ of a function f, analytic in U, with respect to a function $h \in \Lambda$ is defined as

$$\begin{matrix}\rho(h) \\ \lambda(h)\end{matrix} = \lim_{r\to 1} \begin{matrix}\sup \\ \inf\end{matrix} \frac{h(\log^+ M(r))}{h(\log \frac{1}{1-r})}.$$

22. Suppose that $h \in \Lambda$ and $F(x,c) = h^{-1}(ch(\log x))$, $1 \leq c < \infty$, satisfies

(i) $\frac{dF(x,c)}{dx} \leq \frac{F(x,c)}{x}$ for $x > x_o(c)$

and

(ii) $\frac{F(x,c)}{x} \to 0$ as $x \to \infty$.

Then the generalized order $\rho(h)$ (> 1) and the lower generalized order $\lambda(h)$ (> 1) of a function f, analytic in U, are given by

$$\begin{matrix}\rho(h) \\ \lambda(h)\end{matrix} = \lim_{r\to 1} \begin{matrix}\sup \\ \inf\end{matrix} \frac{h(\log \mu(r))}{h(\log \frac{1}{1-r})}.$$

23. Show that the functions

$$h_1(x) = \log^{[p]} x, \ p \geq 1$$

and

$$h_2(x) = \exp (\log x)^{\alpha}, \ 0 < \alpha < 1$$

are in Λ and satisfy the conditions (i) and (ii) of Exercises 22.

1.6 PROXIMATE ORDER

To obtain a more refined measure of growth of an analytic function than is given by the growth parameters in Section 1.5, we consider a real valued function $\rho(r)$ in $0 < r < 1$ having the following properties:

$\rho(r)$ is positive and differentiable in $0 \leq r_o < r < 1$, (1.6.1)

$\rho(r) \to \rho$ as $r \to 1$, where $0 < \rho < \infty$, (1.6.2)

$\lim_{r\to 1} -\rho'(r)\,(1-r)\log(1-r) = 0$, where $\rho'(r)$ denotes the derivative of $\rho(r)$. (1.6.3)

A function $\rho(r)$ satisfying the conditions (1.6.1) to (1.6.3) is said to be a *proximate order*.

For a function f, analytic in U having nonzero finite order ρ, let

$$\begin{matrix}T^* \\ t^*\end{matrix} = \lim_{r\to 1} \begin{matrix}\sup \\ \inf\end{matrix} \frac{\log M(r)}{(1-r)^{-\rho(r)}} \quad (1.6.4)$$

where $M(r) = \max_{|z|=r} |f(z)|$ and ρ in (1.6.2) is the order of f. The numbers T^* and t^* are said to be *type and lower type of* f *with respect to the proximate order* $\rho(r)$. If T^* is different from zero and infinity, then the function $\rho(r)$ satisfying (1.6.1) to (1.6.3) is called a proximate order of f.

It is easily seen then if $\rho(r)$ is a proximate order of a function f, analytic in $\mathcal{U}$, then $\rho(r) + \log T^*/-\log (1-r)$, $0 < T^* < \infty$, is also a proximate order for f. This shows that the proximate order of an analytic function is not unique.

The following theorem shows that there exists a proximate order for every function, analytic in $\mathcal{U}$ and having nonzero finite order:

Theorem 1.6.1. *Let* f *be analytic in* $\mathcal{U}$ *and have order* ρ, $0 < \rho < \infty$. *Then, for every* T^*, $0 < T^* < \infty$, *there exists a proximate order of* f *satisfying* (1.6.1) *to* (1.6.4).

Proof. Let

$$p(r) = (1-r)^{-\rho} \log M(r)/T^*$$

where $M(r) = \max_{|z|=r} |f(z)|$. If we put $x = -\log (1-r)$ and $p_1(x) = \log p(1-e^{-x})$, then

$$\limsup_{x \to \infty} \frac{p_1(x)}{x} = 0. \tag{1.6.5}$$

First, suppose $\limsup_{x \to \infty} p_1(x) = \infty$. Let $y = q(x)$ be the boundary curve of the smallest convex domain containing the curve $y = p_1(x)$ and the positive ray of x-axis. By making suitable changes in the small neighbourhoods of the vertices in this curve, we may assume without loss of generality that the function $q(x)$ is differentiable in $0 \le x < \infty$. The curve $y = q(x)$ has the following properties:

the curve $y = q(x)$ is concave in the sense that a chord joining any two points of the curve lies below the curve; (1.6.6)

the function $q(x)/x$ is monotonic decreasing and non-negative it follows therefore that this function must tend to a limit as $x \to \infty$. Since the curves $y = q(x)$ and $y = p_1(x)$ have infinitely many common points $\{x_n\}$ such that $x_n \to \infty$, using (1.6.5) we get

$$\lim_{x\to\infty} \frac{q(x)}{x} = 0; \tag{1.6.7}$$

$$p_1(x) \leq q(x) \text{ for all } x \geq 0 . \tag{1.6.8}$$

As a consequence of (1.6.7), we have

$$\lim_{x\to\infty} q'(x) = 0. \tag{1.6.9}$$

Now, (1.6.8) implies

$$\log M(r) \leq T^*(1-r)^{-\rho - \frac{q(-\log(1-r))}{-\log (1-r)}} . \tag{1.6.10}$$

Set

$$\rho(r) = \rho + \frac{q(-\log (1-r)}{-\log (1-r)} . \tag{1.6.11}$$

It is clear that $\rho(r)$ is positive and differentiable in $0 \leq r_o < r < 1$. Using (1.6.7), if follows that $\rho(r) \to \rho$ as $r \to 1$. Further,

$$-(1-r)\, \rho'(r)\log(1-r) = q'(-\log(1-r)) - \frac{q(-\log(1-r))}{-\log (1-r)}$$

so that, in view of (1.6.7) and (1.6.9), $-(1-r)\, \rho'(r)\log(1-r) \to 0$ as $r \to 1$. Finally, by (1.6.10) and (1.6.7), we get

$$\log M(r) \leq T^*(1-r)^{-\rho(r)}$$

for all r in $0 \leq r_o < r < 1$ and that there exists a sequence $r_n \to 1$ as $n \to \infty$ on which $\log M(r_n) = T^*(1-r_n)^{-\rho(r_n)}$. Thus, $\rho(r)$, defined by (1.6.11), is a proximate order of f.

To reduce the general case to the above case we construct a polygonal curve $\bar{q}(x)$ as follows. Let d_1 be the segment of the line $y = -\varepsilon_1 x$, $\varepsilon_1 > 0$, from origin to the point x_1 such that

$$q(x_1) > -\varepsilon_1 x_1 + 1.$$

Choose a positive number $0 < \varepsilon_2 < \varepsilon_1$ and from the point $(x_1, -\varepsilon_1 x_1)$ draw a line segment d_2 of the line

$$y + \varepsilon_1 x_1 = -\varepsilon_2(x_2 - x_1)$$

to a point $x_2 > x_1$ at which

$$q(x_2) > -\varepsilon_1 x_1 - \varepsilon_2(x_2 - x_1) + 2$$

and continue this process. The positive numbers $\{\varepsilon_n\}_{n=1}^{\infty}$ satisfy $\varepsilon_1 > \varepsilon_2 > \ldots > \varepsilon_n > \ldots$ and $\varepsilon_n \to 0$ as $n \to \infty$. The points $\{x_n\}$ are chosen so that

$x_n \to \infty$ as $n \to \infty$.

Clearly, the polygonal curve $\bar{q}_1(x)$ constructed in this way satisfies

$$\lim_{x\to\infty} \frac{\bar{q}_1(x)}{x} = 0.$$

Using the arguments in (1.6.7) and (1.6.9), it is easily seen that the function $q_1(x) = -\bar{q}_1(x)$ has the following properties:

$$\lim_{x\to\infty} \frac{q_1(x)}{x} = 0 ; \tag{1.6.12}$$

$$\lim_{x\to\infty} q_1'(x) = 0 ; \tag{1.6.13}$$

$$\limsup_{x \to \infty} \left[p_1(x) + q_1(x)\right] = \infty . \tag{1.6.14}$$

Now, as in the first case, we construct a concave majorant $q_2(x)$ for $p_1(x) + q_1(x)$. Set, $q(x) = q_2(x)-q_1(x)$ and

$$\rho^*(r) = \rho + \frac{q(-\log (1-r))}{-\log (1-r)} . \tag{1.6.15}$$

Now, using (1.6.12) to (1.6.14) instead of using (1.6.7) to (1.6.9) in deriving the properties of the function $\rho(r)$ as before, we find that the function $\rho^*(r)$ also satisfies the properties (1.6.1) to (1.6.3) and T^* defined by (1.6.4) with $\rho^*(r)$ in place of $\rho(r)$ is nonzero finite. Thus, $\rho^*(r)$ given by (1.6.15) serves as a proximate order of f in the general case. □

In order to study the properties of a proximate order of an analytic function, we need the concept of slowly increasing function. A real valued function $L(r)$, $0 < r < 1$, is said to be *slowly increasing* if, for every k satisfying $1 < k < \infty$,

$$\lim_{r\to 1} \frac{L(r + \frac{1}{k} (1-r))}{L(r)} = 1. \tag{1.6.16}$$

We have the following theorem:

Theorem 1.6.2. *Let* $\rho(r)$ *be a proximate order of a function* f, *analytic in* U *and having order* ρ. *Then,*

$$L(r) = (1-r)^{-\rho(r)+\rho} \text{ is a slowly increasing function of } r \text{ in } 0 < r < 1. \tag{1.6.17}$$

and

$(1-r)^{-\rho(r)}$ *is a monotonically increasing function of* r *in* $0 \le r_o < r < 1$ *and tends to* ∞ *as* $r \to 1$. (1.6.18)

Proof. For $k > 1$, using the mean value theorem and (1.6.3), there exists $k' > 1$ such that for any $\varepsilon > 0$,

$$\log \frac{L(r + \frac{1}{k}(1-r))}{L(r)} = [\rho(r + \frac{1}{k}(1-r))-\rho] - [\log(1 - \frac{1}{k}) - \log(1-r)] + (\rho(r)-\rho)\log(1-r)$$

$$= -[\rho(r + \frac{1}{k}(1-r))-\rho]\log(1 - \frac{1}{k}) - [\rho(r + \frac{1}{k}(1-r))-\rho(r)]\log(1-r)$$

$$< -[\rho(r + \frac{1}{k}(1-r))-\rho]\log(1 - \frac{1}{k}) + \frac{\varepsilon\log(1-r)}{k(1-k')\log((1 - \frac{1}{k'})(1-r))}.$$

It follows now from the above inequality that

$$\lim_{r\to 1} \log \frac{L(r + \frac{1}{k}(1-r))}{L(r)} = 0$$

so that (1.6.17) is proved.

To prove that $(1-r)^{-\rho(r)}$ is monotonically increasing, we have

$$\frac{d}{dr}[(1-r)^{-\rho(r)}] = \rho(r)(1-r)^{-\rho(r)-1} - \rho'(r)(1-r)^{-\rho(r)}\log(1-r)$$

$$> (\rho-\varepsilon)(1-r)^{-\rho(r)-1} > 0$$

since (1.6.3) is satisfied. This proves (1.6.18). □

Theorem 1.6.3 . *For* $\rho - 1 > \lambda \ge -1$ *and* $1 > r > \alpha > 0$,

$$\int_\alpha^r (1-t)^{-\rho(t)+\lambda}dt = \frac{(1-r)^{-\rho(r)+\lambda+1}}{\rho-\lambda-1} + o((1-r)^{-\rho(r)+\lambda+1}) . \quad (1.6.19)$$

Proof.

$$\int_\alpha^r (1-t)^{-\rho(t)+\lambda}\, dt = \int_\alpha^r (1-t)^{\lambda-\rho}\, (1-t)^{\rho-\rho(t)}\, dt$$

$$= \frac{1}{\rho-\lambda-1} \Big[(1-t)^{-\rho(t)+\lambda+1}\Big|_\alpha^r$$

$$- \int_\alpha^r (1-t)^{-\rho(t)+\lambda}\{-(1-t)\rho'(t) \log (1-t) + (\rho(t)-\rho)\}dt\Big] .$$

From (1.6.2) and (1.6.3), we have asymptotically

$$|\rho(t) - \rho| < \varepsilon/2$$

and

$$|-(1-t)\, \rho'(t) \log (1-t)| < \varepsilon/2.$$

Hence

$$\int_\alpha^r (1-t)^{-\rho(t)+\lambda} dt = \frac{(1-r)^{-\rho(r)+\lambda+1}}{\rho-\lambda-1}(1+o(1))-o(1) \int_\alpha^r (1-t)^{-\rho(t)+\lambda} dt.$$

Thus,

$$\int_\alpha^r (1-t)^{-\rho(t)+\lambda} dt = \frac{(1-r)^{-\rho(r)+\lambda+1}}{\rho-\lambda-1} + o((1-r)^{-\rho(r)+\lambda+1}).$$

This gives (1.6.19). □

Let

$$\begin{matrix}\gamma \\ \delta\end{matrix} = \lim_{r\to 1} \begin{matrix}\sup \\ \inf\end{matrix} \frac{\nu(r)}{r(1-r)^{-\rho(r)-1}} . \tag{1.6.20}$$

Then we have the following theorem.

Theorem 1.6.4. *Let* f *be analytic in* U *and have order* $\rho, 0 < \rho < \infty$, *and proximate order* $\rho(r)$. *Let* T^*, t^* *and* γ, δ *be given by* (1.6.4) *and* (1.6.20) *respectively. Then,*

$$\delta \le \rho\Big[\Big(\frac{\gamma\rho+\delta}{\gamma\rho+\gamma}\Big)^{\rho} \Big(\frac{\delta}{\rho} + \frac{\gamma-\delta}{\rho+1}\Big)\Big] \le \rho T^* \le \gamma \tag{1.6.21}$$

and

$$\delta \le \rho t^* \le \rho\delta\Big[\Big(\frac{1+\rho}{\rho}\Big) \Big(\frac{\gamma}{\delta}\Big)^{1/(\rho+1)} -1\Big] \le \gamma . \tag{1.6.22}$$

Further, $0 < \delta \le \gamma < \infty$, *if and only if,* $0 < t^* \le T^* < \infty$.

We need the following lemma:

Lemma 1.6.1. *Let* $f(z) = \Sigma_{n=0}^{\infty} a_n z^n$ *be analytic in* $\mathbb{U}$ *and have order* ρ, $0 < \rho < \infty$, *and proximate order* $\rho(r)$. *Then,*

$$\frac{T^*}{t^*} = \lim_{r\to 1} \frac{\sup}{\inf} \frac{\log \mu(r)}{(1-r)^{-\rho(r)}}$$

where $\mu(r) = \max_{n \geq 0} \{|a_n| r^n\}$.

Proof. From (1.4.11), we have, for $0 < r_o < r < 1$,

$$\log M(r) \leq \log \mu(r) + \log \left[\{1 + 2\nu(r + \frac{1-r}{\nu(r)})\} \frac{1}{1-r}\right].$$

Further, for any $\varepsilon > 0$, (1.5.9) gives that

$$\nu(r) < (1-r)^{-(1+\rho+\varepsilon)}$$

for all r in $0 < r_1 < r < 1$. Let $r' = \max (r_o, r_1)$. Then, for $0 < r' < r < 1$,

$$\log M(r) < \log \mu(r) + (1+\rho+\varepsilon) \log \frac{\nu(r)}{\nu(r)-1} - (2+\rho+\varepsilon)\log(1-r) + O(1).$$

Dividing by $(1-r)^{-\rho(r)}$ and proceeding to limits as $r \to 1$, we have

$$\frac{T^*}{t^*} \leq \lim_{r\to 1} \frac{\sup}{\inf} \frac{\log \mu(r)}{(1-r)^{-\rho(r)}}.$$

The reverse inequalities follow from the relation $\mu(r) \leq M(r)$. □

Proof of Theorem 1.6.4. For given $\varepsilon > 0$, by (1.6.20), we have

$$\nu(r) > (\delta-\varepsilon)\, r(1-r)^{-\rho(r)-1}$$

for all r in $0 < r_o(\varepsilon) < r < 1$. Further, for any $k > 1$,

$$\int_r^{r+\frac{1}{k}(1-r)} \frac{\nu(t)}{t} \, dt > \nu(r) \log(1 + \frac{1}{k}\frac{(1-r)}{r}) > \nu(r)\frac{1}{k}(1-r). \qquad (1.6.23)$$

Thus, (1.4.10) gives

$$\log \mu(r + \frac{1}{k}(1-r)) = \log \mu(r_o) + \int_{r_o}^{r} \frac{\nu(t)}{t}\,dt + \int_{r}^{r+\frac{1}{k}(1-r)} \frac{\nu(t)}{t}\,dt$$

$$> \log \mu(r_o)+(\delta-\varepsilon)\int_{r_o}^{r} (1-t)^{-\rho(t)-1}dt+\nu(r)\frac{1}{k}(1-r).$$

Using (1.6.19), with $\lambda = -1$, we get from the above inequality

$$\log \mu(r + \frac{1}{k}(1-r)) > \log \mu(r_o) + \frac{\delta-\varepsilon}{\rho}(1-r)^{-\rho(r)} + o((1-r)^{-\rho(r)}) + \nu(r)\,\frac{1}{k}\,(1-r) \quad .$$

Dividing by $(k/(k-1))^{\rho(r)}(1-r)^{-\rho(r)}$ and proceeding to limits, we get, with the help of Lemma 1.6.1, that

$$T^* \geq \frac{\delta}{\rho}\left(\frac{k-1}{k}\right)^{\rho} + \frac{\gamma}{k}\left(\frac{k-1}{k}\right)^{\rho} \tag{1.6.24}$$

and

$$t^* \geq \frac{\delta}{\rho}\left(\frac{k-1}{k}\right)^{\rho} + \frac{\delta}{k}\left(\frac{k-1}{k}\right)^{\rho} . \tag{1.6.25}$$

Further, since

$$\int_{r}^{r+\frac{1}{k}(1-r)} \frac{\nu(t)}{t}\,dt \leq \nu(r + \frac{1}{k}(1-r))\log(1 + \frac{1}{k}\,\frac{(1-r)}{r})$$

$$< \nu(r + \frac{1}{k}(1-r))\,\frac{1}{k}\,\frac{(1-r)}{r} \tag{1.6.26}$$

proceeding as above, we have for all r in $0 < r_o(\varepsilon) < r < 1$,

$$\log \mu(r + \frac{1}{k}(1-r)) \leq \log \mu(r_o) + \frac{\gamma+\varepsilon}{\rho}\,(1-r)^{-\rho(r)}+o((1-r)^{-\rho(r)}) + \nu(r + \frac{1}{k}(1-r))\,\frac{1}{k}\,\frac{(1-r)}{r} \quad .$$

This inequality gives that

$$T^* \leq \frac{\gamma}{\rho}\left(\frac{k-1}{k}\right)^{\rho} + \frac{\gamma}{k-1} \tag{1.6.27}$$

and

$$t^* \leq \frac{\gamma}{\rho}\left(\frac{k-1}{k}\right)^{\rho} + \frac{\delta}{k-1} . \tag{1.6.28}$$

Since (1.6.25) and (1.6.27) hold for all $k > 1$, we get $\delta' \leq \rho t^*$ and $\gamma \geq \rho T^*$. With $k = (\gamma(\rho+1))/(\gamma-\delta)$, (1.6.24) gives

$$T^* \geq \left(\frac{\gamma\rho+\delta}{\gamma\rho+\gamma}\right)^{\rho} \left[\frac{\delta}{\rho} + \frac{\gamma-\delta}{\rho+1}\right] \geq \frac{\delta}{\rho} .$$

Putting $(k-1)/k = (\delta/\gamma)^{1/(\rho+1)}$, (1.6.28) gives

$$t^* \leq \frac{\gamma}{\rho} \left(\frac{\delta}{\gamma}\right)^{\rho/(\rho+1)} + \delta\{\left(\frac{\gamma}{\delta}\right)^{1/(\rho+1)} - 1\}$$

$$= \delta \left[\left(\frac{1+\rho}{\rho}\right) \left(\frac{\gamma}{\delta}\right)^{1/(\rho+1)} -1\right] \leq \delta/\rho .$$

This completes the proof of (1.6.21) and (1.6.22).

To prove the remaining part of the theorem, we observe that if $\delta > 0$, then by (1.6.25), $t^* > 0$ and if $\gamma < \infty$, then by (1.6.27), $T^* < \infty$. Also (1.6.24) shows that $T^* < \infty$ implies $\gamma < \infty$. Finally, if $t^* > 0$ then $\delta > 0$ because if $\delta = 0$, then by (1.6.28) we have

$$t^* < \frac{\gamma}{\rho} \left(\frac{k-1}{k}\right)^{\rho} .$$

Since this holds for every $k > 1$, making $k \to 1$, we have $t^* = 0$, a contradiction. □

Theorem 1.6.5. *Let* f *be analytic in* U *and have order* ρ, $0 < \rho < \infty$ *and proximate order* $\rho(r)$. *Then, as* $r \to 1$,

$$\nu(r) \sim \gamma r(1-r)^{-\rho(r)-1} \tag{1.6.29}$$

if and only if

$$\log M(r) \sim \frac{\gamma}{\rho} (1-r)^{-\rho(r)} \tag{1.6.30}$$

where γ *is defined in* (1.6.20).

Proof. First suppose that $\nu(r) \sim \gamma r(1-r)^{-\rho(r)-1}$ as $r \to 1$. Then, it follows immediately from Theorem 1.6.4 that $T^* = t^* = \gamma/\rho$ and consequently (1.6.30) follows.

Next assume that (1.6.30) holds. Then $T^* = t^* = \gamma/\rho$. Using Lemma 1.6.1,

$$\log \mu(r) \sim (1-r)^{-\rho(r)}\, T^*$$

as $r \to 1$. By (1.4.10) and the above asymptotic relation,

$$\int_r^{r+\frac{1}{k}(1-r)} \frac{\nu(t)}{t}\,dt = \log\mu(r+\frac{1}{k}(1-r)) - \log\mu(r)$$

$$\sim (1-r)^{-\rho(r)} \left[\left(\frac{k}{k-1}\right)^{\rho(r)} - 1\right] T^* .$$

The above relation and (1.6.23) give asymptotically

$$\frac{\nu(r)}{r(1-r)^{-\rho(r)-1}} \le \frac{k}{r} \left[\left(\frac{k}{k-1}\right)^{\rho(r)} - 1\right] T^*$$

so that

$$\limsup_{r\to 1} \frac{\nu(r)}{r(1-r)^{-\rho(r)-1}} \le k \left[\left(\frac{k}{k-1}\right)^{\rho} - 1\right] T^* .$$

Letting $k \to \infty$ on the right-hand side of this inequality, we get

$$\gamma \le \rho T^* . \tag{1.6.31}$$

Again, using (1.6.26) and proceeding as above,

$$\nu(r+\frac{1}{k}(1-r))\frac{1}{k}\frac{(1-r)}{r} > (1-r)^{-\rho(r)} \left[\left(\frac{k}{k-1}\right)^{\rho(r)} - 1\right] T^* .$$

Dividing by $(1-r)^{-\rho(r)}(1/k)(k/(k-1))^{\rho+1}$ and taking limit, we have

$$\liminf_{r\to 1} \frac{\nu(r)}{r(1-r)^{-\rho(r)-1}} \ge kT^* \frac{[(k/(k-1))^{\rho} - 1]}{(k/(k-1))^{\rho+1}} .$$

Now, letting $k \to \infty$, we have

$$\delta \ge \rho T^* . \tag{1.6.32}$$

Since $\delta \le \gamma$, (1.6.31) and (1.6.32) imply $\delta = \gamma = \rho T^*$ and consequently (1.6.29) holds. □

EXERCISES 1.6

1. Let $\rho(r)$ be a proximate order and $t = (1-r)^{-\rho(r)-1}$ if and only if $(1-r)^{-1} = \chi(t)$. Show that

$$\lim_{t\to\infty} \frac{d(\log \chi(t))}{d(\log t)} = \frac{1}{\rho+1}$$

and, for $0 < \eta < \infty$,

$$\lim_{t\to\infty} \frac{\chi(\eta t)}{\chi(t)} = \eta^{1/(\rho+1)} .$$

2. Prove that, for any $k > 1$,

$$\lim_{r\to 1} \frac{[(k/(k-1)/(1/(1-r))]^{\rho(r + \frac{1}{k}(1-r))}}{(1/(1-r))^{\rho(r)}} = (\frac{k}{k-1})^{\rho} .$$

(Kapoor [1])

3. Prove that $\log\log \mu(r)/-\log(1-r)$ is a proximate order for those functions analytic in U whose central index $\nu(r)$ satisfies $\nu(r) \sim r(1-r)^{-\rho-1}$ as $r \to 1$.

From Theorem 1.6.4, it follows that $\gamma \leq ((\rho+1)^{\rho+1}/\rho^{\rho})\, T^*$ since, $T^* \geq ((\gamma\rho+\delta)^{\rho+1}/(\gamma^{\rho}\, \rho(\rho+1)^{\rho+1})) \geq \gamma\rho^{\rho}/(\rho+1)^{\rho+1}$. (Also see Kapoor [2] and Bajpai and Tanne [1] for this inequality and other related results.) However, this inequality could be sharpened and something more could be said about the improved inequality.

4. For a function f, analytic in U having nonzero finite order ρ, let T^* be defined by (1.6.4) and γ, δ be defined by (1.6.20). Then, prove that

$$\gamma + \delta \leq \frac{(\rho+1)^{\rho+1}}{\rho^{\rho}} T^*$$

$$\delta \leq \rho T^*$$

and that the equality cannot simultaneously hold in the above two inequalities. (Kapoor [4])

5. For a function f, analytic in U, having nonzero finite order ρ, let T^*, t^* be defined by (1.6.4) and γ, δ be defined by (1.6.20). Then show that

$$\gamma + \delta + t^* \leq \gamma + (\rho+1)\, t^* \leq \frac{(\rho+1)^{\rho+1}}{\rho^{\rho}} T^*$$

$$((\rho+1)\delta/\rho^{\rho+1}) + T^* \geq ((\rho+1)^{\rho+1}/\rho^{\rho}) \quad t^* .$$

Hence show that if equality holds in the first inequality in Exercise 4, then $t^* = 0$. (Kapoor [4])

A function $T(r)$ is said to be a *proximate type* for a function f, analytic in U, having order ρ, $0 < \rho < \infty$ and type T, $0 < T < \infty$, if it satisfies the following properties:

(i) $T(r)$ is real and differentiable in $0 < r_0 < r < 1$.

(ii) $T(r) \to T$ as $r \to 1$

(iii) $(1-r)\, T'(r) \to 0$ as $r \to 1$, where $T'(r)$ denotes the derivative of $T(r)$.

(iv) $\limsup_{r \to 1} [M(r)/\exp\{(1-r)^{-\rho} T(r)\}] = 1$, where $M(r) = \max_{|z|=r} |f(z)|$.

6. Prove that there exists a proximate type for every function f, analytic in U, having nonzero finite order and nonzero finite type.

By Hadamard's three circles theorem, it follows that, if f is analytic in U, then log M(r) is a convex function of log r in $0 < r < 1$. Thus we have

$$\log M(r) = \log M(r_o) + \int_{r_o}^{r} x^{-1} \omega(x)dx, \quad 0 < r_o < r < 1,$$

where $\omega(x)$ is a positive, continuous and piecewise differentiable function of x.

7. Prove that if $\lim_{r\to 1} (1-r)^{\rho+1} \omega(r)$ exists, then $(1-r)^{\rho} \log M(r)$ is a proximate type.

1.7. ANALYTIC FUNCTIONS HAVING PRESCRIBED ASYMPTOTIC GROWTH

There exist functions analytic in U whose upper rates of growth are arbitrarily fast and simultaneously whose lower rates of growth are arbitrarily slow. In this direction we have the following theorems.

Theorem 1.7.1. *Let* $\lambda(r)$ *and* $\mu(r)$ *be positive functions of* r *for* $0 \leq r < 1$ *such that* $\log \lambda(r) \neq O(-\log(1-r))$ *as* $r \to 1^-$. *Then, there exists a function* f, *analytic in* U *with nonnegative coefficients and two sequences* $\{S_n\}$ *and* $\{t_n\}$ *of positive numbers tending monotonically to* 1, *such that, for every positive integer* n, $M(S_n,f) > \mu(S_n)$ *and* $M(t_n,f) < \lambda(t_n)$, *where* $M(r,f) = \max_{|z|=r} |f(z)|$, $0 \leq r < 1$.

Proof. Let $S_1 = 1/e$ and λ_1 be a positive integer such that $2^{\lambda_1} > \mu(S_1)$. Due to the condition on $\lambda(r)$ in the theorem, it is possible to find a positive number $r_1 > S_1$ satisfying

$$\left(\frac{2(1-S_1)}{1-r_1}\right)^{\lambda_1} < \lambda(r_1) - 1.$$

Let $n > 1$. Define S_n inductively by

$$\frac{1}{1-S_n} = 4\,\frac{1}{1-r_{n-1}} .$$

Choose a positive integer $\lambda_n > \lambda_{n-1}$ such that $2^{\lambda_n} > \mu(S_n)$. Let r_n be a positive number greater than S_n satisfying

$$\sum_{m=1}^{n} \left[\frac{2(1-S_m)}{1-r_n}\right]^{\lambda_m} < \lambda(r_n) - 1$$

which is possible since $\log \lambda(r) \neq O(-\log(1-r))$ as $r \to 1$.

Now, define f by

$$f(z) = \sum_{n=1}^{\infty} \left(\frac{2(1-S_n)}{1-z}\right)^{\lambda_n} \qquad z \in U.$$

Then, f is analytic in U and has positive coefficients in its Taylor series expansion. Further, since $M(r,f) = f(r)$,

$$f(S_n) > 2^{\lambda_n} > \mu(S_n)$$

and

$$f(r_n) = \sum_{m=1}^{n} \left(\frac{2(1-S_m)}{1-r_n}\right)^{\lambda_m} + \sum_{m=n+1}^{\infty} \left(\frac{2(1-S_m)}{1-r_n}\right)^{\lambda_m}$$

$$< \lambda(r_n) - 1 + 2^{-\lambda_{n+1}} + 2^{-\lambda_{n+2}} + \ldots$$

$$\leq \lambda(r_n) - 1 + 2^{-1} + 2^{-2} + \ldots$$

$$= \lambda(r_n).$$

This completes the proof of the theorem. □

Theorem 1.7.2. *Let* $\mu(r)$ *be a positive function of* r *in* $0 \leq r < 1$ *such that it is bounded on every subinterval of* $(0,1)$. *Then, there exists a function* f, *analytic in* U, *and having nonnegative coefficients, such that* $M(r,f) > \mu(r)$ *for* $0 < r < 1$, *where* $M(r,f) = \max_{|z|=r} |f(z)|$.

Proof. We may assume without loss of generality that $\mu(r)$ is nondecreasing. Let, for every positive integer n, $\{\lambda_n\}$ be an increasing sequence of positive integers satisfying

$$\left(\frac{n+1}{n}\right)^{\lambda_n} > \mu\left(\frac{n+1}{n+2}\right).$$

Define f by

$$f(z) = 1 + \mu\left(\frac{1}{2}\right) + \sum_{n=1}^{\infty} \left|\frac{1}{n(1-z)}\right|^{\lambda_n}, \qquad z \in U.$$

Then, f is analytic in U and has nonnegative coefficients. Let r satisfy $1/2 < r < 1$ and put

$$m = \left[\frac{1}{1-r}\right] - 1.$$

Then

$$M(r) = f(r) > \left(\frac{1}{m(1-r)}\right)^{\lambda_m} \geq \left(\frac{m+1}{m}\right)^{\lambda_m}$$

$$> \mu\left(\frac{m+1}{m+2}\right)$$

$$= \mu\left(1 - \frac{1}{m+2}\right) \geq \mu(r).$$

For $0 < r \leq \frac{1}{2}$, $M(r) > \mu(\frac{1}{2}) \geq \mu(r)$. □

Theorem 1.7.3. *Let* $\lambda(r)$ *be a positive function of* r *in* $0 \leq r < 1$ *such that*

$$\lim_{r \to 1} \frac{\log \lambda(r)}{-\log (1-r)} = \infty .$$

Then, there exists a function f, *analytic in* U *and having nonnegative coefficients such that* $M(r,f) < \lambda(r)$, $0 < r < 1$, *where* $M(r,f) = \max_{|z|=r} |f(z)|$.

Proof. Let m_n be an increasing sequence of positive numbers satisfying

$$m_n^n > \frac{2^n}{(1-r)^n \lambda(r)}$$

for $n = 1,2,\ldots$ and $0 < r < 1$. This is possible in view of the condition on $\lambda(r)$ in the theorem, since $\lambda(r)$ has a positive lower bound. Define f by

$$f(z) = \sum_{n=1}^{\infty} \left(\frac{1}{m_n(1-z)}\right)^n, \qquad z \in U.$$

Then, f is analytic in U and has nonnegative coefficients. Further

$$M(r,f) = f(r) = \sum_{n=1}^{\infty} \left(\frac{1}{m_n(1-r)}\right)^n < \sum_{n=1}^{\infty} \frac{\lambda(r)}{2^n} = \lambda(r). \ \square$$

Our final theorem in this direction shows that there exist functions analytic in U which have arbitrarily prescribed rates of growth on two different although unspecified sequences tending to 1.

Theorem 1.7.4. *Let* $\lambda(r)$ *and* $\mu(r)$ *be positive and continuous functions of* r *for* $0 < r < 1$ *such that* $\log \lambda(r) \neq O(-\log(1-r))$ *and* $\log \mu(r) \neq O(-\log(1-r))$ *as* $r \to 1^-$. *Then, there exists a function* f, *analytic in* U *and having nonnegative coefficients such that* $M(r,f) = \lambda(r)$ *on some sequence of values of* $r \to 1^-$ *and* $M(r,f) = \mu(r)$ *on another such sequence, where* $M(r,f) = \max_{|z|=r} |f(z)|$.

Proof. Let $\alpha(r) = \min(\mu(r), \lambda(r))$ and $\beta(r) = \max(\mu(r), \lambda(r))$. First, let $\log \alpha(r) \neq O(-\log(1-r))$ as $r \to 1$. Then, by Theorem 1.7.1, there exists a function f, analytic in U and having nonnegative coefficients and two sequences $\{S_n\}$ and $\{r_n\}$ of positive numbers tending monotonically to 1 such that, for every positive integer n, $M(S_n,f) > \beta(S_n)$ and $M(r_n,f) < \alpha(r_n)$. Hence, $M(S_n,f) > \mu(S_n)$, $M(r_n,f) < \mu(r_n)$ and $M(S_n,f) > \lambda(S_n)$, $M(r_n,f) < \lambda(r_n)$. Thus, by the continuity of $M(r,f)$, $\lambda(r)$ and $\mu(r)$, it follows that $M(r,f) = \mu(r)$ on one sequence of values tending to 1 and $M(r,f) = \lambda(r)$ on another such sequence.

Next, suppose that $\log \alpha(r) = O(-\log(1-r))$ as $r \to 1^-$. We can find a pair $\mu^*(r)$ and $\lambda^*(r)$ of continuous functions on $0 < r < 1$ which are bounded away from zero, such that

$$\lim_{r\to 1} \frac{\log \mu^*(r)}{-\log(1-r)} = \lim_{r\to 1} \frac{\log \lambda^*(r)}{-\log(1-r)} = \infty$$

and there exist two sequences $\{S_n^*\}$ and $\{r_n^*\}$ tending to 1 such that

$$\mu(S_n^*) = \mu^*(S_n^*) \text{ and } \lambda(r_n^*) = \lambda^*(r_n^*).$$

Let $\delta(r) = \min(\mu^*(r), \lambda^*(r))$. Then by Theorem 1.7.3, there exists a function f, analytic in U and having nonnegative coefficients, such that $M(r,f) < \delta(r)$ for $0 < r < 1$. Now

$$\lim_{n\to\infty} \frac{\log \mu(S_n^*)}{-\log(1-S_n^*)} = \lim_{n\to\infty} \frac{\log \lambda(r_n^*)}{-\log(1-r_n^*)} = \infty .$$

But

$$\frac{\log \alpha(S_n^*)}{-\log(1-S_n^*)} \quad \text{and} \quad \frac{\log \alpha(r_n^*)}{-\log(1-r_n^*)}$$

are bounded as $n \to \infty$. Hence

$$\frac{\log \mu(r_n^*)}{-\log(1-r_n^*)} \quad \text{and} \quad \frac{\log \lambda(S_n^*)}{-\log(1-S_n^*)}$$

are also bounded. Since f is not a rational function,

$$\lim_{n\to\infty} \frac{\log M(S_n^*,f)}{-\log(1-S_n^*)} = \lim_{n\to\infty} \frac{\log M(r_n^*,f)}{-\log(1-r_n^*)} = \infty .$$

Therefore,

$$M(S_n^*,f) > \lambda(S_n^*) \quad \text{and} \quad M(r_n^*,f) > \mu(r_n^*).$$

But

$$M(S_n^*,f) < \delta(S_n^*) \le \mu^*(S_n^*) = \mu(S_n^*)$$

and

$$M(r_n^*,f) < \delta(r_n^*) \le \delta^*(r_n^*) = \delta(r_n^*).$$

Now, using the continuity of the maximum modulus $M(r,f)$ and the functions $\lambda(r)$ and $\mu(r)$, we get $M(r,f) = \mu(r)$ on one sequence tending to 1 and $M(r,f) = \lambda(r)$ on another such sequence. □

In the above theorems we have proved the existence of functions, analytic in U, which can grow arbitrarily fast or arbitrarily slowly on some sequences approaching 1. We now prove the existence of functions, analytic in U, which have arbitrarily prescribed asymptotic growth for *all* r close to 1.

Theorem 1.7.5. *Let* $\phi(r)$ *be increasing and convex in* $\log r$ *for* $0 < r < 1$ *such that*

$$\lim_{r\to 1} \phi(r) = \infty ; \tag{1.7.1}$$

then there exists a function f, *analytic in* U, *such that*

$$\log M(r,f) \sim \phi(r) \text{ as } r \to 1 \tag{1.7.2}$$

where $M(r,f) = \max_{|z|=r} |f(z)|$.

Proof. Since $\phi(r)$ is increasing and convex in $\log r$, one has

$$\phi(r) = \int_0^r \frac{\Psi(t)}{t}\, dt$$

where $\Psi(t)$ is a nondecreasing, unbounded function of t on (0,1). We may assume that $\Psi(t)$ is a continuous, strictly increasing, unbounded function of t satisfying $\Psi(t) = 0$ for $0 \leq t \leq \lambda$ for some $\lambda (0 < \lambda < 1)$. There is no loss of generality in assuming this, since, given a function of this kind, there exists a function $\phi(r)$ satisfying the hypotheses of the theorem to which it is asymptotic as $r \to 1$.

Let $0 < r_1 < r_2 < \ldots < 1$ be the sequence defined by $\Psi(r_n) = n$. Then $r_n \to 1$ as $n \to \infty$. We define F by

$$F(z) = \sum_{n=1}^{\infty} \frac{z^n}{r_1 r_2 \ldots r_n} = \sum_{n=1}^{\infty} a_n z^n, \qquad z \in \mathcal{U}.$$

Since $r_n \to 1$ as $n \to \infty$, it follows that F is analytic in $\mathcal{U}$. If $\mu(r,F)$ denotes the maximum term of F(z) for $|z| = r$, then, for $r_n \leq r \leq r_{n+1}$, we have

$$\mu(r,F) = \frac{r^n}{r_1 r_2 \ldots r_n} .$$

It is easy to check that, for $r > r_1$,

$$A(r_1) + \log \frac{r_1}{r} + \phi(r) \leq \log \mu(r,F) \leq A(r_1) + \phi(r) \tag{1.7.3}$$

where $A(r_1)$ is a constant depending on r_1. Inequality (1.7.3) easily leads to

$$\log \mu(r,F) \sim \phi(r) \quad \text{as } r \to 1. \tag{1.7.4}$$

We now define a sequence of integers $\lambda_1 < \lambda_2 < \ldots$ as follows. Let $\lambda_1 = 1$ and assume that $\lambda_1, \lambda_2, \ldots, \lambda_n$ have been specified. If

$$a_{\lambda_n+1} r_{\lambda_n+1}^{\lambda_n+1} > 2a_{\lambda_n} r_{\lambda_n}^{\lambda_n}$$

take $\lambda_{n+1} = \lambda_n + 1$; otherwise take λ_{n+1} to be the largest integer m for which

$$a_m r_m^m \leq 2a_{\lambda_n} r_{\lambda_n}^{\lambda_n} .$$

Since $\mu(r)$ is a strictly increasing, unbounded function of r, $\mu(r_n) = a_n r_n^n \to \infty$ and so the sequence $\{\lambda_n\}$ is well-defined. Define f by

$$f(z) = \sum_{n=1}^{\infty} \frac{a_{\lambda_n} z^{\lambda_n}}{n^2}, \qquad z \in U. \tag{1.7.5}$$

Since $\lambda_n \geq n$, it follows that f is analytic in U. We show that f is the required function.

We first show that

$$\log n = o(\log (a_{\lambda_n} r_{\lambda_n}^{\lambda_n})) \qquad \text{as } n \to \infty . \tag{1.7.6}$$

From the construction of λ_n it follows that

$$a_{\lambda_{n+2}} r_{\lambda_{n+2}}^{\lambda_{n+2}} > 2a_{\lambda_n} r_{\lambda_n}^{\lambda_n}, \qquad n \geq 1. \tag{1.7.7}$$

In view of the above, it is easy to see that the series $\sum_{n=1}^{\infty} (a_{\lambda_n} r_{\lambda_n}^{\lambda_n})^{-\delta}$ converges for every $\delta > 0$. Since the series consists of positive, decreasing terms, by a theorem of Abel one has

$$\lim_{n\to\infty} n(a_{\lambda_n} r_{\lambda_n}^{\lambda_n})^{-\delta} = 0$$

which gives (1.7.6) immediately.

Now, consider r in the interval $r_\nu \leq r \leq r_{\nu+1}$. If, for some n, we have $\nu = \lambda_n$ then

$$\mu(r,F) = a_{\lambda_n} r^{\lambda_n} \geq a_{\lambda_m} r^{\lambda_m} \qquad \text{for } m \geq 1.$$

Thus, by (1.7.5),

$$\frac{\mu(r,F)}{n^2} \leq \sum_{m=1}^{\infty} \frac{a_{\lambda_m} r^{\lambda_m}}{m^2} = M(r,f) \leq \mu(r,F) \sum_{m=1}^{\infty} \frac{1}{m^2}$$

i.e.,

$$\log \mu(r,F) - 2 \log n \leq \log M(r,f) \leq \log \mu(r,F) + O(1). \tag{1.7.8}$$

However, (1.7.6) yields

$$\log n = o (\log \mu(r,F)), \qquad r_\nu \leq r \leq r_{\nu+1}, \ r \to 1.$$

This, coupled with (1.7.8), gives

$$\log \mu(r,F) \sim \log M(r,f) \qquad \text{as } r \to 1. \tag{1.7.9}$$

If there is no K such that $\lambda_K = \nu$, let λ_n be the largest λ_K such that $\lambda_K < \nu$, then $\lambda_{n+1} \geq \nu+1$ and so

$$a_{\nu+1} r_{\nu+1}^{\nu+1} \leq 2a_{\lambda_n} r_{\lambda_n}^{\lambda_n}.$$

Thus,

$$a_{\lambda_n} r_{\lambda_n}^{\lambda_n} \leq \mu(r,F) \leq \mu(r_{\nu+1},F) \leq 2a_{\lambda_n} r_{\lambda_n}^{\lambda_n} \qquad \text{for } r_\nu \leq r \leq r_{\nu+1}.$$

This implies

$$\frac{\mu(r,F)}{2n^2} \leq \frac{a_{\lambda_n} r_{\lambda_n}^{\lambda_n}}{n^2} \leq \frac{a_{\lambda_n} r^{\lambda_n}}{n^2} \leq \Sigma_{m=1}^{\infty} \frac{a_{\lambda_m} r^{\lambda_m}}{m^2} = M(r,f)$$

$$\leq \mu(r,F) \sum_{m=1}^{\infty} \frac{1}{m^2}$$

i.e.,

$$\log \mu(r,F) - 2 \log n + O(1) \leq \log M(r,f) \leq \log \mu(r,F) + O(1).$$

Again using (1.7.6), we have

$$\log M(r,f) \sim \log \mu(r,F) \qquad \text{as } r \to 1. \tag{1.7.10}$$

(1.7.4), (1.7.9) and (1.7.10) give (1.7.2). □

EXERCISES 1.7.

1. Let $\lambda_1(r)$ and $\mu_1(r)$ be positive functions of r for $1 < r < \infty$ such that $\log \lambda_1(r) \neq O(-\log(r-1))$ as $r \to 1^+$. Then, show that there exist a function G analytic in $|z| > 1$ with nonnegative coefficients, and two sequences $\{S'_n\}$ and $\{t'_n\}$ of positive numbers decreasing monotonically to 1 such that for every positive integer n,

$$M(s'_n,G) > \mu_1(s'_n) \text{ and } M(t'_n,G) < \lambda_1(t'_n).$$

(Kapoor [5])

2. Let $\mu_1(r)$ be a positive function of r for $1 < r < \infty$ such that it is bounded in every finite subinterval of $(1,\infty)$. Then show that there exists a function G analytic in $|z| > 1$, with nonnegative coefficients such that $M(r,G) > \mu_1(r)$ for $1 < r < \infty$. (Kapoor [5])

3. Let $\lambda_1(r)$ be a positive function of r for $1 < r < \infty$ such that

$$\lim_{r\to 1} \frac{\log \lambda_1(r)}{-\log(r-1)} = \infty.$$

Then there exists a function G, analytic $|z| > 1$ with nonnegative coefficients, such that $M(r,G) < \lambda_1(r)$ for $1 < r < \infty$. (Kapoor [5])

4. Let $\lambda_1(r)$ and $\mu_1(r)$ be positive and continuous functions of r for $1 < r < \infty$ such that $\log \lambda_1(r) \neq O(-\log (r-1))$ and $\log \mu_1(r) \neq O(-\log (r-1))$ as $r \to 1^+$. Then there exists a function G, analytic in $|z| > 1$ with nonnegative coefficients, such that $M(r,G) = \lambda_1(r)$ on some sequence of values of r decreasing to 1 and $M(r,G) = \mu_1(r)$ on another such sequence. (Kapoor [5])

*5. Let $\lambda(r)$ be nonnegative, increasing, convex with respect to log r in $0 < r < 1$ and satisfy

$$\lim_{r\to 1} \frac{\lambda(r)}{-\log (1-r)} = \infty .$$

Does there exists a function f, analytic in U, such that as $r \to 1$,

$$\log f(r) = \log M(r,f) \sim T(r,f) \sim \lambda(r)?$$

6. Let $\lambda(r)$ and $\mu(r)$ be nonnegative, increasing, convex with respect to log r in $0 < r < 1$ and satisfy

(i) $\lim_{r\to 1} \frac{\lambda(r)}{-\log (1-r)} = \infty, \quad \lim_{r\to 1} \frac{\mu(r)}{-\log (1-r)} = \infty$;

(ii) $\mu(r) = o(\lambda(r))$ as $r \to 1$;

(iii) $\lambda(r)(1-r) + \int_0^r \lambda(t)dt = o(\mu(r))$ as $r \to 1$;

(iv) $\lambda'(r)(1-r)$ is monotonic increasing in $0 < r < 1$.

Then show that there exists a function f, analytic in $|z| < 1$, such that

$$\log M(r,f) \sim \lambda(r), \quad T(r,f) \sim \mu(r)$$

as $r \to 1$. (Linden [5])

(For a different technique to obtain a result similar to that in Exercise 6, see Shea [1]).

REFERENCES

Bajpai and Tanne [1]

Beuermann [1]

Bogda and Shankar [1], [2]

Duren [1]

Gopal [1]

Hayman [2]

Heins [1]

Hoffman [1]

Juneja [1]

Kapoor [1], [2], [4], [5]

Kapoor and Gopal [1], [3]

Kapoor and Nautiyal [2]

Kovari [1]

Linden [1], [3], [5], [7]

MacLane [1]

Nautiyal [1]

Nevanlinna [1]

Polya and Szego [1]

Porceilli [1]

Rosenbloom [1]

Rudin [1]

Shapiro and Shields [1]

Shea [1]

Sons [1], [2], [3], [4]

Srivastava and Juneja [1]

Tsuji [1], [2]

Valiron [1]

Yamashita [1], [2]

2 Growth measurement by Taylor series coefficients

2.1. INTRODUCTION

Every function f, analytic in the unit disc U, is given by its Taylor series around the origin; therefore, the growth parameters of f, i.e., the order, the lower order, the type, the lower type and proximate order, defined in Sections 1.5 and 1.6, should be determinable by the Taylor series coefficients of f. This becomes of added importance when the computation of the maximum modulus function M(r) of f is complicated (which is indeed the case in many situations), and so evaluation of these growth parameters directly from their definitions presents a difficulty. The present chapter is devoted to the development of results in this direction.

The Taylor series expansion (1.4.1) of a function f, analytic in U, may have some of the coefficients a_n zero. If b_n, n = 0,1,2,..., denote the nonzero coefficients in (1.4.1), then the function f is represented in U by the gap Taylor series

$$f(z) = \Sigma_{n=0}^{\infty} b_n z^{\lambda_n}. \tag{2.1.1}$$

In Section 2.2 we find the characterizations of the order ρ and the type T in terms of the coefficients b_n and exponents λ_n, that are applicable to every function f, analytic in U. The characterization of the lower order λ and the lower type t, in terms of the coefficients b_n and the exponents λ_n, is found in Section 2.3. These cover useful subclasses of functions analytic in U. The relations involving the aforesaid growth parameters and the ratio of consecutive coefficients $|b_{n-1}/b_n|^{1/(\lambda_n-\lambda_{n-1})}$ are developed in Section 2.4. With the help of results obtained in Section 2.2, the decomposition theorems for functions, analytic in U, that are either not of regular growth or not of perfectly regular growth (cf. Section 1.5) are found in Section 2.5. Finally, in Section 2.6, a formula for computing a proximate order of f from the coefficients a_n in (1.4.1) is derived.

2.2. COEFFICIENT CHARACTERIZATIONS OF ORDER AND TYPE

The following lemmas are instrumental in finding the coefficient characterizations of order and type of a function, analytic in $\mathcal{U}$.

Lemma 2.2.1. *Let the maximum modulus* $M(r)$ *of a function*

$f(z) = \Sigma_{n=0}^{\infty} b_n z^{\lambda_n}$, *analytic in* $\mathcal{U}$, *satisfy*

$$\log M(r) < A(1-r)^{-B}, \qquad 0 < A,\ B < \infty, \tag{2.2.1}$$

for all r *such that* $r_o(A,B) < r < 1$. *Then, for all* $n > n_o(A,B)$,

$$\log |b_n| < S(A,B)\ \lambda_n^{B/(B+1)} \tag{2.2.2}$$

where, $S(A,B) = (A/B^B)^{1/(B+1)}\ (B+1+o(1))$.

Proof. Define a sequence $\{r_n\}$ by

$$(1-r_n)^{-1} = \left(\frac{\lambda_n}{AB}\right)^{1/(B+1)} .$$

Then, $r_n \to 1$ as $n \to \infty$. By Cauchy's inequality and (2.2.1), for all $n > n_o(A,B)^*$,

$$\begin{aligned}\log |b_n| &< \log M(r_n) - \lambda_n \log r_n \\ &< A(1-r_n)^{-B} + \lambda_n (1-r_n)\ (1+o(1)) \\ &= A\left(\frac{\lambda_n}{AB}\right)^{B/(B+1)} + \lambda_n \left(\frac{AB}{\lambda_n}\right)^{1/(B+1)}\ (1+o(1)) \\ &= S(A,B)\ \lambda_n^{B/(B+1)} . \quad \square\end{aligned}$$

Lemma 2.2.2. *Let* $f(z) = \Sigma_{n=0}^{\infty} b_n z^{\lambda_n}$, *analytic in* $\mathcal{U}$, *satisfy*

$$\log |b_n| < C\lambda_n^D,\ 0 < C < \infty,\quad 0 < D < 1, \tag{2.2.3}$$

for all $n > n_o(C,D)$. *Then, for all* r *such that* $r_o(C,D) < r < 1$,

$$\log M(r) < L(C,D)\ (1-r)^{-D/(1-D)} \tag{2.2.4}$$

where, $L(C,D) = (1-D)\ C^{1/(1-D)}\ D^{D/(1-D)} + o(1)$ *and* $M(r)$ *is the maximum modulus of* $f(z)$ *on* $|z| = r$.

$^*n_o$ need not be same at each occurrence in the sequel.

Proof. For $|z| = r < 1$,

$$M(r) < \Sigma_{n=0}^{\infty} |b_n| r^{\lambda_n} < K(n_o) + \Sigma_{n=n_o+1}^{\infty} \exp(C\lambda_n^{B/(B+1)}) r^{\lambda_n}$$

where $B = D/(1-D)$.

Choose

$$N \equiv N(r) = \left[(2C/(\log \tfrac{1}{r}))^{B+1}\right]$$

where $[x]$ denotes the greatest integer not greater than x. Then, $N(r) \to \infty$ as $r \to 1$.

The above estimate of M(r), for all r sufficiently close to 1, gives

$$M(r) < K(n_o) + NH(r) + \Sigma_{n=N+1}^{\infty} r^{\lambda_n/2} \tag{2.2.5}$$

where, $H(r) = \max_n \{\exp(C\lambda_n^{B/(B+1)}) r^{\lambda_n}\}$; for, if $n \geq N+1$, then

$$\lambda_n \geq N+1 > \left(\frac{2C}{\log(1/r)}\right)^{B+1}$$

and so,

$$\exp(C\lambda_n^{B/(B+1)}) r^{\lambda_n} < r^{\lambda_n/2}.$$

The infinite series in (2.2.5) is bounded by $r^{(N+1)/2}/(1-r^{1/2})$. Since $B > 0$, we have

$$-\left(\frac{N+1}{2}\right)\log\frac{1}{r} - \log(1-r^{1/2}) < -(2C)^{B+1}\left(\log\frac{1}{r}\right)^{-B} - \log(1-r) + \log(1+r^{\frac{1}{2}})$$

$$\to -\infty \text{ as } r \to 1.$$

Thus, for r sufficiently close to 1,

$$\Sigma_{n=N+1}^{\infty} r^{\lambda_n/2} = o(1). \tag{2.2.6}$$

Further,

$$\log H(r) \leq \log F(r) = \max_{0 \leq x < \infty} \{Cx^{B/(B+1)} - x\log\frac{1}{r}\}.$$

The maximum of the expression on the right-hand side is assumed at the point

$$x_o \equiv x_o(r) = \left(\frac{BC}{(B+1)\log(1/r)}\right)^{B+1}$$

and the maximum value of this expression is

$$\frac{B^B (C/(B+1))^{B+1}}{(\log (1/r))^B} . \tag{2.2.7}$$

Now, in view of (2.2.6) and (2.2.7), inequality (2.2.5), for all r such that* $r_o(C,D) < r < 1$, becomes

$$\begin{aligned}\log M(r) &< \log N(r) + \log H(r) + o(1)\\ &< -(B+1) \log\log \frac{1}{r} + \frac{B^B C^{B+1}}{(B+1)^{B+1}} (\log \frac{1}{r})^{-B} + o(1)\\ &= \frac{B^B C^{B+1}}{(B+1)^{B+1}} (\log \frac{1}{r})^{-B} (1+o(1))\\ &= L(C,D)\ (1-r)^{-D/(1-D)} . \ \square\end{aligned}$$

The following result gives a characterization of order of a function f, analytic in U, in terms of the coefficients b_n and exponents λ_n as in (2.1.1).

Theorem 2.2.1. *Let* $f(z) = \Sigma_{n=0}^{\infty} b_n z^{\lambda_n}$ *be analytic in* U *and have order* $\rho (0 \leq \rho \leq \infty)$. *Then,*

$$\frac{\rho}{1+\rho} = \limsup_{n \to \infty} \frac{\log^+ \log^+ |b_n|}{\log \lambda_n} \tag{2.2.8}$$

where, the left- hand side in (2.2.8) *is interpreted to be* 1 *when* $\rho = \infty$.

Proof. If $|b_n|$ is bounded by K for all n, then $\Sigma |b_n| r^{\lambda_n}$ is bounded by $K/(1-r)$. Therefore, by Definition (1.5.1) of the order of f, $\rho = 0$ and so (2.2.8) is satisfied. Thus, we need to consider only the case $\limsup_{n \to \infty} |b_n| = \infty$. In this case, all the $\log^+$ in (2.2.8) may be replaced by log.

First, let $0 \leq \rho < \infty$ and $\rho' > \rho$. Then, for r sufficiently close to 1,

$$\log M(r) < (1-r)^{-\rho'} .$$

---

*r_o may not be same at each occurrence in the sequel.

Using Lemma 2.2.1 with $A = 1$ and $B = \rho'$, it follows from the above inequality that, for $n > n_o(\rho')$,

$$\log |b_n| < (\rho') \lambda_n^{\rho'/(\rho'+1)} .$$

Consequently, $\limsup_{n \to \infty} (\log \log |b_n|)/\log \lambda_n \leq \rho'/(\rho'+1)$. Since, $\rho' > \rho$ is arbitrary, this inequality gives

$$\limsup_{n \to \infty} \frac{\log \log |b_n|}{\log \lambda_n} \leq \frac{\rho}{\rho+1} . \tag{2.2.9}$$

Since, f is analytic in U, the above inequality is trivially true if $\rho = \infty$ and the right-hand side is interpreted as 1 in this case.

Conversely, if

$$\theta = \limsup_{n \to \infty} \frac{\log \log |b_n|}{\log \lambda_n}$$

then $0 \leq \theta \leq 1$. First, let $\theta < 1$ and choose θ' such that $\theta < \theta' < 1$. Then, for all sufficiently large n,

$$\log |b_n| < \lambda_n^{\theta'} .$$

Using Lemma 2.2.2, with $C = 1$ and $D = \theta'$, it follows from the above inequality that, for all r such that $r_o(\theta') < r < 1$,

$$\log M(r) < ((1-\theta') \theta'^{\theta'/(1-\theta')} + o(1))(1-r)^{-\theta'/(1-\theta')} .$$

Consequently, $\rho \leq \theta'/(1-\theta')$. Since, $\theta' > \theta$ is arbitrary, it follows that

$$\frac{\rho}{1+\rho} \leq \theta = \limsup_{n \to \infty} \frac{\log \log |b_n|}{\log \lambda_n} \tag{2.2.10}$$

If $\theta = 1$, the above inequality is obviously true.

Inequalities (2.2.9) and (2.2.10) together give (2.2.8) when $\limsup_{n \to \infty} |b_n| = \infty$. □

A function f, analytic in U, is said to have *growth* (ρ, T), $0 < \rho < \infty$, if it is of order not exceeding ρ and of type not exceeding T, if of order ρ. The function f is said to have *growth not less than* (ρ, T), if it is of order not less than ρ and of type not less than T, if of order ρ.

Our next result gives the characterization of the type of a function f, analytic in $\mathcal{U}$, in terms of the coefficients b_n and exponents λ_n, as in (2.1.1).

Theorem 2.2.2. For a function $f(z) = \sum_{n=0}^{\infty} b_n z^{\lambda_n}$, *analytic in* $\mathcal{U}$, *set*

$$v \equiv \limsup_{n \to \infty} \frac{(\log^+ |b_n|)^{\rho+1}}{\lambda_n^{\rho}}, \quad 0 < \rho < \infty. \tag{2.2.11}$$

If $0 < v < \infty$, *the function* f *is of order* ρ *and type* T *if and only if*

$$\frac{(\rho+1)^{\rho+1}}{\rho^{\rho}} T = v. \tag{2.2.12}$$

If $v = 0$ *or* ∞, *the function* f *is respectively of growth* $(\rho,0)$ *or of growth not less than* (ρ,∞), *and conversely.*

Proof. Let $v < \infty$. For any $v' > v$ and for all sufficiently large n, (2.2.11) gives,

$$\log^+ |b_n| < v'^{1/(\rho+1)} \lambda_n^{\rho/(\rho+1)} .$$

Therefore,

$$\frac{\log^+ \log^+ |b_n|}{\log \lambda_n} < \frac{\log v'}{(\rho+1) \log \lambda_n} + \frac{\rho}{\rho+1} .$$

Now, using Theorem 2.2.1, it follows that f is of order at most ρ. Similarly, if $v > 0$, one can prove that f is of order at least ρ. Thus, if $0 < v < \infty$, f is of order ρ. For the first part of the theorem, it remains to be proved that, in case $0 < v < \infty$, f is of type T if and only if (2.2.12) holds.

Let $0 \leq T < \infty$. If $T' > T$, then it follows from the definition of type (cf. Section 1.5) that for all r such that $1 > r > r_o(T',\rho)$,

$$\log M(r) < T'(1-r)^{-\rho}.$$

Using Lemma 2.2.1, with $A = T'$ and $B = \rho$, the above inequality implies that for all $n > n_o(T',\rho)$,

$$\log^+ |b_n| < \left(\frac{T'}{\rho^{\rho}}\right)^{1/(\rho+1)} (\rho+1+o(1)) \lambda_n^{\rho/(\rho+1)}.$$

Therefore, for all sufficiently large n,

$$\frac{(\log^+ |b_n|)^{\rho+1}}{\lambda_n^\rho} < \frac{(\rho+1+o(1))^{\rho+1}}{\rho^\rho} T'.$$

Now, on proceeding to limits, it follows that $v \leq ((\rho+1)^{\rho+1}/\rho^\rho)T'$. Since $T' > T$ is arbitrary, we get

$$v \leq \frac{(\rho+1)^{\rho+1}}{\rho^\rho} T. \tag{2.2.13}$$

Inequality (2.2.13) is obviously true when $T = \infty$.

Next, suppose that $0 \leq v < \infty$. If $v' > v$, then for all $n > n_o(v')$,

$$\log^+ |b_n| < v'^{1/(\rho+1)} \lambda_n^{\rho/(\rho+1)}.$$

Using Lemma 2.2.2, with $D = \rho/(1+\rho)$ and $C = v'^{1/(\rho+1)}$, we have, for all r sufficiently close to 1,

$$\log M(r) < \left(\frac{\rho^\rho}{(\rho+1)^{\rho+1}} v' + o(1)\right) (1-r)^{-\rho}.$$

Thus,

$$T = \limsup_{r \to 1} \frac{\log M(r)}{(1-r)^{-\rho}} \leq \frac{\rho^\rho}{(\rho+1)^{\rho+1}} v'.$$

Since $v' > v$ is arbitrary it follows that

$$\frac{(\rho+1)^{\rho+1}}{\rho^\rho} T \leq v. \tag{2.2.14}$$

On combining (2.2.13) and (2.2.14) and using the fact proved in the beginning that, for $0 < v < \infty$, f is of order ρ, we get that if $0 < v < \infty$, then f is of order ρ and type T if and only if

$$\frac{(\rho+1)^{\rho+1}}{\rho^\rho} T = v.$$

Now, if $v = 0$, then f is of order at most ρ and (2.2.14) gives that, if f is of order ρ, then its type is zero. Therefore, if $v = 0$, then f is of growth $(\rho, 0)$.

Similarly, if $v = \infty$, then f is of order at least ρ and (2.2.13) shows that, if f is of order ρ, then $T = \infty$. Hence, if $v = \infty$, then f is of growth not less than (ρ,∞).

The converse also follows in a similar manner. □

EXERCISES 2.2.

1. Show that the q-order (cf. Exercises 1.5) $\rho(q)$ of a function $f(z) = \Sigma_{n=0}^{\infty} b_n z^{\lambda_n}$, analytic in U, is given by

$$\rho(q)+A(q) = \limsup_{n\to\infty} \frac{\log^{[q-1]} \lambda_n}{\log \lambda_n - \log^+ \log^+ |b_n|}$$

where $A(q) = 1$ if $q = 2$ and $A(q) = 0$ if $q = 3,4,\ldots$.

(Kapoor and Gopal [1])

2. If $\rho(q) > 0$, then show that the q-type (cf. Exercises 1.5.) $T(q)$ of a function $f(z) = \Sigma_{n=0}^{\infty} b_n z^{\lambda_n}$, analytic in U, is given by

$$B(q)T(q) = \limsup_{n\to\infty} (\log^{[q-2]} \lambda_n)\left(\frac{\log^+ |b_n|}{\lambda_n}\right)^{\rho(q)+A(q)}$$

where $A(q)$ is as in Exercise 1, $B(q) = (\rho(q)+1)^{\rho(q)+1}/\rho(q)^{\rho(q)}$, if $q = 2$ and $B(q) = 1$, if $q = 3,4,\ldots$. (Kapoor and Gopal [1])

3. Show that the logarithmic order (cf. Exercises 1.5) ρ_o of a function $f(z) = \Sigma_{n=0}^{\infty} b_n z^{\lambda_n}$, analytic in U, satisfies

$$\beta_o \leq \rho_o \leq \max(1,\beta_o)$$

where

$$\beta_o = \limsup_{n\to\infty} \frac{\log^+ \log^+ |b_n|}{\log\log \lambda_n} .$$

(Kapoor and Gopal [2])

4. If $f(z) = \Sigma_{n=0}^{\infty} b_n z^{\lambda_n}$ is analytic in U and has logarithmic order ρ_o $(1 \leq \rho_o < \infty)$ and logarithmic type (cf. Exercises 1.5) T_o then, show that

$$T_o = \limsup_{n\to\infty} \frac{\log^+ |b_n|}{(\log \lambda_n)^{\rho_o}} .$$

(Kapoor and Gopal [2])

5. Let $f(z) = \Sigma_{n=0}^{\infty} b_n z^{\lambda_n}$, analytic in U, be of (α,β)-order $\rho(\alpha,\beta)$ (cf. Exercises 1.5). Prove that, if $\alpha \not\equiv \beta$, then

$$\rho(\alpha,\beta) = \limsup_{n \to \infty} \{\frac{\alpha(\lambda_n)}{\beta(\lambda_n/\log^+ |b_n|)}\}.$$

6. If f is same as in Exercise 5 and $\alpha(x) \equiv \beta(x) \neq \log x$, then show that, for $\rho(\alpha,\alpha) \geq 1$,

$$\rho(\alpha,\alpha) = \max (1, \limsup_{n \to \infty} \{\frac{\alpha(\lambda_n)}{\alpha(\lambda_n/\log^+ |b_n|)}\}).$$

7. Let $h \in \Lambda$ satisfy conditions (i) and (ii) of Exercise 1.5.22. Prove that the generalized order $\rho(h)$, $1 \leq \rho(h) \leq \infty$, (cf. Exercises 1.5), of a function $f(z) = \Sigma_{n=0}^{\infty} b_n z^{\lambda_n}$, analytic in U and having unbounded coefficients $\{b_n\}$, is given by

$$\rho(h) = \max (1, \limsup_{n \to \infty} \{\frac{h(\log |b_n|)}{h(\log \lambda_n)}\}).$$

(Kapoor and Nautiyal [1])

The rational order $\bar{\rho}$ of a function f, analytic in U, is defined as

$$\bar{\rho} = \limsup_{r \to 1} \frac{\log^+ M(r)}{-\log (1-r)}.$$

8. Determine the rational order $\bar{\rho}$ of a function $f(z) = \Sigma_{n=0}^{\infty} b_n z^{\lambda_n}$, analytic in U, in terms of the coefficients $\{b_n\}$ and exponents $\{\lambda_n\}$.

2.3. COEFFICIENT CHARACTERIZATIONS OF LOWER ORDER AND LOWER TYPE

Coefficient characterizations analogous to those obtained in Section 2.2 for the order and the type do not necessarily hold for the lower order and the lower type, respectively, of a function analytic in U. For consider the function

$$f_1(z) = \Sigma_{n=0}^{\infty} \exp (\lambda_{n-1}) z^{\lambda_n}$$

where $\lambda_n = 2^{2^n}$. Then, the function f_1 is analytic in U. Proceeding as in the case of function f_o considered after Theorem 1.5.2 and applying Theorems 2.2.1, 1.5.1 and Corollary 2 of Theorem 1.5.3, it follows that f_1 has order 1, lower order 1/2 and lower type zero. However, in this case

$$\lim_{n\to\infty} \frac{\log\log|b_n|}{\log \lambda_n} = 2$$

and

$$\lim_{n\to\infty} \frac{(\log|b_n|)^2}{\lambda_n} = 1$$

so that analogues of Theorems 2.2.1 and 2.2.2 for lower order and lower type do not hold in case of the function f_1.

We first find lower bounds for the lower order and lower type in terms of the coefficients b_n and exponents λ_n in (2.1.1), that holds for every function f analytic in U. This inequality is used in later theorems of this section for finding coefficient characterizations of lower order for certain subclasses of functions analytic in U.

Theorem 2.3.1. *Let* $f(z) = \Sigma_{n=0}^{\infty} b_n z^{\lambda_n}$ *be analytic in* U *and have lower order* λ $(0 \leq \lambda \leq \infty)$. *Then, for any increasing sequence* $\{n_k\}$ *of natural numbers*

$$1 + \lambda \geq \liminf_{k\to\infty} \left\{\frac{\log \lambda_{n_{k-1}}}{\log \lambda_{n_k} - \log^+ \log^+ |b_{n_k}|}\right\}. \tag{2.3.1}$$

Further, if f *has order* $\rho(0 < \rho < \infty)$ *and lower type* $t(0 \leq t \leq \infty)$, *then*

$$\frac{(\rho+1)^{\rho+1}}{\rho^{\rho}} t \geq \liminf_{k\to\infty} \{\lambda_{n_{k-1}} (\log^+ |b_{n_k}|^{1/\lambda_{n_k}})^{\rho+1}\}. \tag{2.3.2}$$

Proof. Let the right-hand side expression in (2.3.1) be denoted by θ and set $\alpha = 1/\theta$. Then, $0 \leq \alpha \leq \infty$. We need only consider the case $0 \leq \alpha < 1$. For any ε such that $0 < \varepsilon < 1-\alpha$, there exists an integer $K \equiv K(\varepsilon)$ such that, for $k > K$,

$$\log|b_{n_k}| > \lambda_{n_k} \lambda_{n_{k-1}}^{-(\alpha+\varepsilon)}. \tag{2.3.3}$$

Define the sequence $\{r_k\}$, $k = 2,3,\ldots$, by

$$-\log r_k = \frac{1}{e} \lambda_{n_{k-1}}^{-(\alpha+\varepsilon)} . \tag{2.3.4}$$

Then, $r_k \to 1$ as $k \to \infty$. If $k > K$ and $r_k \leq r \leq r_{k+1}$, then by (2.3.3) and Cauchy's inequality

$$\begin{aligned}\log M(r,f) &\geq \log |b_{n_k}| + \lambda_{n_k} \log r \\ &\geq \log |b_{n_k}| + \lambda_{n_k} \log r_k \\ &> (1 - \frac{1}{e}) \lambda_{n_k} \lambda_{n_{k-1}}^{-(\alpha+\varepsilon)} .\end{aligned}$$

Therefore,

$$\begin{aligned}\log \log M(r,f) &> \log (1 - \frac{1}{e}) + \log \lambda_{n_k} - (\alpha+\varepsilon) \log \lambda_{n_{k-1}} \\ &= \log (1 - \frac{1}{e}) + (1 - \frac{1}{\alpha+\varepsilon}) - \frac{\log \log (\frac{1}{r_{k+1}})}{\alpha+\varepsilon} + \log \log (\frac{1}{r_k}) \\ &\geq \log (1 - \frac{1}{e}) + (1 - \frac{1}{\alpha+\varepsilon}) + (1 - \frac{1}{\alpha+\varepsilon}) \log \log (\frac{1}{r}) + o(1)\end{aligned}$$

and so,

$$\lambda = \liminf_{r \to 1} \frac{\log \log M(r,f)}{-\log (1-r)} \geq (1 - \frac{1}{\alpha+\varepsilon}) \lim_{r\to 1} \frac{\log \log (\frac{1}{r})}{-\log (1-r)} .$$

Since $-\log \log \frac{1}{r} \sim -\log (1-r)$ as $r \to 1$ and $\varepsilon > 0$ is arbitrary

$$\lambda \geq \frac{1}{\alpha} - 1 \text{ if } \alpha \neq 0 \text{ and } \lambda = \infty \text{ if } \alpha = 0.$$

This establishes (2.3.2).

Next, set

$$\liminf_{k \to \infty} \{ (\frac{\lambda_{n_{k-1}}}{B^*}) (\frac{\log^+ |b_{n_k}|}{\lambda_{n_k}})^{1+\rho} \} \equiv \beta$$

where, $B^* = \frac{(\rho+1)^{\rho+1}}{\rho^\rho}$. The same technique as adopted in proving (2.3.1) gives $t \geq \beta$; the only difference here being that instead of choosing the sequence $\{r_k\}$, defined by (2.3.4) earlier, we now choose the sequence $\{r_k^*\}$ defined as

$$\frac{1}{r_k^*} = \exp\left\{\frac{(\beta-\varepsilon)\rho}{\lambda_{n_{k-1}}}\right\}^{1/(1+\rho)}, \quad k = 2,3,\ldots .$$

Thus, (2.3.2) follows. □

Corollary 1. *Let* $f(z) = \Sigma_{n=0}^{\infty} b_n z^{\lambda_n}$ *be analytic in* U, *have order* $\rho(0 \leq \rho < \infty)$ *and lower order* $\lambda(0 \leq \lambda < \infty)$. *If*

(i) $$\lim_{n\to\infty} \frac{\log \lambda_{n-1}}{\log \lambda_n} = 1$$

and

(ii) $$S = \lim_{n\to\infty} \left\{\frac{\log \lambda_n}{\log \lambda_n - \log^+ \log^+ |b_n|}\right\}$$ *exists*,

then f *is of regular growth* (cf. Section 1.5) *and of order* S-1.

Proof. By Theorems 2.2.1 and 2.3.1, and Conditions (i) and (ii),

$$1+\rho = \lim_{n\to\infty} \left\{\frac{\log \lambda_n}{\log \lambda_n - \log^+ \log^+ |b_n|}\right\}$$

$$\leq \left(\lim_{n\to\infty} \frac{\log \lambda_n}{\log \lambda_{n-1}}\right) \left(\liminf_{n\to\infty} \left\{\frac{\log \lambda_{n-1}}{\log \lambda_n - \log^+ \log^+ |b_n|}\right\}\right)$$

$$\leq 1+\lambda.$$

Thus, $0 \leq \lambda = \rho < \infty$. □

Corollary 2. *Let* $f(z) = \Sigma_{n=0}^{\infty} b_n z^{\lambda_n}$, *analytic in* U, *have order* $\rho(0 < \rho < \infty)$.

Let,

(i) $$\lim_{n\to\infty} \frac{\lambda_{n-1}}{\lambda_n} = 1$$

and suppose

(ii) $$\beta_o = \lim_{n\to\infty} \lambda_n (\log^+ |b_n|^{1/\lambda_n})^{1+\rho}$$ *exists and* $0 < \beta_o < \infty$.

Then, f *is of perfectly regular growth* (cf. Section 1.5), *and the type* T *and lower type* t *of* f *are given by*

$$BT = Bt = \beta_o$$

where, $B = (\rho+1)^{\rho+1}/\rho^{\rho}$.

It is to be noted, however, that Conditions (i) and (ii) in Corollary 1 are not necessary for a function to be of regular growth (Exercise 2.3.1). Similarly, Conditions (i) and (ii) in Corollary 2 are not necessary for a function to be of perfectly regular growth (Exercise 2.3.2).

The following lemma is needed for finding upper bounds on the lower order and lower type of f, in terms of the coefficients b_n and exponents λ_n as in (2.1.1), for a subclass of functions analytic in U.

Lemma 2.3.1. *Let* $f(z) = \Sigma_{n=0}^{\infty} b_n z^{\lambda_n}$ *be analytic in* U *and have nonzero order. Let* $\Psi(n) \equiv |b_{n-1}/b_n|^{1/(\lambda_n - \lambda_{n-1})}$ *form a nondecreasing function of* n *for* $n > n_o$. *If the maximum term* (cf. Section 1.4) $\mu(r)$ *of* f, *for* $|z| = r$, *satisfies*

$$\log \mu(r) > a(1-r)^{-b}, \quad 0 < a, b < \infty \tag{2.3.5}$$

for all r *such that* $r_o(a,b) < r < 1$, *then*

$$\log |b_n| > s(a,b)\, \lambda_n^{b/(b+1)} \tag{2.3.6}$$

for all $n > n_o(a,b)$, *where* $s(a,b) = ((b+1)/b^{b/(b+1)})\, a^{1/(b+1)}$.

Proof. Since, by hypothesis, $\Psi(n)$ forms a nondecreasing function of n for $n > n_o$, $\Psi(n+1) > \Psi(n)$ for an infinity of n; for otherwise the order of f is zero. Further, $\Psi(n) \to 1$ as $n \to \infty$ (cf. Section 1.4). When $\Psi(n+1) > \Psi(n)$, the term $b_n z^{\lambda_n}$ becomes the maximum term (cf. Section 1.4) for $\Psi(n) \leq r < \Psi(n+1)$; so that for these values of r,

$$\log \mu(r) = \log |b_n| + \lambda_n \log r.$$

Let $b_{n_1} z^{\lambda_{n_1}}$ and $b_{n_2} z^{\lambda_{n_2}}$ be two consecutive maximum terms of f so that $n_1 \leq n_2 - 1$. Then, by (2.3.5) for $\Psi(n_2) \leq r < \Psi(n_2+1)$,

$$\log |b_{n_2}| + \lambda_{n_2} \log r > a(1-r)^{-b}.$$

Let $n_1 \leq n \leq n_2 - 1$. It follows easily that (cf. Section 1.4)

$$\Psi(n_1+1) = \Psi(n_1+2) = \ldots = \Psi(n+1) = \ldots = \Psi(n_2)$$

and

$$|b_n| \{\Psi(n+1)\}^{\lambda_n} = |b_{n_2}| \{\Psi(n_2)\}^{\lambda_{n_2}}.$$

Therefore, for sufficiently large n,

$$\log |b_n| + \lambda_n \log \Psi(n+1) > a(1-r)^{-b}.$$

Since, the above inequality holds for all r such that $\Psi(n_2) \leq r < \Psi(n_2+1)$,

$$\log |b_n| + \lambda_n \log \Psi(n+1) > a(1-\Psi(n_2))^{-b} = a(1-\Psi(n+1))^{-b}$$

which, in view of the inequality $-\log x \geq 1-x$ for $x > 0$, gives that, for all sufficiently large n,

$$\log |b_n| > a(1-\Psi(n+1))^{-\rho} + \lambda_n(1 - \Psi(n+1)). \tag{2.3.7}$$

The minimum value of the function

$$F(x,\lambda_n) = a(1-x)^{-b} + \lambda_n(1-x)$$

is easily seen to be

$$s(a,b)\, \lambda_n^{b/(b+1)},$$

being attained to the point

$$x_o \equiv x_o(\lambda_n) = 1 - (\frac{ab}{\lambda_n})^{1/(b+1)}.$$

Therefore, by (2.3.7), for all $n > n_o(a,b)$,

$$\log |b_n| > s(a,b)\, \lambda_n^{b/(b+1)}. \quad \square$$

Now, we find an upper bound on the lower order of f, in terms of the coefficients b_n and exponents λ_n as in (2.1.1), when $\Psi(n) = |b_{n-1}/b_n|^{1/(\lambda_n-\lambda_{n-1})}$ is a nondecreasing function of n for $n > n_o$.

Theorem 2.3.2. *Let* $f(z) = \Sigma_{n=0}^{\infty} b_n z^{\lambda_n}$ *be analytic in* U, *have order* $\rho(0 < \rho \leq \infty)$ *and lower order* $\lambda(0 \leq \lambda \leq \infty)$. *Further, let* $\Psi(n) = |b_{n-1}/b_n|^{1/(\lambda_n-\lambda_{n-1})}$ *form a nondecreasing function of* n *for* $n > n_o$. *Then,*

$$1 + \lambda \leq \liminf_{n \to \infty} \{\frac{\log \lambda_n}{\log \lambda_n - \log^+ \log^+ |b_n|}\}. \tag{2.3.8}$$

Proof. First, let $0 < \lambda < \infty$. Then, by Theorem 1.5.1 for any λ' such that $0 < \lambda' < \lambda$, and for all r sufficiently close to 1,

$$\log \mu(r) > (1-r)^{-\lambda'} .$$

Using Lemma 2.3.1, with a = 1 and b = λ', it follows from the above inequality that for all sufficiently large n,

$$\log^+ |b_n| > \frac{\lambda'+1}{\lambda'^{\lambda'/(\lambda'+1)}} \lambda_n^{\lambda'/(\lambda'+1)} .$$

Thus, for n sufficiently large

$$\log \lambda_n - \log^+ \log^+ |b_n| < \frac{1}{\lambda'+1} \log \lambda_n + o(1)$$

and it follows that

$$1 + \lambda' \leq \liminf_{n \to \infty} \left\{\frac{\log \lambda_n}{\log \lambda_n - \log^+ \log^+ |b_n|}\right\} .$$

Since $\lambda' < \lambda$ is arbitrary, the above inequality easily gives (2.3.8), in case $0 < \lambda < \infty$. Inequality (2.3.8) is trivially true if $\lambda = 0$ and, if $\lambda = \infty$, the above arguments with an arbitrary large number in place of λ' give that lim inf on the right-hand side of (2.3.8) must be ∞. □

Theorems 2.3.1 and 2.3.2 together give the following characterization of the lower order that is valid for a subclass of functions analytic in U.

Theorem 2.3.3. *Let* $f(z) = \Sigma_{n=0}^{\infty} b_n z^{\lambda_n}$ *be analytic in* U, *have order* $\rho (0 < \rho \leq \infty)$ *and lower order* $\lambda (0 \leq \lambda \leq \infty)$. *Further, let*

(i) $\Psi(n) \equiv |b_{n-1}/b_n|^{1/(\lambda_n - \lambda_{n-1})}$ *form a nondecreasing function of* n *for* $n > n_o$

and

(ii) $\log \lambda_{n-1} \sim \log \lambda_n$ *as* $n \to \infty$.

Then,

$$1 + \lambda = \liminf_{n \to \infty} \left\{\frac{\log \lambda_{n-1}}{\log \lambda_n - \log^+ \log^+ |b_n|}\right\}. \tag{2.3.9}$$

The coefficient characterization (2.3.9) for the lower order may not hold for a function f analytic in U, if the exponents λ_n do not satisfy the condition $\log \lambda_{n-1} \sim \log \lambda_n$ as $n \to \infty$. This can be illustrated by the function f_1 considered in the beginning of this section. For this function, $\rho = 1$, $\lambda = \frac{1}{2}$,

$$\liminf_{n \to \infty} \frac{\log \lambda_n}{\log \lambda_{n-1}} = 2$$

and

$$\liminf_{n \to \infty} \left\{\frac{\log \lambda_{n-1}}{\log \lambda_n - \log^+ \log^+ |b_n|}\right\} = 1$$

so that (2.3.9) is not satisfied for f_1.

The upper bound on the lower order found in Theorem 2.3.2 and the coefficient characterization (2.3.9) may also fail to hold if the condition '$\Psi(n) = |b_{n-1}/b_n|^{1/(\lambda_n - \lambda_{n-1})}$ forms a nondecreasing function of n for $n > n_o$' is dropped. To see this consider

$$f_2(z) = \Sigma_{n=0}^{\infty} a_n z^n$$

$$= \left(\sum_{n=4^m} p(2n+1)z^{2n+1} + \sum_{n\neq 4^m} z^{2n+1}\right) + \Sigma_{n=0}^{\infty} p(2n)z^{2n}, \quad m=1,2,\ldots$$

$$= f_{2,1}(z) + f_{2,2}(z) \quad \text{(say)},$$

where $p(n) = \exp(n^{\zeta/(1+\zeta)})$, $0 < \zeta < \infty$. Then, $f_2(z)$ is analytic in U. Since the coefficients in the series defining $f_{2,1}(z)$ and $f_{2,2}(z)$ are positive,

$$M(r,f_2) \le M(r,f_{2,1}+f_{2,2}) = M(r,f_{2,1}) + M(r,f_{2,2}) \le 2M(r,f_{2,2}).$$

Thus, $\log M(r,f_2) \sim \log M(r,f_{2,2})$ as $r \to 1$ and therefore $f_2(z)$ and $f_{2,2}(z)$ have the same order and same lower order. Set

$$B(n) = \frac{\log n}{\log n - \log^+ \log^+ |a_n|}.$$

Then,

$$\lim_{n\to\infty} B(2n) = \lim_{n\to\infty} \left\{\frac{\log 2n}{\log 2n - \log^+ \log^+ p(2n)}\right\}$$

$$= 1+\zeta .$$

Since the hypotheses of the Corollary of Theorem 2.3.1 are satisfied for $f_{2,2}(z)$, it is of regular growth. Using Theorem 2.2.1, it follows from the above evaluation of $\lim_{n\to\infty} B(2n)$ that both order and lower order of $f_{2,2}(z)$ and, therefore, both order and lower order of $f_2(z)$ are equal to $\zeta > 0$. But it is easily seen that $\liminf_{n\to\infty} B(n) = 1$ so that (2.3.8) or (2.3.9) do not hold for $f_2(z)$, the reason being that $\Psi(n) = |a_{n-1}/a_n|$ is not a nondecreasing sequence for this function.

A function f, analytic in U, is said to be *admissible* if its lower order λ satisfies

$$1 + \lambda = \liminf_{r\to 1} \frac{\log^+ \nu(r)}{-\log (1-r)}$$

where $\nu(r)$ is the central index (cf. Section 1.4) of f for $|z| = r$. Our next result gives a coefficient characterization for the lower order of an admissible function f, given by (2.1.1), for which $\Psi(n) = |b_{n-1}/b_n|^{1/(\lambda_n-\lambda_{n-1})}$ is a nondecreasing function of n for $n > n_o$. We note that Condition (ii) in Theorem 2.3.3 is no longer needed for admissible functions.

Theorem 2.3.4. *Let* $f(z) = \Sigma_{n=0}^{\infty} b_n z^{\lambda_n}$ *be an admissible function having order* $\rho(0 < \rho \leq \infty)$ *and lower order* $\lambda(0 \leq \lambda \leq \infty)$. *Further, let* $\Psi(n) = |b_{n-1}/b_n|^{1/(\lambda_n-\lambda_{n-1})}$ *form a nondecreasing function of* n *for* $n > n_o$. *Then,*

$$1 + \lambda = \liminf_{n\to\infty} \left\{\frac{\log \lambda_{n-1}}{\log \lambda_n - \log^+ \log^+ |b_n|}\right\}. \tag{2.3.10}$$

Proof. Since $\Psi(n)$ forms a nondecreasing function of n for $n > n_o$ and $\rho > 0$, $\Psi(n+1) > \Psi(n)$ for an infinity of n. Further, $\Psi(n) \to 1$ as $n \to \infty$ (cf. Section 1.4). When $\Psi(n+1) > \Psi(n)$, the maximum term $\mu(r)$ and central index $\nu(r)$ of f, for $\Psi(n) \leq r < \Psi(n+1)$, are given by

$$\mu(r) = |b_n| \, r^{\lambda_n} \quad \text{and} \quad \nu(r) = \lambda_n .$$

Since f is admissible and $\rho > 0$, for any $\varepsilon > 0$, there exists an $r_o \equiv r_o(\varepsilon)$ such that, for $0 < r_o < r < 1$,

$$\nu(r) > \left(\frac{1}{1-r}\right)^{1+\lambda-\varepsilon} .$$

Let $b_{m_1} z^{\lambda_{m_1}}$ and $b_{m_2} z^{\lambda_{m_2}}$ be two consecutive maximum terms of f, so that $m_1 \leq m_2-1$. Let $m_1 < n \leq m_2$. Since $b_{m_1} z^{\lambda_{m_1}}$ is a maximum term of f,

$$\nu(r) = \lambda_{m_1} \text{ for } \Psi(m_1) \leq r < \Psi(m_1+1).$$

Thus, for every r satisfying $\Psi(m_1) \leq r < \Psi(m_1+1)$,

$$\lambda_{m_1} = \nu(r) > \left(\frac{1}{1-r}\right)^{1+\lambda-\varepsilon} .$$

In particular, this inequality gives

$$\lambda_{n-1} \geq \lambda_{m_1} \geq \left(\frac{1}{1-\Psi(m_1+1)}\right)^{1+\lambda-\varepsilon} .$$

Further, in view of

$$\Psi(m_1+1) = \Psi(m_1+2) = \ldots = \Psi(n),$$

the above inequality becomes

$$\lambda_{n-1} \geq \left(\frac{1}{1-\Psi(n)}\right)^{1+\lambda-\varepsilon} . \tag{2.3.11}$$

Since

$$|b_n| = |b_N| \left(\frac{1}{\Psi(N+1)}\right)^{(\lambda_{N+1}-\lambda_N)} \cdots \left(\frac{1}{\Psi(n)}\right)^{(\lambda_n-\lambda_{n-1})}$$

it follows that

$$|b_n| \geq |b_N| \left(\frac{1}{\Psi(n)}\right)^{\lambda_n-\lambda_N} .$$

This inequality yields, for n sufficiently large,

$$\log^+ |b_n| \geq \log |b_N| + (\lambda_n - \lambda_N) \log \frac{1}{\Psi(n)}$$

$$\simeq \log |b_N| + (\lambda_n - \lambda_N)\ (1 - \Psi(n)).$$

Thus, for sufficiently large n,

$$\log^+ \log^+ |b_n| \geq o(1) + \log (\lambda_n - \lambda_N) - \log \left(\frac{1}{1-\Psi(n)}\right).$$

Now, using (2.3.11), it follows that

$$1 + \lambda \leq \liminf_{n \to \infty} \left[\frac{\log \lambda_{n-1}}{\log \lambda_n - \log^+ \log^+ |b_n|}\right].$$

The above inequality and Theorem 2.3.1 immediately give (2.3.10). □

We note that Theorems 2.3.3 and 2.3.4 give coefficient characterization for the lower order λ of a function f analytic in $\mathcal{U}$, under the condition that $\Psi(n+1) = |b_n/b_{n+1}|^{1/(\lambda_{n+1} - \lambda_n)}$ is a nondecreasing function of n for $n > n_o$. We now find the coefficient characterization of the lower order of functions, analytic in $\mathcal{U}$, which holds without this condition.

Theorems 2.3.5. *Let* $f(z) = \Sigma_{n=0}^{\infty} b_n z^{\lambda_n}$ *be analytic in* $\mathcal{U}$, *have order* $\rho (0 < \rho \leq \infty)$ *and lower order* $\lambda (0 \leq \lambda \leq \infty)$. *Let the elements* $\{\lambda_{p_k}\}$ *in the range set of the central index of* f *satisfy* $\log \lambda_{p_{k-1}} \sim \log \lambda_{p_k}$ *as* $k \to \infty$. *Then*

$$1 + \lambda = \max_{\{n_k\}} \left\{\liminf_{k \to \infty} \frac{\log \lambda_{n_{k-1}}}{\log \lambda_{n_k} - \log^+ \log^+ |b_{n_k}|}\right\} \tag{2.3.12}$$

where the maximum in (2.3.10) *is taken over all increasing sequences* $\{n_k\}$ *of natural numbers.*

Proof. Define $S(Z) = \Sigma_{k=0}^{\infty} b_{p_k} z^{\lambda_{p_k}}$, $|z| < 1$, where $b_{p_k} \neq 0$ for all k. It is easily seen that S is analytic in U and, for every $z \in U$, the functions f and S have the same maximum term. Thus, by Theorem 1.5.1, S is of lower order λ. Further, let

$$R(p_k) = \max \{r : \nu(r) = \lambda_{p_k}\}$$

where $\nu(r)$ is the central index of f. It follows that (cf. Section 1.4)

$$R(p_{k+1}) = \Psi(p_{k+1}) = |b_{p_k}/b_{p_{k+1}}|^{1/(\lambda_{p_{k+1}} - \lambda_{p_k})}$$

and that $R(p_{k+1})$ is an increasing function of k. Therefore, in view of the condition $\log \lambda_{p_{k-1}} \sim \log \lambda_{p_k}$ as $k \to \infty$, the hypotheses of Theorem 2.3.3 are satisfied for the function S, so that

$$1 + \lambda = \liminf_{k \to \infty} \left\{\frac{\log \lambda_{p_{k-1}}}{\log \lambda_{p_k} - \log^+ \log^+ |b_{p_k}|}\right\}. \tag{2.3.13}$$

On the other hand, by Theorem 2.3.1,

$$1 + \lambda \geq \max_{\{n_h\}} \left\{\liminf_{h \to \infty} \frac{\log \lambda_{n_{h-1}}}{\log \lambda_{n_h} - \log^+ \log^+ |b_{n_h}|}\right\} \tag{2.3.14}$$

On combining (2.3.13) and (2.3.14), we get (2.3.12). □

It is seen in the next theorem that to obtain the coefficient characterization (2.3.12) for lower order of an admissible function f, we do not need the condition $\log \lambda_{p_{k-1}} \sim \log \lambda_{p_k}$ as $k \to \infty$ on the sequence $\{\lambda_{p_k}\}$ of principal indices (cf. Section 1.4) of f.

Theorem 2.3.6. *Let* $f(z) = \Sigma_{n=0}^{\infty} b_n z^{\lambda_n}$, *analytic in* U, *be an admissible function having order* $\rho (0 < \rho \leq \infty)$ *and lower order* $\lambda (0 \leq \lambda \leq \infty)$. *Then,*

$$1 + \lambda = \max_{\{n_k\}} \left\{\liminf_{k \to \infty} \frac{\log \lambda_{n_{k-1}}}{\log \lambda_{n_k} - \log^+ \log^+ |b_{n_k}|}\right\} \tag{2.3.15}$$

where the maximum in (2.3.15) *is taken over all increasing sequences* $\{n_k\}$ *of natural numbers.*

Proof. The proof of this theorem is similar to that of Theorem 2.3.5, the only change being that instead of using Theorem 2.3.3 we now use Theorem 2.3.4. □

A necessary condition for a function f, analytic in U, to be of regular growth may be found from the following theorem, which we derive by an application of Theorem 2.3.6.

Theorem 2.3.7. *Let* $f(z) = \Sigma_{n=0}^{\infty} b_n z^{\lambda_n}$, *analytic in* U, *be an admissible function having order* $\rho(0 < \rho \leq \infty)$ *and lower order* $\lambda(0 \leq \lambda \leq \infty)$. *Then,*

$$1 + \lambda \leq (1+\rho) \liminf_{k \to \infty} \frac{\log \lambda_{k-1}}{\log \lambda_k} . \qquad (2.3.16)$$

Proof. By Theorem 2.3.6,

$$1 + \lambda \leq \max_{\{n_k\}} \{\limsup_{k \to \infty} \frac{\log \lambda_{n_k}}{\log \lambda_{n_k} - \log^+ \log^+ |b_{n_k}|}\}$$

$$\times \max_{\{n_k\}} \{\liminf_{k \to \infty} \frac{\log \lambda_{n_{k-1}}}{\log \lambda_{n_k}}\}$$

$$= \limsup_{k \to \infty} \{\frac{\log \lambda_k}{\log \lambda_k - \log^+ \log^+ |b_k|}\}$$

$$\times \max_{\{n_k\}} \{\liminf_{k \to \infty} \frac{\log \lambda_{n_{k-1}}}{\log \lambda_{n_k}}\}.$$

In view of Theorem 2.3.1, this gives

$$1 + \lambda \leq (1+\rho) \max_{\{n_k\}} \{\liminf_{k \to \infty} \frac{\log \lambda_{n_{k-1}}}{\log \lambda_{n_k}}\} . \qquad (2.3.17)$$

Now, for any arbitrary sequence $\{n_k\}$ of natural numbers, let

$$\xi(m) = \frac{\log \lambda_{n_{k-1}}}{\log \lambda_{n_k}} \qquad \text{for } n_{k-1} < m \leq n_k, \quad k = 2,3,\ldots$$

and

$$\omega(m) = \frac{\log \lambda_{m-1}}{\log \lambda_m}, \quad m = 2,3,\ldots .$$

It is easily seen that $\xi(m) \leq \omega(m)$ for all m. Therefore,

$$\liminf_{k \to \infty} \frac{\log \lambda_{n_{k-1}}}{\log \lambda_{n_k}} = \liminf_{m \to \infty} \xi(m)$$

$$\leq \liminf_{m \to \infty} \omega(m) = \liminf_{k \to \infty} \frac{\log \lambda_{k-1}}{\log \lambda_k} .$$

Since, the sequence $\{n_k\}$ is arbitrary, this gives

$$\sup_{\{n_k\}} \{\liminf_{k \to \infty} \frac{\log \lambda_{n_{k-1}}}{\log \lambda_{n_k}}\} \leq \liminf_{k \to \infty} \frac{\log \lambda_{k-1}}{\log \lambda_k} .$$

In fact, equality holds in the above inequality since the reverse inequality is obviously true. Thus, (2.3.16) follows from (2.3.17). □

Corollary. Let $f(z) = \Sigma_{n=0}^{\infty} b_n z^{\lambda_n}$, *analytic in* U, *be an admissible function of regular growth having nonzero order. Then* $\log \lambda_{k-1} \sim \log \lambda_k$ *as* $k \to \infty$.

In the rest of this section, for subclasses of functions f, analytic in U, we find characterizations of the lower type in terms of the coefficients b_n and exponents λ_n as in (2.1.1).

Our next result gives an upper bound on the lower type of a function analytic in U, in terms of the coefficients b_n and exponents λ_n as in (2.1.1), in the case $\Psi(n) = |b_{n-1}/b_n|^{1/(\lambda_n - \lambda_{n-1})}$ forms a nondecreasing function of n for $n > n_o$.

Theorem 2.3.8. *Let* $f(z) = \Sigma_{n=0}^{\infty} b_n z^{\lambda_n}$, *analytic in* U, *have order* $\rho (0 < \rho < \infty)$ *and lower type* $t (0 \leq t \leq \infty)$. *If* $\Psi(n) = |b_{n-1}/b_n|^{1/(\lambda_n - \lambda_{n-1})}$ *forms a nondecreasing function of* n *for* $n > n_o$, *then*

$$\frac{(\rho+1)^{\rho+1}}{\rho^{\rho}} t \leq \liminf_{n \to \infty} \{\lambda_n (\log^+ |b_n|^{1/\lambda_n})^{\rho+1}\} \tag{2.3.18}$$

$$\leq L \liminf_{n \to \infty} \{\lambda_{n-1} (\log^+ |b_n|^{1/\lambda_n})^{\rho+1}\} \tag{2.3.19}$$

where

$$L = \limsup_{n \to \infty} \frac{\lambda_n}{\lambda_{n-1}} .$$

Inequality (2.3.19) *holds whenever the product on the right-hand side is well defined.*

Proof. First, let $0 < t < \infty$. If t' is such that $0 < t' < t$, then by Theorem 1.5.3 it follows that, for all r such that $r_o(t',\rho) < r < 1$,

$$\log \mu(r) > t'(1-r)^{-\rho}.$$

Using Lemma 2.3.1, with $a = t'$ and $b = \rho$, we get that for sufficiently large n,

$$\log |b_n| > \frac{(\rho+1)}{\rho^{\rho/(\rho+1)}} t'^{1/(\rho+1)} \lambda_n^{\rho/(\rho+1)} .$$

Thus,

$$\frac{(\rho+1)^{\rho+1}}{\rho^{\rho}} t' \leq \liminf_{n \to \infty} \{\lambda_n (\log^+ |b_n|^{1/\lambda_n})^{\rho+1}\}.$$

Since $t' < t$ is arbitrary, (2.3.18) follows easily from the above inequality in the case $0 < t < \infty$. If $t = 0$, (2.3.18) trivially holds and if $t = \infty$, the above arguments with an arbitrary large number in place of t' give that lim inf on the right-hand side is infinity.

Inequality (2.3.19) is an easy consequence of (2.3.18) whenever the product on its right-hand side is well defined. □

Now we are able to find the coefficient characterization for the lower type of a function f, analytic in U and given by (2.1.1), for which $\Psi(n)$ forms a nondecreasing sequence of n for $n > n_o$ and $\lambda_{n-1} \sim \lambda_n$ as $n \to \infty$.

Theorem 2.3.9. *Let* $f(z) = \Sigma_{n=0}^{\infty} b_n z^{\lambda_n}$, *analytic in* U, *have order* $\rho(0 < \rho < \infty)$ *and lower type* $t(0 \leq t \leq \infty)$. *Further, let*

(i) $\Psi(n) = |b_{n-1}/b_n|^{1/(\lambda_n - \lambda_{n-1})}$ *form a nondecreasing function of* n

for $n > n_o$

and

(ii) $\lambda_{n-1} \sim \lambda_n$ *as* $n \to \infty$.

Then,

$$\frac{(\rho+1)^{\rho+1}}{\rho^{\rho}} t = \liminf_{n \to \infty} \{\lambda_{n-1} (\log^+ |b_n|^{1/\lambda_n})^{1+\rho}\} . \tag{2.3.20}$$

Proof. Combining Theorem 2.3.1 and the inequality (2.3.19), and using Condition (ii), (2.3.20) follows immediately. □

We close this section by finding a characterization of lower type of a function f, analytic in U, in terms of the coefficients b_n and exponents λ_n in (2.1.1), where the nondecreasing nature of $\psi(n)$ is no longer needed.

Theorem 2.3.10. *Let* $f(z) = \Sigma_{n=0}^{\infty} b_n z^{\lambda_n}$, *analytic in* U, *have order* $\rho(0 < \rho < \infty)$ *and lower type* $t(0 \leq t \leq \infty)$. *Further, let the sequence* $\{\lambda_{p_k}\}$ *of the principal indices* (cf. Section 1.4) *satisfy* $\lambda_{p_{k-1}} \sim \lambda_{p_k}$ *as* $k \to \infty$. *Then,*

$$\frac{(\rho+1)^{\rho+1}}{\rho^{\rho}} t = \max_{\{n_k\}} \{\liminf_{k \to \infty} \lambda_{n_{k-1}} (\log^+ |b_{n_k}|^{1/\lambda_{n_k}})^{1+\rho}\} \tag{2.3.21}$$

where the maximum in (2.3.21) *is taken over all increasing sequences* $\{n_k\}$ *of natural numbers.*

Proof. Consider the function $G(z) = \Sigma_{k=0}^{\infty} b_{p_k} z^{\lambda_{p_k}}$. It is easily seen that G is analytic in U. Further, for any z in U, f and G have the same maximum term. Thus, in view of Theorem 1.5.1 and Corollary 2 of Theorem 1.5.3, f and G are of the same order and of the same lower type, so that G is of order ρ and lower type t. Also G satisfies the hypotheses of Theorem 2.3.9. Therefore,

$$\frac{(\rho+1)^{\rho+1}}{\rho^{\rho}} t = \liminf_{k \to \infty} \{\lambda_{p_{k-1}} (\log^{+} |b_{p_k}|^{1/\lambda_{p_k}})^{1+\rho}\}. \tag{2.3.22}$$

But by Theorem 2.3.1, for any arbitrary increasing sequence $\{n_h\}$ of natural numbers,

$$\frac{(\rho+1)^{\rho+1}}{\rho^{\rho}} t \geq \max_{\{n_h\}} \{\liminf_{h \to \infty} \lambda_{n_{h-1}} (\log^{+} |b_{n_h}|^{1/\lambda_{n_h}})^{1+\rho}\}.$$

Combining the above inequality with (2.3.22), we get (2.3.21). □

EXERCISES 2.3

1. Show that there exists a function f, analytic in U and having regular growth, such that conditions (i) and (ii) in Corollary 1 of Theorem 2.3.1 are not satisfied.
2. Show that there exists a function f, analytic in U and having perfectly regular growth, such that conditions (i) and (ii) in Corollary 2 of Theorem 2.3.1 are not satisfied.
3. Construct a function f, analytic in U, for which strict inequality holds in (2.3.1). Also, construct a function, analytic in U, to show that strict inequality may hold in (2.3.2).
4. Let $n_1 = 4$, $n_{k+1} = n_k^4$, $k = 1,2,\ldots$, $r_1 = r_2 = r_3 = e$,

$$r_m = \exp(m^{-1/2}) \text{ if } n_k \leq m < n_k^2$$

$$= \exp(n_{k+1}^{-1/2}) + \frac{\exp(m^{-1})-\exp(n_{k+1}^{-1})}{n_{k+1}} \text{ if } n_k^2 \leq m < n_{k+1}.$$

Let

$$g(z) = 1 + \Sigma_{k=1}^{\infty} (r_1 r_2 \cdots r_k)\, z^k .$$

Then, show that g satisfies the conditions of Theorem 2.3.3 but is not of regular growth. Also show that the order of g is 1 and the lower order of g is zero.

5. Does there exist an admissible function $f(z) = \Sigma_{n=0}^{\infty} b_n z^{\lambda_n}$ such that $\log \lambda_n \not\sim \log \lambda_{n+1}$ as $n \to \infty$?

6. Let $f(z) = \Sigma_{n=0}^{\infty} b_n z^{\lambda_n}$ be analytic in U and have q-order $\rho(q) > 0$ and lower q-order $\lambda(q)$, q=3,4,... (cf. Exercises 1.5). Show that, if $\psi(n) = |b_{n-1}/b_n|^{1/(\lambda_n - \lambda_{n-1})}$ forms a nondecreasing function of n for $n > n_o$, then

$$\lambda(q) = \liminf_{n \to \infty} \left\{\frac{\log^{[q-1]} \lambda_{n-1}}{\log \lambda_n - \log^+ \log^+ |b_n|}\right\} .$$

(Kapoor and Gopal [1])

Further, if f has lower q-type $t(q)$ (cf. Exercises 1.5) and satisfies $\log^{[q-2]} \lambda_{n-1} \sim \log^{[q-2]} \lambda_n$ as $n \to \infty$, $q = 3,4,\ldots$, then show that

$$t(q) = \liminf_{n \to \infty} \left\{(\log^{[q-2]} \lambda_{n-1}) \left(\frac{\log^+ |b_n|}{\lambda_n}\right)^{\rho(q)}\right\} .$$

(Kapoor and Gopal [1])

7. Let $f(z) = \Sigma_{n=0}^{\infty} b_n z^{\lambda_n}$, analytic in U, have lower logarithmic order $\lambda_o > 1$ (cf. Exercises 1.5). Let $\psi(n)$ be as in Exercise 6 and $\log \log \lambda_{n-1} \sim \log \log \lambda_n$ as $n \to \infty$. Then show that

$$\lambda_o = \liminf_{n \to \infty} \left\{\frac{\log^+ \log^+ |b_n|}{\log \log \lambda_n}\right\}.$$

Gopal [1])

8. Let $f(z) = \Sigma_{n=0}^{\infty} b_n z^{\lambda_n}$ be analytic in U, have logarithmic order ρ_o, $1 < \rho_o < \infty$, and lower logarithmic type t_o, $0 \le t_o \le \infty$, (cf. Exercises 1.5). If $\psi(n)$ is as in Exercise 6 and $\log \lambda_n \sim \log \lambda_{n+1}$ as $n \to \infty$, then show that

$$t_o = \liminf_{n \to \infty} \left\{\frac{\log^+|b_n|}{(\log \lambda_n)^{\rho_o}}\right\} \qquad \text{(Gopal [1])}$$

9. Find the coefficient characterizations analogous to that of Theorem 2.3.5 for lower q-order, q = 3,4,..., and lower logarithmic order of a function analytic in U.

10. Find the coefficient characterizations analogous to that of Theorem 2.3.10 for lower q-type, q = 3,4,..., and lower logarithmic type of a function analytic in U.

11. Let $f(z) = \Sigma_{n=0}^{\infty} b_n z^{\lambda_n}$, analytic in U, be of lower (α,β)-order $\lambda(\alpha,\beta)$ (cf. Exercises 1.5). Assume that $\psi(n)$ be as in Exercise 6. Prove that, if, $\alpha \not\equiv \beta$, then

$$\lambda(\alpha,\beta) = \liminf_{n \to \infty} \left\{\frac{\alpha(\lambda_{n-1})}{\beta(\lambda_n/\log^+|b_n|)}\right\}.$$

12. If f is the same as in Exercise 11, and $\alpha(x) \equiv \beta(x) \neq \log x$, then show that, for $\lambda(\alpha,\alpha) \ge 1$,

$$\lambda(\alpha,\alpha) = \max\left(1, \liminf_{n \to \infty} \left\{\frac{\alpha(\lambda_{n-1})}{\alpha(\lambda_n/\log^+|b_n|)}\right\}\right).$$

13. Let $f(z) = \Sigma_{n=0}^{\infty} b_n z^{\lambda_n}$, analytic in U, be of lower generalized order $\lambda(h) > 1$ (cf. Exercises 1.5). Assume that h satisfies Conditions (i) and (ii) of Exercise 1.5.22 and $F(x,c) = h^{-1}(ch(\log x))$, $1 < c < \infty$, is such that

$$x^2 \frac{dF(x,c)}{dx} \to \infty \text{ as } x \to \infty.$$

If $\psi(n)$ is the same as in Exercise 6 and $h(\log \lambda_{n-1}) \sim h(\log \lambda_n)$ as $n \to \infty$, then show that

$$\lambda(h) = \liminf_{n \to \infty} \frac{h(\log^+ |b_n|)}{h(\log \lambda_n)} . \qquad \text{(Nautiyal [1])}$$

14. Find coefficient characterizations analogous to Theorem 2.3.5 for lower (α,β)-order and lower generalized order of a function analytic in U (cf. Exercises 1.5).

The *lower rational order* $\bar{\lambda}$ of a function f, analytic in U, is defined as

$$\bar{\lambda} = \liminf_{r \to 1} \frac{\log^+ M(r)}{-\log (1-r)} .$$

15. Find coefficient characterizations analogous to those in Theorems 2.3.4 and 2.3.5 for the lower rational order of a function analytic in U.

2.4. GROWTH PARAMETERS AND RATIO OF CONSECUTIVE COEFFICIENTS

In certain cases, it is easier to find the growth parameters of f by using the ratio $|b_{n-1}/b_n|$ instead of $|b_n|$. The present section is devoted to the development of such results.

In the following theorem we find an interrelation between the order and the ratio $|b_{n-1}/b_n|$ of a function f, analytic in U and given by (2.1.1).

Theorem 2.4.1. *Let* $f(z) = \Sigma_{n=0}^{\infty} b_n z^{\lambda_n}$, *analytic in* U, *have order* $\rho (0 \le \rho \le \infty)$. *If* $\Psi(n) = |b_{n-1}/b_n|^{1/(\lambda_n - \lambda_{n-1})} \ge 1/e$ *for all* $n > n_1$, *then*

$$1 + \rho \le \max (1,\theta) \qquad (2.4.1)$$

where

$$\theta = \limsup_{n \to \infty} \{\frac{\log \lambda_n}{\log ((\lambda_n - \lambda_{n-1})/(\log^+ |b_n/b_{n-1}|))}\} . \qquad (2.4.2)$$

The quotient occurring on the right-hand side of (2.4.2) *is to be interpreted to be zero for the values of* k *for which* $|b_k/b_{k-1}| \le 1$.

Further, if $\Psi(n)$ *is a nondecreasing function of* n *for* $n > n_o$, *then equality holds in* (2.4.1).

Proof. Since $\Psi(n) \geq 1/e$ for all $n > n_1$, we get $0 \leq \theta \leq \infty$. First, let $\theta < \infty$. For any β, such that $\theta < \beta < \infty$ and all $n > N \equiv N(\beta)$, it follows from the definition of θ that

$$\log^+ |b_n/b_{n-1}| < (\lambda_n - \lambda_{n-1})\, \lambda_n^{-1/\beta} .$$

Therefore, for all $n > \max(N, n_1)$,

$$\begin{aligned}
\log |b_n| &= \log |b_N| + \log |b_{N+1}/b_N| + \ldots + \log |b_n/b_{n-1}| \\
&< \log |b_N| + (\lambda_{N+1} - \lambda_N)\lambda_{N+1}^{-1/\beta} + \ldots + (\lambda_n - \lambda_{n-1})\, \lambda_n^{-1/\beta} \\
&= \log |b_N| + \lambda_n^{(\beta-1)/\beta} - \Sigma_{m=N+1}^{n-1} \lambda_m (\lambda_{m+1}^{-1/\beta} - \lambda_m^{-1/\beta}) - \lambda_N \lambda_{N+1}^{-\beta} \\
&= \log |b_N| + \lambda_n^{(\beta-1)/\beta} - \int_{\lambda_{N+1}}^{\lambda_n} n(t)\, d(t^{-1/\beta}) - \lambda_N \lambda_{N+1}^{-\beta} ,
\end{aligned} \tag{2.4.3}$$

where $n(t) = \lambda_m$ for $\lambda_m < t \leq \lambda_{m+1}$, $m = N+1, \ldots, n-1$.
Since,

$$\begin{aligned}
\int_{\lambda_{N+1}}^{\lambda_n} n(t)\, d(t^{-1/\beta}) &= -\frac{1}{\beta} \int_{\lambda_{N+1}}^{\lambda_n} \frac{n(t)}{t}\, t^{-1/\beta}\, dt \\
&\geq -\frac{1}{\beta} \int_{\lambda_{N+1}}^{\lambda_n} t^{-1/\beta}\, dt \\
&= -\frac{1}{\beta-1} \{\lambda_n^{(\beta-1)/\beta} - \lambda_{N+1}^{(\beta-1)/\beta}\} ,
\end{aligned}$$

by (2.4.3),

$$\log |b_n| < \log |b_N| + \frac{\beta}{\beta-1} \lambda_n^{(\beta-1)/\beta} - \frac{1}{\beta-1} \lambda_{N+1}^{(\beta-1)/\beta} - \lambda_N \lambda_{N+1}^{-\beta} . \tag{2.4.4}$$

By Theorem 2.2.1, if $\theta < \beta < 1$, then (2.4.4) gives $\rho = 0$ and so (2.4.1) obviously holds. Hence, suppose that $1 \leq \theta < \beta < \infty$. Inequality (2.4.4) now gives, for all n sufficiently large,

$$\log^+ \log^+ |b_n| < \frac{\beta-1}{\beta} \log \lambda_n + o(1).$$

Using Theorem 2.2.1, the last inequality gives

$$\frac{\rho}{1+\rho} \leq \frac{\beta-1}{\beta} .$$

Since this holds for every β satisfying $1 \leq \theta < \beta < \infty$, we get

$$\frac{\rho}{1+\rho} \leq \frac{\theta-1}{\theta} ,$$

which implies $1+\rho \leq \theta$. If $\theta=\infty$, this inequality is trivially true. This completes the proof of (2.4.1).

Next to prove that $1+\rho \geq \max(1,\theta)$ if $\Psi(n)$ is nondecreasing for $n > n_o$, it is sufficient to consider $1 < \theta < \infty$. Let $N > n_o$. Then, for all $n > N$,

$$\begin{aligned}\log|b_n| &= \log|b_N| + \log|b_{N+1}/b_N| + \ldots + \log|b_n/b_{n-1}| \\ &= \log|b_N| + (\lambda_{N+1}-\lambda_N)\log\frac{1}{\Psi(N+1)} + \ldots + (\lambda_n-\lambda_{n-1})\log\frac{1}{\Psi(n)} \\ &> \log|b_N| + (\lambda_n-\lambda_N)\log\frac{1}{\Psi(n)} ,\end{aligned}$$

which, for all $n > N$, gives

$$\log^+\log^+|b_n| > O(1) + \log(\lambda_n-\lambda_N) + \log\log\frac{1}{\Psi(n)} .$$

Thus, by the definition of θ, for every $\varepsilon > 0$ there is a sequence of values of n tending to infinity for which

$$\begin{aligned}\frac{\log\lambda_n}{\log\lambda_n - \log^+\log^+|b_n|} &> \frac{\log\lambda_n}{\log\left(\log\frac{1}{\Psi(n-1)}\right)^{-1}} + o(1) \\ &> (\theta-\varepsilon) + o(1). \qquad (2.4.5)\end{aligned}$$

Proceeding to limits and using Theorem 2.2.1, we get $1+\rho \geq \theta$. This inequality is true for $\theta=\infty$ also, since in that case $(\theta-\varepsilon)$ in (2.4.5) may be replaced by an arbitrary large number so that $\rho=\infty$. □

Corollary. Let $f(z) = \sum_{n=0}^{\infty} b_n z^{\lambda_n}$, *analytic in* U, *have order* $\rho\,(0 \leq \rho \leq \infty)$. *Let* $\{\lambda_{n_k}\}$ *be the principal indices* (cf. Section 1.4) *and* $R(n_k)$ *be the jump points of the central index* (cf. Section 1.4) *of* f. *Then*,

$$1+\rho = \max(1,V) \qquad (2.4.6)$$

where

$$V = \limsup_{k \to \infty} \frac{\log \lambda_{n_k}}{-\log (1-R(n_k))} .$$

Theorem 2.4.2. *Let* $f(z) = \Sigma_{n=0}^{\infty} b_n z^{\lambda_n}$, *analytic in* U, *have lower order* $\lambda (0 \leq \lambda \leq \infty)$. *Let* $\{n_k\}_{k=1}^{\infty}$ *be any increasing sequence of natural numbers. Then,*

$$1+\lambda \geq \max \{1, \liminf_{k \to \infty} (\log \lambda_{n_{k-1}} / \log \{\frac{\lambda_{n_k} - \lambda_{n_{k-1}}}{\log^+ |a_{n_k} / a_{n_{k-1}}|}\})\} \tag{2.4.7}$$

where the quotient occurring at the right-hand side of (2.4.7) *is interpreted as zero for the values of* k *for which* $|a_{n_k} / a_{n_{k-1}}| \leq 1$.

Proof. Let

$$\limsup_{k \to \infty} (\log \{\frac{\lambda_{n_k} - \lambda_{n_{k-1}}}{\log^+ |a_{n_k} / a_{n_{k-1}}|}\} / \log \lambda_{n_{k-1}}) = \alpha .$$

It is sufficient to consider that $0 \leq \alpha < 1$. For any ε such that $0 < \varepsilon \leq 1-\alpha$, there exists an integer $N = N(\varepsilon)$ such that, for all $m > N$,

$$\log^+ |a_{n_m} / a_{n_{m-1}}| > (\lambda_{n_m} - \lambda_{n_{m-1}}) \lambda_{n_{m-1}}^{-(\alpha+\varepsilon)} . \tag{2.4.8}$$

Since

$$|a_{n_k}| = |a_{n_N}| \Pi_{m=N+1}^{k} |a_{n_m} / a_{n_{m-1}}| ,$$

using (2.4.8) with the fact that the right-hand side of this inequality is positive, we get

$$\log |a_{n_k}| > \log |a_{n_N}| + \Sigma_{m=N+1}^{k} (\lambda_{n_m} - \lambda_{n_{m-1}}) \lambda_{n_{m-1}}^{-(\alpha+\varepsilon)}$$

$$> \log |a_{n_N}| + (\lambda_{n_k} - \lambda_{n_N}) \lambda_{n_{k-1}}^{-(\alpha+\varepsilon)} . \tag{2.4.9}$$

Let the sequence $\{r_k\}$, $k = 2,3,\ldots$ be defined by

$$-\log r_k = \frac{1}{e} \lambda_{n_{k-1}}^{-(\alpha+\varepsilon)} .$$

Obviously, $r_k \to 1$ as $k \to \infty$. If $k > N$ and $r_k \leq r \leq r_{k+1}$, then by (2.4.9) and Cauchy's inequality,

$$\log M(r) \geq \log |a_{n_k}| + \lambda_{n_k} \log r_k$$

$$> \log |a_{n_N}| + (1-e^{-1}) \lambda_{n_k} \lambda_{n_{k-1}}^{-(\alpha+\varepsilon)} - \lambda_{n_N} \lambda_{n_{k-1}}^{-(\alpha+\varepsilon)} .$$

Therefore,

$$\log \log M(r) > \log \lambda_{n_k} - (\alpha+\varepsilon) \log \lambda_{n_{k-1}} + O(1)$$

$$= - \frac{\log \log (1/r_{k+1})}{\alpha+\varepsilon} + \log \log (1/r_k) + O(1)$$

$$\geq (1 - \frac{1}{\alpha+\varepsilon}) \log \log (1/r) + O(1).$$

The above inequality, on dividing both the sides by $-\log (1-r)$ and proceeding to limits, easily gives

$$\lambda \geq \frac{1}{\alpha} - 1 \text{ if } \alpha = 0 \text{ and } \lambda = \infty \text{ if } \alpha = 0. \quad \square$$

Corollary. Let $f(z) = \Sigma_{k=0}^{\infty} b_k z^{\lambda_k}$, *be analytic in* U *and have order* $\rho (0 \leq \rho < \infty)$ *and lower order* $\lambda (0 \leq \lambda < \infty)$. *If*

(i) $\lim_{k\to\infty} \frac{\log \lambda_{k-1}}{\log \lambda_k} = 1$

(ii) $\Psi(k) = |a_{k-1}/a_k|^{1/(\lambda_k - \lambda_{k-1})} \geq \frac{1}{e}$

and

(iii) $S \equiv \lim_{k\to\infty} (\log \lambda_k / \log (\frac{\lambda_k - \lambda_{k-1}}{\log^+ |a_k/a_{k-1}|}))$ *exists*

then f *is of regular growth and*

$$1 + \rho = 1 + \lambda = \max (1,S).$$

In view of (2.4.1), the corollary is an immediate consequence of Theorem 2.4.2.

If f is an admissible function (cf. Section 2.3) of positive order and $\Psi(k) = |b_{k-1}/b_k|^{1/(\lambda_k-\lambda_{k-1})}$ is nondecreasing for $k > k_o$, then the inequality in (2.4.7) can be replaced by equality. More precisely, we have

Theorem 2.4.3. *Let* $f(z) = \Sigma_{n=0}^{\infty} b_n z^{\lambda_n}$ *be an admissible function having order* $\rho(0 < \rho \leq \infty)$ *and lower order* $\lambda(0 \leq \lambda \leq \infty)$. *Suppose that* $\Psi(n) = |b_{n-1}/b_n|^{1/(\lambda_n-\lambda_{n-1})}$ *forms a nondecreasing function of* n *for* $n > n_o$. *Then,*

$$1+\lambda = \max \{1, \liminf_{n \to \infty} (\log \lambda_{n-1}/\log \{\frac{\lambda_n-\lambda_{n-1}}{\log |b_n/b_{n-1}|}\}) . \quad (2.4.10)$$

Proof. First let $\lambda < \infty$. Then, for every $\varepsilon > 0$ such that $0 < \varepsilon < 1 + \lambda$ and for all $n > n_o$, we have, as in the proof of Theorem 2.3.4,

$$\lambda_{n-1} \geq (\frac{1}{1-\Psi(n)})^{1+\lambda-\varepsilon} .$$

Therefore, on proceeding to limits after some simple calculations, it follows that

$$1 + \lambda \leq \liminf_{n \to \infty} \frac{\log \lambda_{n-1}}{-\log(1-\Psi(n))}$$

$$= \liminf_{n \to \infty} (\log \lambda_{n-1}/\log(\frac{\lambda_n-\lambda_{n-1}}{\log |a_n/a_{n-1}|})) .$$

If $\lambda = \infty$, then proceeding as above with an arbitrary large number in place of $(1+\lambda-\varepsilon)$, we get

$$\liminf_{n \to \infty} (\log \lambda_{n-1}/\log (\frac{\lambda_n-\lambda_{n-1}}{\log |a_n/a_{n-1}|})) = \infty .$$

Thus, in view of Theorem 2.4.2, (2.4.10) follows. □

Corollary. Let $f(z) = \Sigma_{n=0}^{\infty} b_n z^{\lambda_n}$ *be an admissible function having order* $\rho(0 < \rho \leq \infty)$ *and lower order* $\lambda(0 \leq \lambda \leq \infty)$. *Let* $\{\lambda_{n_k}\}_{k=0}^{\infty}$ *be the principal indices and* $R(n_k)$ *be the jump points* (cf. Section 1.4) *of the central index of* f, *then*

$$1 + \lambda = \max \{1, \liminf_{k \to \infty} \frac{\log \lambda_{n_{k-1}}}{-\log(1-R(n_k))}\} .$$

We now obtain another coefficient characterization of the lower order of an admissible function where the condition that $\Psi(n)$ is nondecreasing for $n > n_o$ is not needed.

Theorem 2.4.4. *Let* $f(z) = \Sigma_{n=0}^{\infty} b_n z^{\lambda_n}$ *be an admissible function having order* $\rho(0 < \rho \leq \infty)$ *and lower order* $\lambda(0 \leq \lambda \leq \infty)$. *Then*

$$1 + \lambda = \max_{\{n_k\}} \{\liminf_{k \to \infty} (\log\lambda_{n_{k-1}} / \log \{\frac{\lambda_{n_k} - \lambda_{n_{k-1}}}{\log^+ |a_{n_k} / a_{n_{k-1}}|}\})\} \qquad (2.4.11)$$

where $\{n_k\}_{k=0}^{\infty}$ *is any increasing sequence of natural numbers and the quotient occurring on the right-hand side is interpreted as zero for those values of* k *for which* $|a_{n_k} / a_{n_{k-1}}| \leq 1$.

Proof. Let $\{\lambda_{n_k}\}$ be the principal indices and $R(n_k)$ be the jump points of f. It follows from the discussions in Section 1.4 that

$$R(n_k) \to 1 \text{ as } k \to \infty ,$$

$$0 \leq R(n_k) < R(n_{k+1}) < 1$$

and that the central index $\nu(r)$ of f is given by

$$\nu(r) = \lambda_{n_{k-1}} \quad \text{for } R(n_{k-1}) \leq r < R_{n_k} .$$

Further,

$$R(n_k) = |b_{n_{k-1}} / b_{n_k}|^{1/(\lambda_{n_k} - \lambda_{n_{k-1}})} .$$

Therefore, since f is admissible and $\rho > 0$,

$$1 + \lambda = \liminf_{k \to \infty} \left(\log \lambda_{n_{k-1}} / \log \left\{\frac{\lambda_{n_k} - \lambda_{n_{k-1}}}{\log^+ |b_{n_k}/b_{n_{k-1}}|}\right\}\right). \tag{2.4.12}$$

But by Theorem 2.4.2,

$$1 + \lambda \geq \max_{\{n_m\}} \left\{ \left(\log \lambda_{n_{m-1}} / \log\left\{\frac{\lambda_{n_m} - \lambda_{n_{m-1}}}{\log^+ |a_{n_m}/a_{n_{m-1}}|}\right\}\right)\right\}. \tag{2.4.13}$$

On comparing (2.4.12) and (2.4.13), we get (2.4.11). □

The interrelations among type, lower type and the ratio $|b_n/b_{n-1}|$ for a function f, analytic in U and given by (2.1.1), can also be obtained as follows.

Theorem 2.4.5. *Let* $f(z) = \Sigma_{n=0}^{\infty} b_n z^{\lambda_n}$ *be analytic in* U *and have order* $\rho (0 < \rho < \infty)$, *type* T *and lower type* t $(0 \leq t \leq T < \infty)$. *Then,*

$$\frac{1}{\rho} \left(\frac{\rho+\beta}{\rho+1}\right)^{\rho+1} R \leq t \leq T \leq \frac{Q}{\rho} \tag{2.4.14}$$

where

$$R = \liminf_{n \to \infty} \left\{\lambda_{n-1} \left(\frac{1}{\lambda_n - \lambda_{n-1}} \log^+ |b_n/b_{n-1}|\right)^{\rho+1}\right\}$$

$$Q = \limsup_{n \to \infty} \left\{\lambda_n \left(\frac{1}{\lambda_n - \lambda_{n-1}} \log^+ |b_n/b_{n-1}|\right)^{\rho+1}\right\}$$

and

$$\beta = \liminf_{n \to \infty} \frac{\lambda_{n-1}}{\lambda_n}.$$

Proof. First assume that $0 < R < \infty$. For any ε such that $R > \varepsilon > 0$,

$$\log^+ |b_m/b_{m-1}| > (R-\varepsilon)^{1/(\rho+1)} \lambda_{m-1}^{-1/(\rho+1)} (\lambda_m - \lambda_{m-1})$$

for all $m > M = M(\varepsilon)$.

Writing the above inequality for m=M+1,...,n and adding all such inequalities,

$$\log^+ |b_n| > (R-\varepsilon)^{1/(\rho+1)}\{\lambda_n \lambda_{n-1}^{-1/(\rho+1)} - \Sigma_{m=M+1}^{n-1} \lambda_m (\lambda_m^{-1/(\rho+1)} - \lambda_{m-1}^{-1/(\rho+1)})$$

$$- \lambda_M^{\rho/(\rho+1)}\} + \log |a_M|$$

$$= (R-\varepsilon)^{1/(\rho+1)}\{\lambda_n \lambda_{n-1}^{-1/(\rho+1)} - \int_{\lambda_M}^{\lambda_{n-1}} n(t)d(t^{-1/(\rho+1)})$$

$$- \lambda_M^{\rho/(\rho+1)}\} + \log |a_M|$$

where $n(t) = \lambda_m$ for $\lambda_{m-1} \leq t < \lambda_m$. Therefore,

$$\log^+ |b_n| > (R-\varepsilon)^{1/(\rho+1)}\{\lambda_n \lambda_{n-1}^{-1/(\rho+1)} + \frac{1}{\rho+1} \int_{\lambda_M}^{\lambda_{n-1}} \frac{n(t)}{t} \times t^{-1/(\rho+1)} dt$$

$$- \lambda_M^{\rho/(\rho+1)}\} + \log |a_M|$$

$$> (R-\varepsilon)^{1/(\rho+1)}\{\lambda_n \lambda_{n-1}^{-1/(\rho+1)} + \frac{1}{\rho} \lambda_{n-1}^{\rho/(\rho+1)} - \frac{\rho+1}{\rho} \lambda_M^{\rho/(\rho+1)}\}$$

$$+ \log |a_M|$$

which gives

$$\lambda_{n-1}(\log^+ |b_n|^{1/\lambda_n})^{\rho+1} > (R-\varepsilon) \{1 + \frac{1}{\rho} \frac{\lambda_{n-1}}{\lambda_n} - o(1)\}^{\rho+1} + o(1).$$

Thus, on proceeding to limits and using (2.3.14),

$$\frac{(\rho+1)^{\rho+1}}{\rho^\rho} t \geq R \left(\frac{\rho+\alpha}{\rho}\right)^{\rho+1}.$$

This inequality is obviously true if $R = 0$. If $R = \infty$, the above arguments with an arbitrary large number in place of $(R-\varepsilon)$ give $t = \infty$. Thus, the first inequality in (2.4.14) follows. The inequality $t \leq T$ is obviously true.

The last inequality in (2.4.14) is obviously true if $Q = \infty$, so we assume that $Q < \infty$. For any $\varepsilon > 0$ and for all $m > M = M(\varepsilon)$, the definition of Q gives that

$$\log^+ |b_m/b_{m-1}| < (Q+\varepsilon)^{1/(\rho+1)} \lambda_m^{-1/(\rho+1)} (\lambda_m - \lambda_{m-1}).$$

Writing the above inequality for m = N+1, N+2,...,n and adding all such inequalities, it follows that

$$\log^+ |b_n| < (Q+\varepsilon)^{1/(\rho+1)} \{\lambda_n^{\rho/(\rho+1)} - \Sigma_{n=M+1}^{n-1} \lambda_m (\lambda_{m+1}^{-1/(\rho+1)} - \lambda_m^{-1/(\rho+1)})$$

$$- \lambda_M \lambda_{M+1}^{-1/(\rho+1)}\} + \log |a_M|$$

$$= (Q+\varepsilon)^{1/(\rho+1)} \{\lambda_n^{\rho/(\rho+1)} - \int_{\lambda_{M+1}}^{\lambda_n} n(t) d(t^{-1/(\rho+1)})$$

$$- \lambda_M \lambda_{M+1}^{-1/(\rho+1)}\} + \log |a_M|$$

where $n(t) = \lambda_m$ for $\lambda_m \leq t < \lambda_{m+1}$. This gives

$$\log^+ |b_n| < (Q+\varepsilon)^{1/(\rho+1)} \{\lambda_n^{\rho/(\rho+1)} + \frac{1}{\rho+1} \int_{\lambda_{M+1}}^{\lambda_n} t^{-1/(\rho+1)} dt$$

$$- \lambda_M \lambda_{M+1}^{-1/(\rho+1)}\} + \log |a_M|$$

$$= (Q+\varepsilon)^{1/(\rho+1)} (\frac{\rho+1}{\rho}) \lambda_n^{\rho/(\rho+1)} (1-o(1)) + o(1).$$

Therefore,

$$\lambda_n (\log^+ |b_n|^{1/\lambda_n})^{\rho+1} < (Q+\varepsilon) (\frac{\rho+1}{\rho})^{\rho+1} .$$

On proceeding to limits in the above inequality and using Theorem 2.2.2, we get

$$\frac{(\rho+1)^{\rho+1}}{\rho^\rho} T \leq Q \frac{(\rho+1)^{\rho+1}}{\rho^\rho} .$$

This proves the last inequality in (2.4.14). □

EXERCISES 2.4

1. Let $f(z) = \Sigma_{n=0}^{\infty} b_n z^{\lambda_n}$ be analytic in $\mathcal{U}$ and have order $\rho (0 < \rho < \infty)$, type T and lower type t. If

   (i) $\lambda_{n-1} \sim \lambda_n$ as $n \to \infty$

   and

   (ii) $\omega = \lim_{n\to\infty} \lambda_n (\frac{1}{\lambda_n - \lambda_{n-1}} \log^+ |b_n/b_{n-1}|)^{\rho+1}$ exists,

   prove that f is of perfectly regular growth and

   $$T = t = \frac{\omega}{\rho} .$$

2. Let

   $$F_1(z) = \Sigma_{k=0}^{\infty} \exp (2k+1)^{1/2} z^{2k+1} + \Sigma_{k=0}^{\infty} \exp (k^{1/2}) z^k .$$

   Show that F_1 is analytic in $\mathcal{U}$ and is of perfectly regular growth, while $R = 0$ and $Q = \infty$, where R and Q are as in Theorem 2.4.5. In fact, show that the order ρ, lower order λ, type T and lower type t of F_1 are

   $$\rho = \lambda = 1; \; T = t = \frac{1}{4} .$$

3. Let $f(z) = \Sigma_{n=0}^{\infty} b_n z^{\lambda_n}$, analytic in $\mathcal{U}$, have order $\rho (0 < \rho < \infty)$ and type T. If $\Psi(n) = |b_{n-1}/b_n|^{1/(\lambda_n - \lambda_{n-1})}$ is a nondecreasing function of n for $n > n_o$, show that

   $$\rho T \leq Q \leq \frac{(\rho+1)^{\rho+1}}{\rho^\rho} T$$

   where Q is as in Theorem 2.4.5.

4. Show that, for the function

   $$F_2(z) = 1 + \Sigma_{k=1}^{\infty} (r_1 \ldots r_k) z^k ,$$

   where $r_k = \exp (\rho T/k)^{1/(\rho+1)}$, $k = 1,2,\ldots$, $0 < \rho, T < \infty$, the following holds:

   (i) $F_2(z)$ is analytic in $\mathcal{U}$.

   (ii) $\Psi(k)$, as defined as Exercise 3, is an increasing function of k.

   (iii) $F_2(z)$ is of order ρ and type T.

   (iv) $\rho T = Q$, where Q is as in Theorem 2.4.5.

5. For the function g in Exercise 2.3.4, show that

 (i) The type of g is 1/4.

 (ii) Q = 1, where Q is as in Theorem 2.4.5.

Thus deduce that $(\rho+1)^{\rho+1}T/\rho^{\rho} = Q$ holds for g.

6. Let $f(z) = \Sigma_{n=0}^{\infty} b_n z^{\lambda_n}$, analytic in U, be of (α,β)-order $\rho(\alpha,\beta)$ (cf. Exercises 1.5). Assume that $\Psi(n)$ is as in Exercise 3. Then prove that, if $\alpha \not\equiv \beta$, then

$$\rho(\alpha,\beta) = \limsup_{n \to \infty} \{\alpha(\lambda_n)/(\beta(\frac{\lambda_n - \lambda_{n-1}}{\log^+ |b_n/b_{n-1}|}))\} .$$

7. If f is the same as in Exercise 6 and $\alpha(x) \equiv \beta(x) \not\equiv \log x$, prove that, for $\rho(\alpha,\alpha) \geq 1$,

$$\rho(\alpha,\alpha) = \max (1, \limsup_{n \to \infty} \{\alpha(\lambda_n)/(\alpha(\frac{\lambda_n - \lambda_{n-1}}{\log^+ |b_n/b_{n-1}|}))\}).$$

8. Let $f(z) = \Sigma_{n=0}^{\infty} b_n z^{\lambda_n}$, analytic in U, be of lower (α,β)-order $\lambda(\alpha,\beta)$ (cf. Exercises 1.5). Assume $\Psi(n)$ is as in Exercise 3. Then show that, if $\alpha \not\equiv \beta$,

$$\lambda(\alpha,\beta) = \liminf_{n \to \infty} \{\alpha(\lambda_{n-1})/(\beta(\frac{\lambda_n - \lambda_{n-1}}{\log^+ |b_n/b_{n-1}|}))\}. \qquad \text{(Nautiyal [1])}$$

9. Let f be the same as in Exercise 8 and $\alpha(x) \equiv \beta(x) \not\equiv \log x$, then show that, for $\lambda(\alpha,\alpha) \geq 1$,

$$\lambda(\alpha,\alpha) = \max (1, \liminf_{n \to \infty} \{\alpha(\lambda_{n-1})/(\alpha(\frac{\lambda_n - \lambda_{n-1}}{\log^+ |b_n/b_{n-1}|}))\}).$$

(Nautiyal [1])

10. By analogy to Theorem 2.4.4, find a charcterization for the lower (α,β)-order of a function $f(z) = \Sigma_{n=0}^{\infty} b_n z^{\lambda_n}$, analytic in U, in terms of the ratio of the consecutive coefficients $|b_n/b_{n-1}|$.

11. Find interrelations of the following growth parameters (cf. Exercises 1.5) of a function $f(z) = \Sigma_{n=0}^{\infty} b_n z^{\lambda_n}$, analytic in U, with the ratio of the consecutive coefficients $|b_n/b_{n-1}|$:

(i) q-order and lower q-order;

(ii) logarithmic order and lower logarithmic order;

(iii) generalized order and lower generalized order;

(iv) rational order and lower rational order.

2.5. DECOMPOSITION THEOREMS

The results obtained in this section are concerned with the decomposition of those functions, analytic in U, that are either not of regular growth or not of perfectly regular growth (cf. Section 1.5). The following lemmas are instrumental in the proofs of these results.

Lemma 2.5.1. *Let the maximum modulus* $M(r,f)$ *of a function* $f(z) = \Sigma_{n=0}^{\infty} b_n z^{\lambda_n}$ *analytic in* U, *satisfy*

$$(1-r)^{-b} < \log M(r,f) < (1-r)^{-B}, \quad 0 < b < B < \infty, \tag{2.5.1}$$

for all r *in* $r_o(b,B) < r < 1$. *Then, for any* μ, $b < \mu < B$,

$$f(z) = g_\mu(z) + h_\mu(z)$$

where g_μ *is analytic in* U *and for* r *sufficiently near to* 1,

$$\log M(r,g_\mu) \leq \left(\frac{\mu^\mu}{(1+\mu)^{(1+\mu)}} + o(1)\right) (1-r)^{-\mu} \tag{2.5.2}$$

and $h_\mu(z) = \Sigma_{n=0}^{\infty} b_{m_n} z^{\lambda_{m_n}}$ *is such that, for* r *sufficiently near to* 1,

$$\log M(r,f) \geq \frac{\mu^{\mu\beta}}{(1+\mu)^{1+\mu\beta}} (\log \frac{1}{r})^{-\mu\beta}, \tag{2.5.3}$$

the number β *being defined by*

$$\beta = \liminf_{n \to \infty} \frac{\log \lambda_{m_n}}{\log \lambda_{m_{n+1}}}. \tag{2.5.4}$$

Proof. Let $g_\mu(z) = \Sigma_{n=0}^{\infty} a_n z^{\lambda_n}$, where

$$a_n = b_n, \text{ if } \log^+ |b_n| < \lambda_n^{\mu/(1+\mu)}$$

$$= 0, \text{ otherwise.}$$

Then, g_μ is analytic in U, and it follows by Lemma 2.2.2 that, for r sufficiently near to 1,

$$\log M(r,g_\mu) \leq \left(\frac{\mu^\mu}{(1+\mu)^{1+\mu}} + o(1)\right) (1-r)^{-\mu}.$$

Let

$$h_\mu(z) = f(z) - g_\mu(z) = \Sigma_{n=0}^{\infty} b_{m_n} z^{\lambda_{m_n}}$$

and set $A_{m_n} = |b_{m_n}|$. Then,

$$\log^+ A_{m_n} \geq \lambda_{m_n}^{\mu/(1+\mu)} .$$

Now define the sequence $\{r_{m_n}\}$ by

$$\log \frac{1}{r_{m_n}} = \frac{\mu}{1+\mu} \lambda_{m_n}^{-1/(1+\mu)} .$$

Then $r_{m_n} \to 1$ as $n \to \infty$. Choose r such that $r_{m_n} \leq r < r_{m_{n+1}}$. In view of the above estimate of $\log^+ A_{m_n}$ and by Cauchy inequality, for r sufficiently near to 1,

$$\begin{aligned} \log M(r,f) &\geq \log A_{m_n} - \lambda_{m_n} \log \frac{1}{r} \\ &\geq \log A_{m_n} - \lambda_{m_n} \log \frac{1}{r_{m_n}} \\ &\geq \left(\frac{1}{1+\mu}\right) \lambda_{m_n}^{\mu/(1+\mu)} . \end{aligned} \qquad (2.5.5)$$

Since (2.5.3) trivially holds if $\beta = 0$, assume that $\beta > 0$. Then, by (2.5.4), for any $\beta' < \beta$ and n sufficiently large,

$$\lambda_{m_n} > \lambda_{m_{n+1}}^{\beta'} .$$

Therefore, by (2.5.5), for r sufficiently near to 1,

$$\log M(r,f) > \left(\frac{1}{1+\mu}\right) \lambda_{m_{n+1}}^{\beta'\mu/(1+\mu)}$$

$$> \frac{\mu^{\beta'\mu}}{(1+\mu)^{1+\beta'\mu}} \left(\log \frac{1}{r}\right)^{-\beta'\mu} .$$

The inequality (2.5.3) now follows from the above inequality, since $\beta' < \beta$ is arbitrary. □

Lemma 2.5.2. *Let the maximum modulus* $M(r,f)$ *of a function* $f(z) = \Sigma_{n=0}^{\infty} b_n z^{\lambda_n}$, *analytic in* U, *satisfy*

$$c(1-r)^{-B} < \log M(r,f) < C(1-r)^{-B}, \quad 0 < c < C < \infty, \ 0 < B < \infty, \tag{2.5.6}$$

for all r *in* $r_o(c,C,B) < r < 1$. *Then, for any* ζ, $c < \zeta < C$,

$$f(z) = g_\zeta(z) + h_\zeta(z)$$

where $g_\zeta(z)$ *is analytic in* U *and, for* r *sufficiently near to* 1,

$$\log M(r,g_\zeta) \le (\zeta+o(1))(1-r)^{-B} \tag{2.5.7}$$

and $h_\zeta(z) = \Sigma_{n=0}^{\infty} b_{k_n} z^{\lambda_{k_n}}$ *is such that, for* r *sufficiently near to* 1,

$$\log M(r,f) \ge \delta\zeta\left(\log \frac{1}{r}\right)^{-B} \tag{2.5.8}$$

the number δ *being given by*

$$\delta = \liminf_{n \to \infty} \left(\frac{\lambda_{k_n}}{\lambda_{k_{n+1}}}\right)^{B/(1+B)} . \tag{2.5.9}$$

Proof. Let $g_\zeta(z) = \Sigma_{n=0}^{\infty} a_n z^{\lambda_n}$, where

$$a_n = b_n, \text{ if } \log^+ |b_n| < K\lambda_n^{B/(1+B)}$$

$$= 0, \text{ otherwise,}$$

and $K = \zeta^{1/(1+B)} (1+B) B^{-B/(1+B)}$. Then, g_ζ is analytic in U. Applying Lemma 2.2.2 with $C = K$ and $D = B/(1+B)$, it follows that

$$\log M(r, g_\zeta) \leq (\zeta + o(1)) (1-r)^{-B}.$$

Now, setting

$$h_\zeta(z) = f(z) - g_\zeta(z) = \Sigma_{n=0}^{\infty} b_{k_n} z^{\lambda_{k_n}}$$

and proceeding exactly as in Lemma 2.5.1, (2.5.8) follows, the only change here being that we work with the sequence $\{r_{k_n}\}$ defined by

$$\log \frac{1}{r_{k_n}} = \frac{KB}{1+B} \lambda_{k_n}^{-1/(1+B)}$$

and the number δ given by (2.5.9) instead of the sequence $\{r_{m_n}\}$ and the number β as in Lemma 2.5.1. □

Theorem 2.5.1. *Let* $f(z) = \Sigma_{n=0}^{\infty} b_n z^{\lambda_n}$, *analytic in* U, *have order* $\rho (0 < \rho < \infty)$ *and lower order* $\lambda (0 < \lambda < \infty)$ *satisfying* $\lambda < \rho$. *Let* μ *be such that* $\lambda < \mu < \rho$. *Then,*

$$f(z) = g_\mu(z) + h_\mu(z)$$

where g_μ, *analytic in* U, *has order less than or equal to* μ *and*

$h_\mu(z) = \Sigma_{n=0}^{\infty} b_{m_n} z^{\lambda_{m_n}}$ *is such that*

$$\lambda \geq \mu \liminf_{n \to \infty} \frac{\log \lambda_{m_n}}{\log \lambda_{m_{n+1}}}. \tag{2.5.10}$$

Proof. Let ε be such that $\lambda > \varepsilon > 0$. Then, by the definition of the order and the lower order of f, it follows that, for all r sufficiently near to 1,

$$(1-r)^{-(\lambda-\varepsilon)} < \log M(r,f) < (1-r)^{-(\rho+\varepsilon)}.$$

Now, applying Lemma 2.5.1, with $b = \lambda-\varepsilon$ and $B = \rho+\varepsilon$, it follows that

$$f(z) = g_\mu(z) + h_\mu(z)$$

where g_μ is analytic in U, and for r sufficiently near to 1,

$$\log M(r,g_\mu) \leq \left(\frac{\mu^\mu}{(1+\mu)^{1+\mu}} + o(1)\right) (1-r)^{-\mu}.$$

The above inequality immediately gives that the order of g_μ does not exceed μ.

Further, (2.5.10) follows easily from (2.5.3). □

Theorem 2.5.2. Let $f(z) = \Sigma_{n=0}^{\infty} b_n z^{\lambda_n}$, *analytic in* U, *have order* $\rho (0 < \rho < \infty)$, *type* T *and lower type* t $(0 < t < T < \infty)$. *Let* ζ *be such that* $t < \zeta < T$. *Then,*

$$f(z) = g_\zeta(z) + h_\zeta(z)$$

where $g_\zeta(z)$, *analytic in* U, *is of growth* (ρ,ζ) (cf. Section 2.2) *and* $h_\zeta(z) = \Sigma_{n=0}^{\infty} b_{k_n} z^{\lambda_{k_n}}$ *satisfies*

$$t \geq \zeta \liminf_{n \to \infty} \left(\frac{\lambda_{k_n}}{\lambda_{k_{n+1}}}\right)^{\rho/(1+\rho)} . \tag{2.5.10}$$

Proof. Let ε be such that $t > \varepsilon > 0$. Then, by the definition of the type and the lower type of f, it follows that, for r sufficiently near to 1,

$$(t-\varepsilon)(1-r)^{-\rho} < \log M(r,f) < (T+\varepsilon)(1-r)^{-\rho}.$$

Applying Lemma 2.5.2, with $B = \rho$, $c = (t-\varepsilon)$ and $C = (T+\varepsilon)$, we get that

$$f(z) = g_\zeta(z) + h_\zeta(z)$$

where g_ζ is analytic in U and, for r sufficiently near to 1,

$$\log M(r,g_\zeta) \leq (\zeta+o(1))\ (1-r)^{-\rho}.$$

The above inequality gives that g_ζ is of growth (ρ,ζ).

Further, (2.5.10) follows from (2.5.8). □

EXERCISES 2.5

1. Let $f(z) = \Sigma_{n=0}^{\infty} b_n z^{\lambda_n}$ be analytic in U and have q-order $\rho(q)$ and lower q-order $\lambda(q)$, $0 < \lambda(q) < \rho(q)$, $q = 2,3,\ldots$, (cf. Exercises 1.5). Let $\mu(q)$ be such that $\lambda(q) < \mu(q) < \rho(q)$. Then, prove that

$$f(z) = g_{\mu(q)}(z) + h_{\mu(q)}(z)$$

where $g_{\mu(q)}$, analytic in U, has q-order not greater than $\mu(q)$ and $h(z) = \Sigma_{n=0}^{\infty} b_{m_n} z^{\lambda_{m_n}}$ satisfies

$$\lambda(q) \geq \mu(q) \liminf_{n \to \infty} \left\{\frac{\log^{[q-1]} \lambda_{m_n}}{\log^{[q-1]} \lambda_{m_{n+1}}}\right\}.$$

(Kapoor and Gopal [1])

2. Let $f(z) = \Sigma_{n=0}^{\infty} b_n z^{\lambda_n}$ be analytic in U and have q-order $\rho(q)$ (> 0), q type $T(q)$ and lower q-type $t(q)$, $0 < t(q) < T(q) < \infty$, $q = 2,3,\ldots$, (cf. Exercises 1.5). Let $\zeta(q)$ be such that $t(q) < \zeta(q) < T(q)$. Then prove that

$$f(z) = g_{\zeta(q)}(z) + h_{\zeta(q)}(z)$$

where $g_{\zeta(q)}$, analytic in U, has q-order not greater than $\rho(q)$ and q-type not greater than $\zeta(q)$, if of order $\rho(q)$. Further, $h_{\zeta(q)}(z) = \Sigma_{n=0}^{\infty} b_{k_n} z^{\lambda_{k_n}}$ satisfies

$$t(q) \geq \zeta(q) \liminf_{n \to \infty} \left\{\frac{\log^{[q-2]} \lambda_{k_n}}{\log^{[q-2]} \lambda_{k_{n+1}}}\right\}^{\alpha(q)}$$

where $\alpha(q) = \rho(2)/(1+\rho(2))$ if $q = 2$, and $\alpha(q) = 1$ if $q = 3,4,\ldots$.

(Kapoor and Gopal [1])

3. Let $f(z) = \Sigma_{n=0}^{\infty} b_n z^{\lambda_n}$ be analytic in U and have logarithmic order ρ_o and lower logarithmic order λ_o (cf. Exercises 1.5) such that $0 < \lambda_o < \mu_o < \rho_o < \infty$. Then show that

$$f(z) = g_{\mu_o}(z) + h_{\mu_o}(z)$$

where g_{μ_o}, analytic in U, has logarithmic order less than or equal to $\max(1,\mu_o)$ and $h_{\mu_o}(z) = \Sigma_{n=0}^{\infty} b_{m_n} z^{\lambda_{m_n}}$ satisfies

$$\lambda_o \geq \mu_o \liminf_{n \to \infty} \left\{\frac{\log\log \lambda_{m_n}}{\log\log \lambda_{m_{n+1}}}\right\}$$

(Kapoor and Gopal [2])

4. Let $f(z) = \Sigma_{n=0}^{\infty} b_n z^{\lambda_n}$ be analytic in U and have logarithmic order $\rho_o (1 < \rho_o < \infty)$, logarithmic type T_o and lower logarithmic type t_o (cf. Exercises 1.5) such that $0 < t_o < \zeta_o < T_o < \infty$. Then prove that

$$f(z) = g_{\zeta_o}(z) + h_{\zeta_o}(z)$$

where g_{ζ_o}, is analytic in U, has logarithmic order less than or equal to ρ_o and logarithmic type less than or equal to T_o, if of logarithmic order ρ_o. Further,

$$h_{\zeta_o}(z) = \Sigma_{n=0}^{\infty} b_{k_n} z^{\lambda_{k_n}}$$

satisfies

$$t_o \geq \zeta_o \liminf_{n \to \infty} \left\{\frac{\log \lambda_{k_n}}{\log \lambda_{k_{n+1}}}\right\}^{\rho} . \qquad \text{(Kapoor and Gopal [2])}$$

5. Let $f(z) = \Sigma_{n=0}^{\infty} b_n z^{\lambda_n}$, analytic in U, be of (α,β)-order $\rho(\alpha,\beta)$ and lower (α,β)-order $\lambda(\alpha,\beta)$ (cf. Exercises 1.5) such that $\sigma(\alpha,\beta) < \lambda(\alpha,\beta) < \mu(\alpha,\beta) < \rho(\alpha,\beta) < \infty$, where $\sigma(\alpha,\beta) = 0$, if $\alpha \not\equiv \beta$ and $\sigma(\alpha,\beta) = 1$, if $\alpha \equiv \beta$. Then show that

$$f(z) = g_{\mu(\alpha,\beta)}(z) + g^*_{\mu(\alpha,\beta)}(z) ,$$

where (α,β)-order of $g_{\mu(\alpha,\beta)}$ is at most $\mu(\alpha,\beta)$ and $h_{\mu(\alpha,\beta)}(z) = \Sigma_{n=0}^{\infty} b_{m_n} z^{\lambda_{m_n}}$ satisfies

$$\lambda(\alpha,\beta) \geq \mu(\alpha,\beta) \liminf_{n \to \infty} \left\{\frac{\alpha(\lambda_{n_m})}{\alpha(\lambda_{n_{m+1}})}\right\} . \qquad \text{(Nautiyal [1])}$$

6. Let $f(z)$, analytic in U, be of generalized order $\rho(h)$ and lower generalized order $\lambda(h)$ (cf. Exercises 1.5), such that $0 < \lambda(h) < \mu(h) < \rho(h)$. Then prove that

$$f(z) = g_{\mu(h)}(z) + g^*_{\mu(h)}(z)$$

where $g_{\mu(h)}$, analytic in U, has generalized order less than or equal to

$\max(1,\mu(h))$ and $g^*_{\mu(h)}(z) = \sum_{n=0}^{\infty} b_{m_n} z^{\lambda_{m_n}}$ satisfies

$$\lambda(h) \geq \mu(h) \liminf_{n \to \infty} \left\{\frac{h(\log \lambda_{m_n})}{h(\log \lambda_{m_{n+1}})}\right\}. \qquad \text{(Nautiyal [1])}$$

2.6. PROXIMATE ORDER AND COEFFICIENTS

We recall that a proximate order of a function f, analytic in $\mathcal{U}$ and having order $\rho(0 < \rho < \infty)$, is a real valued function $\rho(r)$ on (0,1), satisfying (1.6.1) - (1.6.4). In the theorem below we find the type T* of f with respect to a proximate order $\rho(r)$, defined by (1.6.4), in terms of the coefficients a_n in the Taylor series (1.4.1) of f.

Since (cf. Theorem 1.6.2), $(1-r)^{-\rho(r)}$ is a monotonically increasing function of r for $0 < r_o < r < 1$, a single valued function $\chi(x)$ can be defined for $x > x_o$, such that

$$x = (1-r)^{-\rho(r)-1} \text{ if and only if } (1-r)^{-1} = \chi(x).$$

Theorem 2.6.1. *Let* $f(z) = \sum_{n=0}^{\infty} a_n z^n$, *analytic in* $\mathcal{U}$, *have order* $\rho(0 < \rho < \infty)$ *and a proximate order* $\rho(r)$. *Then the type* T* *of* f *with respect to the proximate order* $\rho(r)$, (cf. Section 1.6.4), *is given by*

$$\frac{(\rho+1)^{\rho+1}}{\rho^{\rho}} T^* = \limsup_{n \to \infty} \left\{\frac{\chi(n) \log^+ |a_n|}{n}\right\}^{\rho+1} \qquad (2.6.1)$$

Proof. From (1.6.4), for every $\varepsilon > 0$ and for all r sufficiently close to 1,

$$\log M(r,f) < (T^*+\varepsilon)(1-r)^{-\rho(r)}.$$

Using Cauchy's estimate, this gives for all r sufficiently close to 1,

$$\log^+ |a_n| < (T^*+\varepsilon)(1-r)^{-\rho(r)} - n \log r. \qquad (2.6.2)$$

Now let

$$(1-r)^{-\rho(r)-1} = \frac{n}{\rho(T^*+\varepsilon)}$$

and

$$H(n) = 1 - \frac{1}{\chi(n/\rho(T^*+\varepsilon))} .$$

Then $H(n) \to 1$ as $n \to \infty$. Put

$$P(n) = \frac{1}{1 + \rho H(n)} .$$

Now, for all sufficiently large n, by (2.6.2),

$$\log^+ |a_n| < \frac{(T^*+\varepsilon)^{P(n)} n^{1-P(n)}}{1-P(n)} - n \log H(n).$$

Therefore

$$\frac{\chi(n) \log^+|a_n|}{n} < \frac{(T^*+\varepsilon)^{P(n)}}{\rho^{1-P(n)}} \frac{\chi(n)}{n^{P(n)}} \{1 - \frac{\rho n^{P(n)} \log H(n)}{(\rho(T^*+\varepsilon))^{P(n)}}\} . \tag{2.6.3}$$

Since

$$\lim_{n\to\infty} \frac{\chi(n)}{n^{P(n)}} = 1, \lim_{n\to\infty} \frac{n^{P(n)} \log H(n)}{(\rho(T^*+\varepsilon))^{P(n)}} = -1$$

and $\rho(r) \to \rho$ as $r \to 1$, (2.6.3) gives that

$$\frac{(\rho+1)^{\rho+1}}{\rho^{\rho}} T^* \geq \limsup_{n \to \infty} \{\frac{\chi(n) \log^+ |a_n|}{n}\}^{\rho+1} . \tag{2.6.4}$$

To prove the reverse inequality, let α be defined by the equation

$$\limsup_{n \to \infty} \{\frac{\chi(n) \log^+ |a_n|}{n}\}^{\rho+1} = \frac{(\rho+1)^{\rho+1}}{\rho^{\rho}} \alpha.$$

Then, for every $\beta > \alpha$ and for all r sufficiently close to 1,

$$|a_n| r^n < \exp \{\frac{n(1+\rho)}{\rho^{\rho/(1+\rho)}} \frac{1/(1+\rho)}{\chi(n)} + n \log r\}.$$

Using Exercise 1 of Section 1.6, for sufficiently large n and for all r sufficiently close to 1,

$$|a_n| r^n < \exp \{\frac{n(1+\rho)}{\rho\chi(n/\beta\rho)} - k(1-r)\}.$$

Thus, for r sufficiently close to 1, the maximum term $\mu(r)$ of f for $|z| = r$ satisfies

$$\log \mu(r) < \max_{n\geq 0} \{\frac{n(1+\rho)}{\rho\chi(n/\beta\rho)} - n(1-r)\} .$$

Another use of Exercise 1 of Section 1.6 gives that the maximum value of the right-hand side in the above inequality is attained for

$$n = [\beta\rho(1-r)^{-\rho(r)-1}]$$

so that it follows that, for r sufficiently close to 1,

$$\frac{\log \mu(r)}{(1-r)^{-\rho(r)}} < \beta .$$

Proceeding to limit and using Corollary 2 of Theorem 1.5.3, this gives $T^* \leq \beta$. Since the last inequality holds for all $\beta > \alpha$, we have $T^* \leq \alpha$. This and (2.6.4) together prove (2.6.1). □

In the end, we obtain the lower type t^* of a function f, analytic in $\mathcal{U}$, with respect to a proximate order (cf. Section 1.6), in terms of the coefficients a_n in the Taylor series (1.4.1) of f.

Theorem 2.6.2. *Let* $f(z) = \Sigma_{n=0}^{\infty} a_n z^n$, *analytic in* $\mathcal{U}$, *having order* ρ $(0 < \rho < \infty)$ *and proximate order* $\rho(r)$ *be such that* $\Psi(n) = |a_{n-1}/a_n|$ *forms a nondecreasing function of* n *for* $n > n_o$. *Then the lower type* t^* *of* f, *with respect to the proximate order* $\rho(r)$, (cf. (1.6.4)), *is given by*

$$\frac{(\rho+1)^{\rho+1}}{\rho^{\rho}} t^* = \liminf_{n \to \infty} \left\{\frac{\chi(n) \log^+ |a_n|}{n}\right\}^{\rho+1} \tag{2.6.5}$$

where $\chi(n)$ *is defined in the beginning of this section.*

Proof. Proceeding as in Theorem 2.3.9, it can be seen that $\Psi(n+1) > \Psi(n)$ for infinitely many values of n and $\Psi(n) \to 1$ as $n \to \infty$. When $\Psi(n+1) > \Psi(n)$, the maximum term $\mu(r)$ and the central index $\nu(r)$ of f, for $\Psi(n) \leq r < \Psi(n+1)$, are given by

$$\mu(r) = |a_n| r^n \text{ and } \nu(r) = n.$$

First, let $0 < t^* < \infty$. In view of Corollary 2 of Theorem 1.5.3, for given $\varepsilon > 0$ and for all r sufficiently close to 1,

$$\log \mu(r) > (t^*-\varepsilon)(1-r)^{-\rho(r)} .$$

Then, as in Theorem 2.3.9, if $a_{n_1} z^{n_1}$ and $a_{n_2} z^{n_2}$ are consecutive maximum terms of f, and $n_1 \leq n \leq n_2-1$,

$$\Psi(n_1+1) = \Psi(n_1+2) = \dots = \Psi(n+1) = \dots = \Psi(n_2)$$

and

$$|a_n| \, r^n = |a_{n_2}| \, r^{n_2} \quad \text{for } r = \Psi(n+1).$$

Therefore,

$$\log^+ |a_n| + n \log \Psi(n+1) > (t^*-\varepsilon) \{1-\Psi(n+1)\}^{-\rho\Psi(n+1)}.$$

Since $-\log x \geq 1-x$ for $x > 0$,

$$\frac{\chi(n) \log^+|a_n|}{n} > \frac{(t^*-\varepsilon)\ \chi(n)}{n} \{(1-\Psi(n+1))^{-\rho\Psi(n+1)}$$

$$- \frac{n}{t^*-\varepsilon} (1-\Psi(n+1))\}. \qquad (2.6.6)$$

Let

$$h(x) = (1-x)^{-\rho(x)} + \frac{n}{t^*-\varepsilon} (1-x).$$

The minimum value of the function $h(x)$ occurs at a point $x_1 = x_1(n)$ given by, for n sufficiently large,

$$(1-x)^{-\rho(x)-1} = \frac{n}{(t^*-\varepsilon)\ (\rho+o(1))}.$$

Using the definition of $\chi(x)$, we get

$$(1-x)^{-1} = \chi(\frac{n}{(t^*-\varepsilon)\ (\rho+o(1))}).$$

Thus, for n sufficiently large,

$$\inf_{0<x<1} h(x) = \{n/((t^*-\varepsilon)(\rho+o(1))\ \chi(\frac{n}{(t^*-\varepsilon)(\rho+o(1))})\}$$

$$+\{\frac{n}{(t^*-\varepsilon)\chi(n/(t^*-\varepsilon)(\rho+o(1)))}\}$$

$$=\{n/((t^*-\varepsilon)\ \chi(\frac{n}{(t^*-\varepsilon)(\rho+o(1))}))\ (\frac{1+\rho+o(1)}{\rho+o(1)})$$

Therefore, by using (2.6.6) and Exercise 1 of Section 1.6,

$$\liminf_{n\to\infty} \{\frac{\chi(n) \log^+ |a_n|}{n}\}^{\rho+1} \geq \frac{(\rho+1)^{\rho+1}}{\rho^\rho} t^*. \qquad (2.6.7)$$

Inequality (2.6.7) obviously holds if $t^* = 0$.

We now prove that strict inequality cannot hold in (2.6.7). For, if it holds, then there exists a number $\delta > t^*$, such that

$$\liminf_{n \to \infty} \left\{\frac{\chi(n) \log^+ |a_n|}{n}\right\}^{\rho+1} = \frac{(\rho+1)^{\rho+1}}{\rho^\rho} \delta.$$

Let δ_1 satisfy $\delta > \delta_1 > t^*$. Then, for all n sufficiently large

$$\log^+ |a_n| > \frac{n}{\chi(n)} \frac{1+\rho}{\rho^{\rho/(1+\rho)}} \delta_1^{1/(1+\rho)}.$$

Therefore, for sufficiently large n and all r sufficiently close to 1,

$$\log M(r,f) > \frac{n}{\chi(n)} \frac{(1+\rho)}{\rho^{\rho/(1+\rho)}} \delta_1^{1/(1+\rho)} + n \log r$$

$$\simeq \frac{n}{\chi(n)} \frac{(1+\rho)}{\rho^{\rho/(1+\rho)}} \delta_1^{1/(1+\rho)} - n(1-r).$$

Let

$$n = \left[\delta_1 \rho(1-r)^{-\rho(r)-1}\right].$$

Then, by the definition of $\chi(n)$, for sufficiently large n,

$$\log M(r,f) > \frac{n}{\rho\chi(n/\delta_1\rho)} = \delta_1(1-r)^{-\rho(r)}$$

so that $t^* \geq \delta_1$, which is a contradiction. □

EXERCISES 2.6

1. Let $f(z) = \Sigma_{n=0}^{\infty} b_n z^{\lambda_n}$, analytic in $\mathcal{U}$, have order $\rho(0 < \rho < \infty)$ and proximate order $\rho(r)$. Then, by analogy with Theorems 2.6.1 and 2.6.2, determine the type T* and lower type t* of f with respect to proximate order $\rho(r)$, in terms of the coefficients b_n and exponents λ_n.

2. Construct a function $F(z) = \Sigma_{n=0}^{\infty} b_n z^{\lambda_n}$, analytic in $\mathcal{U}$, having order 1 and proximate order $\rho(r) \not\equiv 1$, such that the type T* and the lower type t* of F with respect to proximate order $\rho(r)$ satisfy $t^* < T^*$.

3. Construct proximate orders $\rho_1(r)$ and $\rho_2(r)$ for the function

$F_1(z) = \Sigma_{n=0}^{\infty} \exp(n^{\alpha})z^n$, $0 < \alpha < 1$, such that

$$T_1^* < T < T_2^* ,$$

where T is the type of F_1 and T_1^* and T_2^* are types of F_1 with respect to proximate orders $\rho_1(r)$ and $\rho_2(r)$ respectively.

4. Construct proximate orders such that the parameter T*, defined by (1.6.4), for the function $F_2(z) = \Sigma_{n=0}^{\infty} \exp(n^{\alpha})z^{2n}$, $0 < \alpha < 1$, is (i) zero, (ii) infinity.

Let f be analytic in U_R, $0 < R < \infty$, and have order $\rho_R (0 < \rho_R < \infty)$, where

$$\rho_R = \limsup_{r \to R} \frac{\log^+ \log^+ M(r)}{\log (Rr/(R-r))} .$$

A function $\rho_R(r)$ is said to be a proximate order of f in U_R, if it satisfies the following conditions:

(i) $\rho_R(r)$ is positive, continuous and piecewise differentiable.

(ii) $\lim_{r\to R} \rho_R(r) = \rho_R$

(iii) $\rho_R'(r)(R-r) \log (\frac{R-r}{R}) \to 0$ as $r \to R$.

(iv) The parameter T_R, defined as

$$T_R = \limsup_{r \to R} \frac{\log M(r)}{(rR/(R-r))^{\rho_R(r)}} ,$$

is nonzero finite.

Then T_R is called the *type of* f *in* U_R *with respect to the proximate order* ρ_R. Similarly, t_R defined as

$$t_R = \liminf_{r \to 1} \frac{\log M(r)}{(rR/(R-r))^{\rho_R(r)}}$$

is said to be the *lower type of* f *in* U_R *with respect to the proximate order* ρ_R, if $0 < t_R < \infty$.

5. Using Theorems 2.6.1 and 2.6.2, determine T_R and t_R in terms of the coefficients a_n of a function $f(z) = \Sigma_{n=0}^{\infty} a_n z^n$, analytic in U_R.

6. Define a q-proximate order $\rho(q,r)$ that is associated with q-order $\rho(q)$, $0 < \rho(q) < \infty$, $q = 2,3,\ldots$, (cf. Exercises 1.5) of a function $f(z) = \Sigma_{n=0}^{\infty} a_n z^n$, analytic in U. Determine q-type $T^*(q)$, $0 < T^*(q) < \infty$, of f with respect to q-proximate order $\rho(q,r)$, in terms of the coefficients a_n, where

$$T^*(q) = \limsup_{r \to 1} \frac{\log^{[q-1]} M(r)}{(1-r)^{-\rho_q(r)}} .$$

7. Determine the proximate type and lower proximate type (cf. Exercises 1.6) of a function $f(z) = \Sigma_{n=0}^{\infty} a_n z^n$, analytic in U, in terms of the coefficients a_n.

REFERENCES

Aborn and Shankar [1] ;
Bajpai, Tanne and Whittier [1] ;
Beuermann [1] ;
Gopal [1] ;
Juneja [2] ;
Kapoor [2] , [3] ;
Kapoor and Gopal [1] , [2] ;
Kapoor and Juneja [1] ;
Kapoor and Nautiyal [1] ;
MacLane [1] ;
Nautiyal [1] ;
Seremeta [1] ;
Sons [1] , [2] .

3 Minimum modulus of analytic functions

3.1. INTRODUCTION

Let f be analytic in $\mathcal{U}$. For $0 \leq r < 1$, set

$$m(r) \equiv m(r,f) = \min_{0 \leq \theta \leq 2\pi} |f(re^{i\theta})| . \qquad (3.1.1)$$

Then, $m(r)$ is called the minimum modulus function of f. If we consider several simple examples, we find considerable diversity in the behaviour of the minimum modulus of analytic functions. When f possesses zeros, $m(r)$ cannot be expected in general to have such properties as monotonicity or convexity as are possessed by the maximum modulus function

$M(r) = \max_{0 \leq \theta \leq 2\pi} |f(re^{i\theta})|$, $\quad 0 < r < 1$. This makes the study of the behaviour of the minimum modulus for analytic functions much more complex. However, the rate of growth of $M(r)$ as $r \to 1$ does affect the rate of growth of $m(r)$ when r approaches 1 through a specified set contained in $(0,1)$. The present chapter is concerned with the study of this aspect of the behaviour of the minimum modulus of analytic functions in $\mathcal{U}$.

We consider bounded analytic functions in $\mathcal{U}$ in Section 3.2. The behaviour of minimum moduli of nonvanishing bounded analytic functions in $\mathcal{U}$ and Blaschke products is first investigated in this section and is followed by similar investigation for any bounded analytic function in $\mathcal{U}$. Section 3.3 deals with the behaviour of minimum modulus of analytic functions in $\mathcal{U}$ having order (cf. Section 1.5) at least one. Finally, in Section 3.4 we concentrate on the behaviour of minimum modului of those analytic functions in $\mathcal{U}$ that have order less than 1.

3.2 MINIMUM MODULUS OF BOUNDED ANALYTIC FUNCTIONS

Suppose $f \not\equiv 0$ is analytic and bounded in $\mathcal{U}$. Without loss of generality, we may assume that $|f(z)| < 1$ for $z \in \mathcal{U}$ and $f(0) \neq 0$. In this case, we have the factorization (cf. Exercises 1.2.12)

$$f(z) = B(z)\, g(z) \qquad (3.2.1)$$

where

$$B(z) = \Pi_{n=1}^{\infty} \frac{\bar{a}_n}{|a_n|} \frac{a_n - z}{1 - z\bar{a}_n}$$

is the Blaschke product formed with the zeros $\{a_n\}$ of f in U, and g, analytic and nonvanishing in U, is such that $|g(z)| < 1$ for z in U. Further, since $1/g$ is analytic and nonvanishing in U, the function $\log (1/g)$ is analytic and has a positive real part in U. Therefore, by the remark following Theorem 1.2.4, there exists a nondecreasing function $\Psi(t)$ on $|0,2\pi|$ such that

$$g(z) = \exp \{\frac{1}{2\pi} \int_0^{2\pi} \frac{e^{it}+z}{e^{it}-z} d\Psi(t) + i\lambda\} \tag{3.2.2}$$

where $\lambda = \text{Im } g(0)$. Let us have the normalization

$$\Psi(t) = \frac{\Psi(t+) - \Psi(t-)}{2}, \quad \Psi(0) = 0 \text{ and } \Psi(t-) = \Psi(t)$$

so that $\Psi(t)$ is uniquely determined. We use the symbols $\Psi(2\pi+)$ and $\Psi(0-)$ for $\Psi(2\pi)$ and $\Psi(0)$ to facilitate the presentation.

A function f, analytic and bounded by 1 in U, is said to be *exponentially dominated* in U, if there exist numbers $\alpha_o > 0$ and t_o, $-\infty < t_o < \infty$, such that, for $z = re^{i\theta}$, $r < 1$,

$$|f(z)| \leq \exp \{-\alpha_o \frac{1-|z|^2}{|e^{it_o}-z|^2}\}$$

$$= \exp \{-\alpha_o P(r,\theta-t_o)\} \tag{3.2.3}$$

where

$$p(r,\theta-t_o) = \frac{1-r^2}{1+r^2-2r \cos (\theta-t_o)}.$$

We note that, if $\theta = t_o$, then $P(r,\theta-t_o) \to \infty$ as $r \to 1$ and if $\theta_o \neq t_o$, then $P(r,\theta-t_o) \to 0$ as $r \to 1$. It is clear that along at least one ray arg $z = t$ (with $t=t_o$) an exponentially dominated function tends to zero as $r \to 1$. It is also clear that if a function f is exponentially dominated in U, then the function g occurring in its factorization (3.2.1) is also exponentially dominated.

The exponential domination of a nonvanishing bounded analytic function g in $\mathcal{U}$ results from discontinuity of the function $\Psi(t)$ occurring in (3.2.2). This fact is observed in the following lemma:

Lemma 3.2.1. *Suppose* g *is analytic and nonvanishing in* $\mathcal{U}$ *and* $|g(z)| < 1$ *for* $z \in \mathcal{U}$. *Then,* g *is exponentially dominated in* $\mathcal{U}$ *if and only if the function* $\Psi(t)$ *in the representation of* g *has at least one discontinuity in* $[0,2\pi)$.

Proof. By (3.2.2), we have

$$|g(z)| = \exp \{ -\frac{1}{2\pi} \int_0^{2\pi} P(r,\theta-t)\, d\Psi(t)\}$$

$$\leq \exp \{-\alpha_o P(r,\theta-t_o)\} \tag{3.2.4}$$

for $\alpha_o > 0$ and $-\infty < t_o < \infty$, if and only if

$$\alpha_o \leq \frac{\frac{1}{2\pi}\int_0^{\pi} P(r,\theta-t)\ \ d\Psi(t)}{P(r,\theta-t_o)}$$

$$= \frac{\Sigma_{k=0}^{\infty} P(r,\theta-t_k^*) \{\frac{\Psi(t_k^*+) - \Psi(t_k^*-)}{2\pi}\}}{P(r,\theta-t_o)} \tag{3.2.5}$$

where, $\{t_k^*\}_{k=0}^{\infty}$ are the discontinuities of Ψ in $[0,2\pi)$. Letting $r \to 1$ in (3.2.5), it follows that (3.2.4) holds, if and only if

$$\alpha_o \leq \max_k \{\frac{\Psi(t_k^*+) - \Psi(t_k^*-)}{2\pi}\}$$

$$= \max_{0\leq t\leq 2\pi} \{\frac{\Psi(t+) - \Psi(t-)}{2\pi}\} \equiv \beta_o \quad \text{(say)}. \tag{3.2.6}$$

The assertion of the lemma now follows from (3.2.4) and (3.2.6). □

Now, we first find the behaviour of the minimum modulus of a bounded, nonvanishing analytic function.

Theorem 3.2.1. *Let* $g(z)$ *be analytic and nonvanishing in* $\mathcal{U}$. *Suppose* $|g(z)| \leq 1$ *for* $z \in \mathcal{U}$. *Then,*

$$\lim_{r\to 1} (1-r) \log m(r) = -2\beta_o \tag{3.2.7}$$

where β_o *is defined by* (3.2.6) *and* $m(r)$ *is the minimum modulus of* g *on* $|z| = r$.

Proof. It is clear from the proof of Lemma 3.2.1 that if β_o is as in (3.2.6) then, for some t_o in $0 \leq t_o \leq 2\pi$,

$$|g(z)| \leq \exp \{-\beta_o P(r,\theta-t_o)\}$$

where $z = re^{i\theta}$, $0 \leq r < 1$. Since

$$P(r,\theta-t_o) \geq \frac{1+r}{1-r}$$

it follows that

$$\log m(r) \leq -\beta_o \frac{1+r}{1-r}$$

and so

$$\limsup_{r \to 1} (1-r) \log m(r) \leq -2\beta_o. \tag{3.2.8}$$

To show that limit exists in (3.2.8) and is equal to $-2\beta_o$ we must show that $\liminf_{r \to 1} (1-r) \log m(r) \geq -2\beta_o$. To this end, we divide the interval $[0,2\pi]$ into n subintervals I_k, $k = 1,2,\ldots,n$, where

$$I_k = [\frac{2\pi(k-1)}{n}, \frac{2\pi k}{n}] .$$

For any $\varepsilon > 0$, it follows from the definition of β_o that there exists a positive integer n so large that the total variation of $\Psi(\theta)/(2\pi)$ over any three successive intervals I_{k_1}, I_{k_2}, I_{k_3}, where $k_1 = k-1 \pmod n$, $k_2 = k$, $k_3 = k+1 \pmod n$, is less than $\beta_o+\varepsilon$. Let J_k denote the closure of the subset of I interval complementary to $I_{k_1} \cup I_{k_2} \cup I_{k_3}$. Then

$$\log |g(re^{i\theta})| = -\frac{1}{2\pi} \left[(\int_{J_k} + \int_{I_{k_1} \cup I_{k_2} \cup I_{k_3}})P(r,\theta-t)\,d\Psi(t)\right]. \tag{3.2.9}$$

If $\theta \in I_k$, the first integral tends uniformly to zero as $r \to 1$, $k = 1,\ldots,n$, and, for $0 < r < 1$,

$$\left| \int_{I_{k_1} \cup I_{k_2} \cup I_{k_3}} P(r,\theta-t)\,d\Psi(t) \right| \le (\beta_o+\varepsilon)\left(\frac{1+r}{1-r}\right) .$$

Therefore, for r sufficiently close to 1 and $\theta \in I_k$, it follows by (3.2.9) that

$$\log |g(re^{i\theta})| > -(\beta_o+\varepsilon)\,\frac{1+r}{1-r} . \tag{3.2.10}$$

Now there are only finitely many intervals I_k, so that (3.2.10) holds uniformly in $[0,2\pi]$ for r sufficiently close to 1. Since ε is arbitrary, it follows that

$$\liminf_{r \to 1} (1-r) \log m(r) \ge -2\beta_o . \tag{3.2.11}$$

Inequalities (3.2.8) and (3.2.11) give (3.2.7). □

Next, we give our attention to the behaviour of the minimum modulus of Blaschke product $B(z)$ (cf.(1.2.1)). In place of $m(r,B) = \min_{|z|=r} |B(z)|$, it is more convenient to consider the function

$$m^*(r,B) = \sup_{r \le \rho < 1} m(\rho,B).$$

We shall also need the function

$$m^*(r,S;B) = \sup_{r \le \rho \le S(r)} m(\rho,B)$$

where $\omega = S(z)$ is the hyperbolic Mobius transformation defined as

$$\frac{\omega-1}{\omega+1} = \lambda\,\frac{z-1}{z+1}, \quad 0 < \lambda < 1. \tag{3.2.12}$$

With $B(z)$, given by (1.2.1), we associate the convergent product

$$b(z) = \Pi_{k=1}^{\infty} \frac{|a_k|+z}{1+|a_k|z}, \quad |z| < 1. \tag{3.2.13}$$

It is clear that

$$b(-r) = m(r,b) \le m(r,B). \tag{3.2.14}$$

Further, it follows from the above inequality that

$$m^*(r,b) \le m^*(r,B) \tag{3.2.15}$$

and

$$m^*(r,S,b) \leq m^*(r,S,B). \tag{3.2.16}$$

The following theorem gives the behaviour of the functions $m^*(r,B)$ and $m^*(r,S,B)$ as $r \to 1$.

Theorem 3.2.2. *For a convergent Blaschke product* B, *given by* (1.2.1),

$$\lim_{r\to 1} (1-r) \log m^* (r,S,B) = 0 \tag{3.2.17}$$

and, a fortiori,

$$\lim_{r\to 1} (1-r) \log m^*(r,B) = 0. \tag{3.2.18}$$

Proof. First note that, by (3.2.14), for $b(z)$ given by (3.2.13)

$$m^*(r,S,b) = \sup_{r\leq\rho<1} |b(-\rho)| , \quad 0 \leq r < 1.$$

Now, in view of (3.2.16), (3.2.17) is proved if we show that

$$\lim_{r\to 1} (1-r) \log m^*(r,S,b) = 0. \tag{3.2.19}$$

If (3.2.19) does not hold, then

$$\liminf_{r \to 1} (1-r) \log m^*(r,S,b) < 0. \tag{3.2.20}$$

Thus, for every $\eta > 0$ such that $-\eta$ is greater than the left-hand side of (3.2.20), there exists an increasing sequence $\{r_k\}$ of numbers in (0,1) such that $\lim_{k\to\infty} r_k = 1$ and

$$(1-r_k) \log m^*(r_k,S,b) < - \eta, \; k = 1,2,\dots . \tag{3.2.21}$$

Now let $\omega(z,r)$, $0 \leq r < 1$, be the unique bounded harmonic function defined on

$$G_r = U - \{x: -S(r) \leq x \leq -r\},$$

such that

$$\omega(z,r) = 0 \text{ on } |z| = 1$$

and

$$\omega(z,r) = 1 \text{ on } -S(r) \leq z \leq -r.$$

It is easily seen that

$$\omega(z,r) \equiv \omega(T_r(z),0) \tag{3.2.22}$$

where $T_r(z)$ is the unique hyperbolic Mobius transformation defined by

$$\frac{1-T_r(z)}{1+T_r(z)} = \sigma\,\frac{1-z}{1+z}$$

where $\sigma = (1-r)/(1+r)$. By using (3.2.21) and (3.2.22), for any z_o in U and for all $k = 1,2,3,\ldots$,

$$|\log b(z_o)| \le \omega(z_o,r_k)\log m^*(r_k,S,b)$$

$$< -\frac{\eta}{1-r_k}\,\omega(T_{r_k}(z_o),0). \tag{3.2.23}$$

We note that, with $z = x + iy$, $\omega(1,0) = 0$, $\omega_x(1,0) = -\tau$, $\tau > 0$, and $\omega_y(1,0) = 0$. Further $T_{r_k}(z_o)$ lies in an arbitrary small neighbourhood of $z = 1$. By extending $w(z,0)$ in some neighbourhood of the point $z=1$ so that it remains harmonic, we have

$$\omega(T_{r_k}(z_o),0) = \omega(T_{r_k}(z_o,0)) - \omega(1,0)$$

$$= \mathrm{Re}\left[T_{r_k}(z_o)-1\right]\,\omega_x(P_k,0)-\mathrm{Im}\left[T_{r_k}(z_o)\right]\,\omega_y(Q_k,0)$$

where P_k and Q_k are in U and tend to 1 as $k \to \infty$.

From the definition of the transformation $T_r(z)$ it therefore follows that

$$\lim_{k\to\infty}\frac{\omega(T_{r_k}(z_o),0)}{1-r_k} = \tau\,\mathrm{Re}\left(\frac{1-z_o}{1+z_o}\right).$$

Thus, by (3.2.23)

$$\log|b(z_o)| \le -\eta\tau\;\mathrm{Re}\left(\frac{1-z_o}{1+z_o}\right).$$

Since η and τ are independent of z, and $z_o \in U$ is arbitrary,

$$\log|b(z)| \le -\eta\tau\;\mathrm{Re}\left(\frac{1-z}{1+z}\right),$$

and this gives that $b(z)$ is exponentially dominated, which is a contradiction. This proves (3.2.19) and hence (3.2.17) follows. That the limit in (3.2.18) is zero follows trivially from (3.2.17). □

It is to be noted in the context of Theorem 3.2.2 that there are Blaschke products for which $\lim_{r\to 1} (1-r) \log m(r,b)$ does not exist (cf. Exercise 3.2.8).

Theorems 3.2.1 and 3.2.2 lead to the following result concerning the behaviour of the minimum modulus of any bounded analytic function in U.

Theorem 3.2.3. *Let* $f \not\equiv 0$ *be analytic and* $|f| < 1$ *in* U. *Then,*

$$\lim_{r\to 1} (1-r) \log m^*(r,f) = -2\beta_o \tag{3.2.24}$$

where β_o *is defined by* (3.2.6).

Proof. Without loss of generality we may consider $f(0) = 1$. Thus, by (3.2.1),

$$f(z) = B(z)g(z)$$

where $B(z)$ is the Blaschke product formed with zeros of f in U and $g(z) \neq 0$ in U. Since

$$m(r,g) \geq m^*(r,f) \geq m(r,f) \geq m(r,g)\ m(r,B) \tag{3.2.25}$$

it follows, on using Theorem 3.2.1, that

$$\limsup_{r\to 1} (1-r) \log m^*(r,f) \leq \lim_{r\to 1} (1-r) \log m(r,g) = -2\beta_o . \tag{3.2.26}$$

Now suppose it possible that

$$\liminf_{r \to 1} (1-r) \log m^*(r,f) < -2\beta_o .$$

Then, there exists a positive number ε and an increasing sequence $\{r_k\}$ in $(0,1)$ with $r_k \to 1$ as $k \to 1$ such that

$$\log m^*(r_k,f) < -(\beta_o+2\varepsilon) \left(\frac{1+r_k}{1-r_k}\right) , \quad k = 1,2,\ldots .$$

It follows from Theorem 3.2.1 that, for r sufficiently close to 1,

$$\log m(r,g) > - (\beta_o + \varepsilon) \frac{1+r}{1-r} .$$

Therefore, on using (3.2.25),

$$\log m(r,B) \leq \log m^*(r,f) - \log m(r,g)$$

$$< \log m^*(r_k,f) + (\beta_o+\varepsilon)\left(\frac{1+r}{1-r}\right)$$

$$< -(\beta_o+2\varepsilon)\left(\frac{1+r_k}{1-r_k}\right) + (\beta_o+\varepsilon)\left(\frac{1+r}{1-r}\right) . \qquad (3.2.27)$$

Now let $w(z)$ be the Mobius transformation defined by

$$\frac{1-w(z)}{1+w(z)} = \frac{1}{1+\eta}\,\frac{1-z}{1+z}, \qquad \eta > 0.$$

If r satisfies $r_k \leq r \leq w(r_k)$, then

$$\frac{1+r}{1-r} \leq (1+\eta)\frac{1+r_k}{1-r_k} .$$

Hence, for sufficiently large k, if $r_k \leq r \leq w(r_k)$, it follows that

$$\log m(r,B) \leq -(\beta_o+2\varepsilon)\left(\frac{1+r_k}{1-r_k}\right) + (\beta_o+\varepsilon)(1+\eta)\left(\frac{1+r_k}{1-r_k}\right) . \qquad (3.2.28)$$

If η is chosen sufficiently small, the coefficient of $(1+r_k)/(1-r_k)$ on the right-hand side of (3.2.28) is negative and independent of k. Let us denote this coefficient by $-\gamma$, $\gamma > 0$. Then, for k sufficiently large,

$$\log m^*(r_k,W,B) < -\gamma\,\frac{1+r_k}{1-r_k}$$

and we have

$$\liminf_{r \to 1}\,(1-r)\log m^*(r,W,B) < 0,$$

a contradiction of Theorem 3.2.2. This shows that

$$\liminf_{r \to 1}\,(1-r)\log m^*(r,f) = -2\beta_o .$$

The theorem now follows in view of (3.2.26). □

EXERCISES 3.2.

1. Find the minimum moduli on $|z| = r$ for the following functions and investigate their behaviour as $r \to 1$:

   (a) $\frac{1+z}{1-z}$ (b) $\frac{1-z}{1+z}$ (c) $\exp\left(\frac{1+z}{1-z}\right)$ (d) $\exp\left(-\frac{1+z}{1-z}\right)$ (e) $\sin\left(\frac{1}{1-z}\right)$

   (f) $\sin\left(\frac{1-z}{1+z}\right)$.

2. Let

$$f_o(z) = \exp\left\{-\alpha\left(\frac{1-z}{1+z}\right)^{1-\varepsilon}\right\}$$

   where $\alpha > 0$ and $0 < \varepsilon < 1$. Prove that

$$\lim_{r \to 1} m(r, f_o) = 0.$$

3. If f is analytic and one-to-one in U, then show that Ψ_1 and Ψ_2 defined as

$$\Psi_1(r) = \frac{(1+r)^2}{r} m(r,f) \quad \text{and} \quad \Psi_2(r) = \frac{(1+r)^3}{1-r} m(r,f')$$

   are nondecreasing functions in $[0,1)$. (Dinghas [1])

4. Let $f \not\equiv 1$ be analytic in U and

$$\alpha = \liminf_{r \to 1} m(r,f)$$

$$\beta = \limsup_{r \to 1} m(r,f).$$

   (a) If $\alpha = \beta = 1$, prove that

$$f(z) = C \prod_{m=1}^{n} \frac{\bar{z}_m}{|z_m|} \frac{z_m - z}{1 - \bar{z}_m z}$$

   where $|C| = 1$.

   (b) If $\alpha = 0$, $\beta = 1$, prove that

$$f(z) = C \prod_{m=1}^{\infty} \frac{\bar{z}_m}{|z_m|} \frac{z_m - z}{1 - \bar{z}_m z}$$

   where $|C| = 1$; and, *a fortiori*, f has infinitely many zeros in U.

5. If $\alpha = \beta = 0$, where α and β are as in Exercise 4, prove that f is exponentially dominated.

6. If $0 < \alpha = \beta < 1$, where α and β are as in Exercise 4, prove that f has a finite number of zeros in U.

7. Define a sequence $\{x_k\}$ and partial Blaschke products

$$b_n(z) = \Pi_{k=1}^{n} \frac{z+x_k}{1+x_k z}, \quad n = 1,2,\ldots$$

as follows. Let $x_1 = \frac{1}{2}$. Let $x_2 > x_1$ be such that

$$\max_{x_1<x<x_2} |b_2(-x)| = 1 - \frac{1}{2^2}.$$

Continuing this process, define $x_n > x_{n-1}$ such that

$$\max_{x_{n-1}<x<x_n} |b_n(-x)| = 1 - \frac{1}{2^n}.$$

Letting $\xi_n \in (x_{n-1}, x_n)$ satisfy $|b_n(-\xi_n)| = 1 - 1/2^n$, $n = 2,3,\ldots$, show that

(i) as $n \to \infty$, the sequence $\{b_n(z)\}$ converges in $|z| < 1$ to the product

$$b(z) = \Pi_{k=1}^{\infty} \frac{z+x_k}{1+x_k z};$$

(ii) $m(r,b) = b(-r)$, $0 < r < 1$;

(iii) $\lim_{n\to\infty} |b(-\xi_n)| = 1$.

(Heins [1])

8. Let b be the Blaschke product constructed in Exercise 7. Prove that

$$\liminf_{r\to 1} (1-r) \log m(r,b) = -\infty$$

$$\limsup_{r\to 1} (1-r) \log m(r,b) = 0.$$

9. Let f by analytic in U and $|f(z)| < 1$ for z in U. Let $M_o(r) = \sup_{r\leq\rho<1} |f(-\rho)|$ satisfy

$$\liminf_{r\to 1} (1-r) \log M_o(r) < 0.$$

Then, prove that f is exponentially dominated and that, for some $K > 0$,

$$f(z) = \exp\left(-K \frac{1-z}{1+z}\right) f_1(z),$$

where f_1 is analytic in U and satisfies $|f_1(z)| < 1$ in U.

10. Prove that $\lim_{r\to 1} (1-r) \log m^*(r,f)$ in Theorem 3.2.3 is

(i) $-\infty$ if and only if $f \equiv 0$,

(ii) 0 if and only if f is not exponentially dominated,

where $m^*(r,f) = \sup_{r\leq\rho<1} m(\rho,f)$.

11. If $f \not\equiv 0$ is analytic and $|f(z)| < 1$ in U and $m^*(r,f)$ is as in Exercise 10, then show that

$$\limsup_{r\to 1} (1-r) \log m(r,f) = \lim_{r\to 1} (1-r) \log m^*(r,f)$$

and that limit exists on the left-hand side of the above relation if and only if f has a finite set of zeros in U.

3.3. MINIMUM MODULUS OF FUNCTIONS HAVING ORDER GREATER THAN ONE

In the study of the minimum modulus of functions f, analytic in U, and having order $\rho > 1$ (cf. (1.5.1)), the transformation

$$1-z = (1-\omega)^\alpha, \qquad \tfrac{1}{2} \leq \alpha < 1 \tag{3.3.1}$$

plays a central role. We choose that branch of $(1-\omega)^\alpha$ in (3.3.1) which is positive when (1-z) is positive. If, under this transformation, D is the image in the z-plane of $|\omega| \leq 1$, then D is simply connected and each boundary point of D corresponds to some point $\omega = e^{i\Psi}$, $0 \leq \Psi < 2\pi$. For convenience we define the functions S and S^{-1} as

$$S(z) = 1-(1-z)^{1/\alpha}, \qquad S^{-1}(\omega) = 1-(1-\omega)^\alpha, \qquad \tfrac{1}{2} \leq \alpha < 1 \tag{3.3.2}$$

so that we are concerned with the tranformations

$$\omega = S(z) \quad \text{or} \quad z = S^{-1}(\omega).$$

We first show that $D \subset U \cup \{1\}$.

Theorem 3.3.1. $S^{-1}(|\omega| \leq 1) \subset U \cup \{1\}$.

Proof. For $\omega = e^{i\Psi}$, $0 \leq \Psi < 2\pi$, let z be any point such that

$$z = 1 - (1-e^{i\Psi})^\alpha$$
$$= 1 - e^{\frac{1}{2} i\alpha(\Psi-\pi)} \left(2 \sin \frac{\Psi}{2}\right)^\alpha.$$

Then

$$|z|^2 = 1-2 \cos \frac{\alpha(\Psi-\pi)}{2} \left(2 \sin \frac{\Psi}{2}\right)^\alpha + \left(2 \sin \frac{\Psi}{2}\right)^{2\alpha}. \tag{3.3.3}$$

Set

$$J(\Psi) \equiv 2 \cos \frac{\alpha(\Psi-\pi)}{2} (2 \sin \frac{\Psi}{2})^{\alpha} .$$

It is easily seen that the function $J(\Psi)$ is continuous in $(0,2\pi)$ and, since $\frac{1}{2} \leq \alpha < 1$,

$$J(0) = J(2\pi) = 2 \cos \frac{\pi\alpha}{2} = 2 \sin \frac{\pi(1-\alpha)}{2} \geq 2(1-\alpha).$$

Further, $J(\Psi)$ is minimum at $\Psi = \Psi^*$ satisfying $dJ/d\psi = 0$. The values of ψ^* are given by

$$(2 \sin \frac{\Psi^*}{2})^{\alpha-1} = - \frac{\sin(\alpha(\Psi^* - \pi)/2)}{\cos (\Psi^*/2)} . \tag{3.3.4}$$

Therefore, for the values of Ψ^* in $(0,2\pi)$ satisfying (3.3.4),

$$J(\Psi^*) = 2 \cos \frac{\alpha(\Psi^*-\pi)}{2} + \frac{2 \sin (\Psi^*/2) \sin (\alpha(\Psi^*-\pi)/2)}{\cos (\Psi^*/2)}$$

$$= \frac{2 \cos (\{\Psi^*-(\Psi^*-\pi)\alpha\}/2)}{\cos (\Psi^*/2)} = 2 \frac{\sin ((1-\alpha)(\pi-\Psi^*)/2)}{\sin ((\pi-\Psi^*)/2)} \geq 2(1-\alpha).$$

Thus, $J(\Psi) \geq 2(1-\alpha)$ for all Ψ in $[0,2\pi]$. Now, using (3.3.3), for all Ψ in $[0,2\pi]$,

$$|z|^2 \leq 1 - 2(1-\alpha) (2 \sin \frac{\Psi}{2})^{\alpha} \leq 1$$

with equality in the last inequality only for $\Psi = 0$. This shows $S^{-1}(|\omega| = 1) \subset \mathcal{U} \cup \{1\}$ and consequently the theorem follows. □

The next theorem gives the size of image of $|z-r| \leq h$, $0 < h < \frac{1}{4}(1-r)$, under the transformation $S(z)$.

Theorem 3.3.2. *Suppose* $0 < r < 1$ *and* $s = S(r)$, *where* S *is defined by* (3.3.2). *Let* $0 < h < \frac{1}{4}(1-r)$ *and*

$$h_1 = \tfrac{1}{2}h(1-r)^{\frac{1}{\alpha}-1} , \qquad h_2 = 3h(1-r)^{\frac{1}{\alpha}-1} .$$

Then,

$$\{|\omega-s| \leq h_1\} \subset S\{|z-r| \leq h\} \subset \{|\omega-s| \leq h_2\}.$$

Proof. Let z be any point on the circle $|z-r| = h$. Then

$$|1-z| \le 1-r + |r-z| \le \frac{5}{4}(1-r)$$

and, since $\frac{1}{2} \le \alpha < 1$,

$$\left|\frac{dS(z)}{dz}\right| = \frac{1}{\alpha}|1-z|^{\frac{1}{\alpha}-1} < 3(1-r)^{\frac{1}{\alpha}-1}.$$

By the definition of S in (3.3.2),

$$S(z) - s = (1-r)^{\frac{1}{\alpha}} - (1-z)^{\frac{1}{\alpha}}$$

so that

$$|S(z)-s| \le |z-r| \max_{|z-r|\le\frac{1}{4}(1-r)} \left|\frac{dS(z)}{dz}\right|$$

$$< 3h(1-r)^{\frac{1}{\alpha}-1} = h_2 ,$$

which shows that $S\{|s-r| = h\} \subset \{|\omega-s| \le h_2\}$ and consequently $S\{|z-r| \le h\} \subset \{|\omega-s| \le h_2\}$.

For the other inclusion relation of the theorem, we prove that $S^{-1}\{|\omega-s| \le h_1\} \subset \{|z-r| \le h\}$. To this end, we note that, by the definition of S^{-1} in (3.3.2),

$$S^{-1}(\omega) - r = (1-s)^{\alpha} - (1-\omega)^{\alpha}$$

and

$$h_1 = \frac{1}{2} h(1-r)^{\frac{1}{\alpha}-1} < \frac{1}{8}(1-s).$$

Therefore it follows that, for ω on the circle $|\omega-z| = h_1$,

$$|1-\omega| \ge 1-s - \omega|s-\omega| > \frac{7}{8}(1-s)$$

and so for these values of ω

$$|S^{-1}(\omega)-r| \le |\omega-s| \max_{|\omega-s| \le \frac{1}{8}(1-s)} \left|\frac{dS^{-1}(\omega)}{d\omega}\right|$$

$$< \frac{8}{7} h_1(1-s)^{\alpha-1} = \frac{4}{7} h < h.$$

The above inequality shows that $S\{|\omega-s| = h_1\} \subset \{|z-r| \leq h\}$ and the desired inclusion relation follows. □

Now we consider a function g, analytic in $\overline{U}$, and its associated function G defined as

$$G(\omega) = g(S^{-1}(\omega)) = g\{1-(1-\omega)^{\alpha}\}, \qquad \tfrac{1}{2} \leq \alpha < 1. \tag{3.3.5}$$

The function G is well defined in view of Theorem 3.3.1. Set

$$\begin{aligned} n_g(z,s) &= n(z,s,g) \\ &= \text{number of zeros of } g \text{ in } \{|\xi-z| \leq s\} \subset \overline{U}, \\ &\quad \text{the multiple zeros being counted according to} \\ &\quad \text{their multiplicity} \end{aligned} \tag{3.3.6}$$

and

$$N_g(z,h) = N(z,h,g) = \int_0^h \frac{n(z,s,g)}{s}\, ds.$$

It is clear that $n_g(z,s)$ is always finite. If on the measurable space $\{\overline{U},\mathcal{B}\}$, where $\mathcal{B}$ is the class of all Borel measurable subsets of $\overline{U}$, a Borel measure μ is defined by

$$\int_{|\xi-z|<s} d\mu = \mu\{|\xi-z| < s\} = n_g(z,s) \tag{3.3.7}$$

then, on integrating by parts, it follows easily that

$$\begin{aligned} N_g(z,h) &= \int_0^h \frac{n_g(z,s)}{s}\, ds \\ &= \int_{|\xi-z|<h} \log \left|\frac{h}{\xi-z}\right| d\mu . \end{aligned} \tag{3.3.8}$$

If $v(z) = \log |g(z)|$ and $V(z) = \log |G(\omega)|$, we define

$$\begin{cases} \omega_g(z,h) = v(z) + N_g(z,h) \\ W_g(\omega,h) = V(\omega) + N_G(\omega,h). \end{cases} \tag{3.3.9}$$

The following result is needed in the sequel.

Theorem 3.3.3. *Suppose* $0 < r < 1$ *and* $s = S(r)$, *where* S *is defined in* (3.3.2). *Let* $0 < h < \frac{1}{4}(1-r)$ *and*

$$h_1 = \tfrac{1}{2} h(1-r)^{\frac{1}{\alpha}-1} \quad \textit{and} \quad h_2 = 3h(1-r)^{\frac{1}{\alpha}-1} .$$

Then, for a function g *analytic in* $\overline{U}$ *and its associated function* G, *given by* (3.3.5),

$$n_G(s,h_1) \leq n_g(r,h) \leq n_G(s,h_2) \tag{3.3.10}$$

and

$$W_G(s,h_1) \leq \omega_g(r,h) \leq W_G(s,h_2) \tag{3.3.11}$$

where $n_f(x,s)$ *denotes the number of zeros of* f *in* $|z-x| \leq s$ *and* ω_g, W_G *are defined as in* (3.3.9).

Proof. We observe that a zero z_o of g corresponds to a zero ω_o of G, where $\omega_o = 1-(1-z_o)^{1/\alpha}$. Therefore the number of zeros of g(z) in $|z-r| \leq h$ is equal to the number of zeros of $G(\omega)$ in $S\{|z-r| \leq h\}$. Since, by Theorem 3.3.2, $S\{|z-r| \leq h\} \subset \{|\omega-s| \leq h_2\}$, it follows that $n_g(r,h) \leq n_G(s,h_2)$. Similarly, the inequality $n_G(s,h_1) \leq n_g(r,h)$ in (3.3.10) is also proved.

To prove (3.3.11), we have from (3.3.10) that for any η, $0 < \eta \leq 1$,

$$n_G(s,\eta h_1) \leq n_g(r,\eta h) \leq n_G(s,\eta h_2).$$

Since,

$$N(z,h) = \int_0^h \frac{n(z,s)}{s}\, ds = \int_0^1 \frac{n(z,\eta h)}{\eta}\, d\eta ,$$

dividing by η in the above inequalities and integrating we get

$$N_G(s,h_1) \leq N_g(r,h) \leq N_G(s,h_2).$$

Since $v(r) = V(s)$ by definition of the function G, (3.3.11) follows from (3.3.9). □

Theorem 3.3.4. *Let* g *be analytic in* $\overline{U}$ *and* G *be its associated function given by* (3.3.5). *If* $v(z) = \log |g(z)|$ *and* $V(\omega) = \log |G(\omega)|$, *then, for every* $\varepsilon > 0$, *there exists an* $r_o = r_o(\varepsilon) < 1$ *such that*

$$\frac{1}{2\pi}\int_0^{2\pi} V^+(e^{i\Psi})\, d\Psi$$

$$\leq \frac{1+\varepsilon}{\alpha\pi} \left(\cos \frac{\pi\alpha}{2}\right)^{-1/\alpha} \int_0^1 M(t,v^+)(1-t)^{1/\alpha-1}\, dt + M(r_o,v^+) \tag{3.3.12}$$

where $p^+(z) = \max(p(z),0)$ *and* $M(t,v^+) = \max_{|z|=t} v^+(z)$.

Proof. Denote $a(\Psi) = 1-(1-e^{i\Psi})^{\alpha}$, $0 \leq \Psi < 2\pi$. For every $\omega = e^{i\Psi}$, let $z = re^{i\theta}$ be a point such that $z = a(\Psi)$. Then

$$\frac{1}{2\pi}\int_0^{2\pi} v^+(e^{i\Psi})\, d\Psi = \int_0^{\pi} \{v^+[a(\Psi)] + v^+[a(2\pi-\Psi)]\}\, d\Psi. \tag{3.3.13}$$

Since, for $0 < \eta \leq \Psi \leq \pi$,

$$|a(\Psi)| = |a(2\pi-\Psi)| \leq r_o = r_o(\alpha,\eta) < 1,$$

it follows that

$$v^+[a(\Psi)] + v^+[a(2\pi-\Psi)] \leq 2M(r_o,v^+).$$

Thus

$$\frac{1}{2\pi}\int_{\eta}^{\pi} \{v^+[a(\Psi)] + v^+[a(2\pi-\Psi)]\}\, d\Psi \leq M(r_o,v^+). \tag{3.3.14}$$

Now it is to be shown that the contribution to the integral on the right-hand side of (3.3.13) in the range $(0,\eta)$ is dominated by the first expression on the right-hand side of (3.3.12)

Using (3.3.3), it is easily seen that

$$1-r^2 = (2\sin\frac{\Psi}{2})^{\alpha}\{2\cos\frac{\alpha(\Psi-\pi)}{2} - (2\sin\frac{\Psi}{2})^{\alpha}\}$$

so that, as $\Psi \to 0$,

$$\frac{d\Psi}{dr} \sim -(\alpha\cos\frac{\pi\alpha}{2})^{-1}(2\sin\frac{\Psi}{2})^{1-\alpha}.$$

Therefore, if $0 \leq \Psi \leq \eta$, where $\eta = \eta(\alpha,\varepsilon)$ is a sufficiently small positive number,

$$\left|\frac{d\Psi}{dr}\right| \leq \frac{1+\varepsilon}{\alpha}(\cos\frac{\pi\alpha}{2})^{-1/\alpha}(1-r)^{\frac{1}{\alpha}-1}.$$

Further,

$$v^+[a(\Psi)] + v^+[a(2\pi-\Psi)] \leq 2M(r,v^+).$$

Thus,

$$\frac{1}{2\pi}\int_0^{\eta}\{v^+\left[a(\Psi)\right] + v^+\left[a(2\pi-\Psi)\right]\}\, d\Psi$$

$$\leq \frac{1+\varepsilon}{\alpha\pi}\left(\cos\frac{\pi\alpha}{2}\right)^{-\frac{1}{\alpha}}\int_{|a(\eta)|}^{1} M(r,v^+)(1-r)^{\frac{1}{\alpha}-1}\, dr$$

$$\leq \frac{1+\varepsilon}{\alpha\pi}\left(\cos\frac{\pi\alpha}{2}\right)^{-\frac{1}{\alpha}}\int_0^1 M(s,v^+)(1-s)^{\frac{1}{\alpha}-1}\, ds. \tag{3.3.15}$$

Inequalities (3.3.14) and (3.3.15) together with (3.3.13) give estimate (3.3.12). □

To obtain a lower bound on $V(\omega) = \log|G(\omega)|$, where G is the associated function given by (3.3.5), we must get a suitable upper bound on $W_G(\omega,h)$ and lower bound on $N_G(\omega,h)$ where $W_G(\omega,h)$ and $N_G(\omega,h)$ are as in (3.3.9). This we do by obtaining the corresponding bounds of $\omega_g(z,h)$ and $N_g(z,h)$ for any function g(z) analytic in $|z| \leq 1$, instead of the particular function $G(\omega)$ which is analytic throughout in $|\omega| \leq 1$, except at the point $z = 1$, where it is bounded. The following Theorems 3.3.5, 3.3.7 and 3.3.8 give the desired estimates.

Theorem 3.3.5. *Let* g *be analytic in* $\overline{U}$ *and* $g(0) = 1$. *Let* $0 < s < 1$ *and* $h = \delta(1-s)$ *where* $0 < \delta \leq \frac{1}{2}$. *If* $v(z) = \log|g(z)|$, *then*

$$|\omega_g(s,h)| \leq \frac{A(\delta)}{1-s}\left\{\int_0^{2\pi} v^+(e^{i\Psi})\, d\Psi\right\} \tag{3.3.16}$$

and

$$n_g(s,h) \leq \frac{1+2\delta}{(1-s)}\left\{\frac{1}{2\pi}\int_0^{2\pi} v^+(e^{i\Psi})\, d\Psi\right\} \tag{3.3.17}$$

where $v^+(e^{i\Psi})=\max(v(e^{i\Psi}),0)$, ω_g *is defined by* (3.3.9) *and* $n_g(s,h)$ *is the number of zeros of* g *in* $|z-s| \leq h$.

Proof. By Poisson-Jensen's formula (cf. Appendix A.10), if g has m zeros at $a_1,\ldots,a_m$ in $\overline{U}$, then for $z = se^{i\theta}$,

$$\log|g(z)| = \frac{1}{2\pi}\int_0^{2\pi} \frac{1-s^2}{1-2s\cos(\theta-\Psi)+s^2} v(e^{i\Psi})\, d\Psi - \Sigma_{n=1}^{m} \log\left|\frac{1-\bar{a}_n z}{z-a_n}\right|$$

$$= \frac{1}{2\pi}\int_0^{2\pi} \frac{1-s^2}{1-2s\cos(\theta-\Psi)+s^2} v(e^{i\Psi})\, d\Psi - \int_{|\xi|<1} \log\left|\frac{1-\bar{\xi}z}{z-\xi}\right| d\mu \tag{3.3.18}$$

where μ is the Borel measure on $\overline{\mathcal{U}}$, defined by (3.3.7). We first prove inequality (3.3.16).

By the definition of ω_g (cf. (3.3.9)) and (3.3.8),

$$\omega_g(s,h) = \log|g(s)| + \int_{|\xi-s|<h} \log\left|\frac{h}{\xi-s}\right| d\mu$$

so that

$$\log|g(s)| + \int_{|\xi|<1} \log\left|\frac{1-s\bar{\xi}}{s-\xi}\right| d\mu = \omega_g(s,h) + \int_{|\xi|<1} K(s,\xi)\, d\mu \tag{3.3.19}$$

where

$$K(s,\xi) = \log\left|\frac{1-s\bar{\xi}}{h}\right| \quad \text{if} \quad |\xi-s| < h$$

$$= \log\left|\frac{1-s\bar{\xi}}{\xi-s}\right| \quad \text{if} \quad |\xi-s| \geq h .$$

If $|\xi-s| < h$, then, since $h = \delta(1-s)$,

$$\frac{1-|\xi|}{1-s} \geq \frac{1-s-h}{1-s} = 1 - \delta \geq \tfrac{1}{2}$$

so, in this case, it follows that

$$K(s,\xi) = \log\left|\frac{1-s\bar{\xi}}{h}\right| = \log\left|\frac{1-s^2+s(s-\bar{\xi})}{h}\right|$$

$$< \log\left(1 + \frac{2(1-s)}{h}\right)$$

$$= \log\left(1 + \frac{2}{\delta}\right)$$

$$< 2\log\left(1 + \frac{2}{\delta}\right)\frac{1-|\xi|}{1-s} .$$

Again, if $|\xi - s| \geq h$, then, since $\log(1+x) \leq x$ for $x \geq 0$,

$$K(s,\xi) = \tfrac{1}{2} \log \left|\frac{1-s\bar{\xi}}{s-\xi}\right|^2$$

$$= \tfrac{1}{2} \log \left(1 + \frac{(1-s^2)(1-|\xi|^2)}{|s-\xi|^2} \right)$$

$$\leq \frac{2(1-s)(1-|\xi|)}{h} = \frac{2}{\delta^2} \frac{1-|\xi|}{1-s} .$$

Thus, in all cases,

$$K(s,\xi) < B(\delta) \frac{1-|\xi|}{1-s}$$

where $B(\delta)$ is a constant depending only on δ. Now (3.3.18) and the above estimate of $K(s,\xi)$ when applied in (3.3.19) give

$$|\omega_g(s,h)| \leq |\log|g(s)|| + \int_{|\xi|<1} \log \left|\frac{1-s\bar{\xi}}{s-\xi}\right| d\mu| + |\int_{|\xi|<1} K(s,\xi)\, d\mu|$$

$$= |\frac{1}{2\pi} \int_0^{2\pi} \frac{1-s^2}{1-2s\cos(\theta-\Psi)+s^2} v(e^{i\Psi})d\Psi| + |\int_{|\xi|<1} K(s,\xi)d\mu|$$

$$\leq \frac{1+s}{1-s} \{\frac{1}{2\pi} \int_0^{2\pi} |v(e^{i\Psi})|\, d\Psi\} + \frac{B(\delta)}{1-s} \int_{|\xi|<1} (1-|\xi|)\, d\mu$$

$$\leq \frac{C(\delta)}{1-s} |\frac{1}{2\pi} \int_0^{2\pi} |v(e^{i\Psi})|d\Psi + \int_{|\xi|<1} \log \frac{1}{|\xi|}\, d\mu|$$

$$= \frac{C(\delta)}{1-s} |\frac{1}{2\pi} \int_0^{2\pi} |v(e^{i\Psi})|\, d\Psi + \frac{1}{2\pi} \int_0^{2\pi} v(e^{i\Psi})\, d\Psi|$$

$$\leq \frac{C(\delta)}{(1-s)2\pi} \int_0^{2\pi} \{|v(e^{i\Psi})| + v(e^{i\Psi})\}\, d\Psi$$

$$\leq \frac{C(\delta)}{(1-s)\pi} \int_0^{2\pi} v^+(e^{i\Psi})\, d\Psi ,$$

since $f(0) = 1$. The above inequality gives (3.3.16).

To get estimate (3.3.17), we have from (3.3.18)

$$\frac{1}{2\pi}\int_0^{2\pi} v(e^{i\Psi})\, d\Psi = \int_{|\xi|<1} \log \frac{1}{|\xi|}\, d\mu \,. \tag{3.3.20}$$

Set $s_1 = s+h = s+\delta(1-s)$. Then, in view of the definition of μ given by (3.3.7),

$$\int_{|\xi|<1} \log \left|\frac{1}{\xi}\right| d\mu = \int_{|\xi|\le 1} \log \left|\frac{1}{\xi}\right| d\mu + \int_{1>|\xi|>s_1} \log \left|\frac{1}{\xi}\right| d\mu$$

$$\geq \int_{|\xi|\le s_1} \log \left|\frac{1}{\xi}\right| d\mu \geq n(0,s_1) \log \frac{1}{s_1} \,.$$

Since $\delta \le \frac{1}{2}$, the above inequality together with (3.3.20) gives

$$n_g(0,s_1) \le \frac{1}{\log\left(\frac{1}{s_1}\right)} \int_{|\xi|<1} \log \left|\frac{1}{\xi}\right| d\mu$$

$$\le \frac{1}{2\pi(1-s_1)} \int_0^{2\pi} v^+(e^{i\Psi})\, d\Psi$$

$$= \frac{1}{2\pi(1-\delta)(1-s)} \int_0^{2\pi} v^+(e^{i\Psi})\, d\Psi$$

$$\le \frac{1+2\delta}{2\pi(1-s)} \int_0^{2\pi} v^+(e^{i\Psi})\, d\Psi \,.$$

Now, the disc $|z-s| < h$ is contained in the disc $|z| < s+h = s_1$, therefore $n_g(s,h) \le n_g(0,s_1)$ and (3.3.17) follows from the last inequality. □

Theorem 3.3.6. *Let* g *be analytic in* $\overline{U}$. *Suppose*

$$\{|\zeta-z_o| \le h\} \subset U$$

and $0 < d < h/2$. *Then there exists a set* S_d *of open discs contained in* $\{|\zeta-z_o| \le h\}$ *and having the sum of radii at most* d *such that, for* $z \in \{|\zeta-z_o| < \frac{h}{2}\} \cap S_d^c$,

$$N_g(z, \tfrac{h}{2}) \le n_g(z_o,h) \log \frac{16h}{d} \tag{3.3.21}$$

where $n_g(z_o,h)$ *is the number of zeros of* g *in* $|z-z_o| < h$, N_g *is defined by* (3.3.8) *and* S_d^c *denotes the complement of* S_d.

Proof. Denote $n_o \equiv n(z_o,h)$. For a positive integer k, choose $a_{k,1}$ to be any point in $|z-z_o| < h/2$ such that

$$n_g(a_k, \frac{d}{2.4^k}) > \frac{n_o}{2^k}.$$

Now, choose $a_{k,2}$ in $|z-z_o| < h/2$ such that $|a_{k,1}-a_{k,2}| \geq d/4^k$ and

$$n_g(a_{k,2}, \frac{d}{2.4^k}) > \frac{n_o}{2^k}.$$

Suppose $a_{k,1},\ldots,a_{k,p_{k-1}}$ have been chosen in this way. Let a_k,p_k be any point in $|z-z_o| < h/2$ such that

$$|a_{k,p_k} - a_{k,i}| \geq \frac{d}{4^k}, \quad i = 1,2,\ldots p_k-1$$

and

$$n_g(a_{k,p_k}, \frac{d}{2.4^k}) > \frac{n_o}{2^k}.$$

Let q_k denote the number of all points $a_{k,i}$ chosen in this way. We prove that $q_k \leq 2^k$.

Since the discs

$$|z-a_{k,i}| < \frac{d}{2.4^k}, \quad i = 1,2,\ldots,q_k$$

are mutually disjoint and are contained in $|z-z_o| < h$, it follows that

$$n_o = n_g(z_o,h) \geq \sum_{i=1}^{q_k} n_g(a_{k,i}, \frac{d}{2.4^k}) \geq \sum_{i=1}^{q_k} \frac{n_o}{2^k} = \frac{n_o q_k}{2^k},$$

which gives $q_k \leq 2^k$. Thus the system of points $\{a_{k,1},\ldots,a_{k,q_k}\}$ is complete in the sense that no additional point can be added to this system.

Let

$$S_d = \{z : |z-a_{k,i}| < \frac{d}{4^k}, 1 \leq i \leq q_k, 1 \leq k < \infty\}.$$

If $z \in \{|\zeta - z_o| < h/2\} \cap S_d^c$, then the point z cannot be any of $a_{k,i}$ and, in fact, for $1 \leq i \leq q_k$, $1 \leq k < \infty$,

$$|z - a_{k,i}| \geq \frac{d}{4^k}.$$

Since the system $\{a_{k,1}, \ldots, a_{k,q_k} : 1 \leq k < \infty\}$ is complete and z is not in this system, we must have

$$n_g(z, \frac{d}{2.4^k}) \leq \frac{n_o}{2^k}.$$

Therefore, for such a point z,

$$N_g(z, \frac{h}{2}) = \int_0^{h/2} \frac{n_g(z,t)}{t} dt = \int_{d/8}^{h/2} \frac{n_g(z,t)}{t} dt + \Sigma_{k=1}^{\infty} \int_{d/(2.4^{k+1})}^{d/(2.4^k)} \frac{n_g(z,t)}{t} dt$$

$$\leq n_g(z_o, \frac{h}{2}) \log \frac{4h}{d} + \Sigma_{k=1}^{\infty} n_g(z, \frac{d}{2.4^k}) \log 4$$

$$\leq n_o \left[\log \frac{4h}{d} + \Sigma_{k=1}^{\infty} \frac{\log 4}{2^k}\right]$$

$$= n_o \log \frac{16h}{d}$$

$$= n_g(z_o, h) \log \frac{16h}{d}.$$

This proves (3.3.21). □

Theorem 3.3.7. *Let* g *be analytic in* $\overline{U}$ *and, for* $|z| = r$, $0 < r < 1$ *and* $0 < h < 1-r$, *let*

$$n_g(z,h) \leq n_o.$$

Then there exists a set $F_r \subset (r, r + \frac{h}{2})$ *such that the Lebesgue measure of* F_r *is at least* $\frac{h}{4}$ *and, for* $|z| \in F_r$,

$$N_g(z, \frac{h}{2}) < n_o \log \frac{B}{h}, \tag{3.3.22}$$

where $n_g(z,h)$ *is the number of zeros of* g *in* $|\zeta - z| \leq h$, N_g *is defined as in* (3.3.8) *and* B *is a constant.*

Proof. If $z = se^{i\theta}$ is any point such that

$$r < s < r + \frac{h}{3}, \qquad |\theta| < \frac{h}{6r} \tag{3.3.23}$$

then

$$|se^{i\theta} - re^{i\theta}| < \frac{h}{2}, \qquad |re^{i\theta} - r| < \frac{h}{6}$$

and so z lies in the disc $|z-r| < \frac{h}{2}$. It follows, therefore, that, if $k = [\frac{6\pi r}{h}] + 1$ so that $\frac{\pi}{k} < \frac{h}{6r}$, the system of k discs

$$D(re^{2\pi ip/k}, \frac{h}{2}) \equiv \{z: |z-re^{2\pi ip/k}| < \frac{h}{2}, \; p = 0,1,\ldots,k-1\}$$

covers the annulus

$$\Omega \equiv \{z : r < |z| < r + \frac{h}{3}\}.$$

Now choose

$$d = \frac{h}{24k}.$$

Since $n_g(re^{i\theta},h) \le N$, it follows by Theorem 3.3.6 that, for every fixed p, there exists a set $S_{d,p}$ of discs, the sum of whose radii is at most d, such that if z lies in $D(re^{2\pi ip/k}, \frac{h}{2}) \cap S^c_{d,p}$, then

$$N_g(z, \frac{h}{2}) \le n_o \log \frac{16h}{d}. \tag{3.3.24}$$

Thus (3.3.22) holds for all $z \in \Omega \cup (\bigcup_{p=0}^{k-1} S^c_{d,p})$.

Now the sum of the radii of the discs $S_{d,p}$, $p = 0,\ldots,k-1$, is at most kd. Since a disc of radius ℓ can meet the circles $|z| = s$ for a set of s having Lebesgue measure at most 2ℓ, it follows that if W is a disc in any $S_{d,p}$, then the Lebesgue measure of the set

$$E = \{s : W \cap \{|z| = s\} \neq \phi\}$$

is at most $2kd = \frac{h}{12}$. Therefore (3.3.22) holds throughout the annulus Ω, except possibly on a set of circles $\{|z| = s\}$, where s runs through intervals $\{I_\alpha\}$ such that Lebesgue measure of $\bigcup_\alpha I_\alpha$ is at most $\frac{h}{12}$. Set

$$F_r = (r, r + \frac{h}{2}) - \bigcup_\alpha I_\alpha.$$

Then (3.3.22) holds for F_r. It is easily seen that the Lebesgue measure of F_r is at least $\frac{h}{3} - \frac{h}{12} = \frac{h}{4}$. This proves the theorem, since

$$\frac{16h}{d} = \frac{16x24hk}{h} \leq \frac{16x24x6\pi}{h}(r+h) < \frac{B}{h}. \qquad \square$$

For the functions f, analytic in U, the following result gives a lower bound on $u(z) = \log |f(z)|$ in terms of the maximum modulus function $M(s,v^+)$.

Theorem 3.3.8. *Let* $f(z)$ *be analytic in* U *and* $f(0) = 1$. *If* $\frac{1}{2} \leq \alpha < 1$ *and* $r \leq R < 1$, *then there exists a set* $F_{r,R} \subset (Rr, R(r + \frac{1}{16}(1-r)))$ *having Lebesque measure at least* $\frac{1}{32}R(1-r)$ *and a number* $r_o = r_o(\alpha) < 1$ *such that, when* $r_o \leq r < 1$ *and* $|z| \in F_{r,R}$,

$$u(z) = \log |f(z)| \geq -\left(\frac{1}{1-r}\right)^{\frac{1}{\alpha}} \log \frac{A(\alpha)}{1-r} \quad x \tag{3.3.25}$$

$$x \left\{\frac{2}{\alpha\pi}\left(R \cos \frac{\pi\alpha}{2}\right)^{-\frac{1}{\alpha}} \int_0^R M(x,u^+)(R-x)^{\frac{1}{\alpha}-1} dx + 2M(r_o,u^+)\right\}$$

where $u^+ = \max(u,0)$, $A(\alpha)$ *is a constant depending on* α, *and* $M(s,u^+) = \max_{|z|=s} u^+(z)$.

Proof. The function

$$g(z) = f(Rz)$$

is analytic in $|z| \leq 1$. We first show that, for every $\varepsilon > 0$, there is an $r_o = r_o(\alpha,\varepsilon) < 1$ such that, with $|z| = r$ and $h = \delta(1-r)$, $0 < \delta < \frac{1}{6}$, the following bounds on the function $\omega_g(z,h)$ (cf.(3.3.9)) and on the number of zeros $n_g(z,h)$ of the function g in the disc $|\zeta-z| \leq h$ hold:

$$\omega_g(z,h) \geq -A(\alpha,\delta)\left(\frac{1}{1-r}\right)^{\frac{1}{\alpha}} \left\{\int_0^1 M(x,u^+)(1-x)^{\frac{1}{\alpha}-1} dx + M(r_o,u^+)\right\} \tag{3.3.26}$$

$$n_g(z,h) \leq (1+6\delta)\left(\frac{1}{1-r}\right)^{\frac{1}{\alpha}} \left\{\frac{1+\varepsilon}{\alpha\pi}\left(\cos \frac{\pi\alpha}{2}\right)^{\frac{1}{\alpha}} \int_0^1 M(x,u^+)(1-x)^{\frac{1}{\alpha}-1} dx + M(r_o,u^+)\right\}. \tag{3.3.27}$$

Let G be the associated function (cf.(3.3.5)) of g. Since G is analytic throughout in $\overline{U}$ except at the point $\omega = 1$, where it is bounded, on applying

Theorems 3.3.5 and 3.3.4 we get, for $0 < s < 1$, $h_1 = \delta_1(1-s)$, $0 < \delta_1 < 1/8$, that there exists $r_o < 1$ such that

$$W_G(s,h_1) \geq -\frac{A(\delta_1)}{1-s}\{\int_0^{2\pi} V^+(e^{i\Psi})\,d\Psi\}$$

$$\geq -\frac{A(\alpha,\delta_1)}{1-s}\{\int_0^1 M(x,v^+)(1-x)^{\frac{1}{\alpha}-1}\,dx + M(r_o,v^+)\}$$

where $W_G(\omega,h)$ is defined by (3.3.9), $v^+(z) = \max(\log|g(z)|, 0)$, $V^+(z) = \max(\log|G(\omega)|, 0)$ and $M(x,v^+) = \max_{|z|=x} v^+(z)$. Now, if $s = S(r) = 1-(1-r)^{1/\alpha}$ and

$$h = 2h_1(1-r)^{1-\frac{1}{\alpha}} = 2\delta_1(1-s)(1-r)^{1-\frac{1}{\alpha}} = 2\delta_1(1-r),$$

then the hypothesis of Theorem 3.3.3 are satisfied with $\delta = 2\delta_1 < \frac{1}{4}$. Therefore it follows by (3.3.11) and the above inequality that, for $h = \delta(1-r)$,

$$\omega_g(r,h) \geq W_G(s,h_1) \geq -A(\alpha,\delta)(\frac{1}{1-r})^{\frac{1}{\alpha}}\{\int_0^1 M(x,v^+)(1-x)^{\frac{1}{\alpha}-1}\,dx+M(r_o,v^+)\}.$$

This proves inequality (3.3.26) by bringing the point $z = re^{i\theta}$ to the real axis by a rotation.

Similarly, by taking $s = S(r)$, $0 < \delta_2 < 1/2$, and

$$h_2 = \delta_2(1-s) < \tfrac{1}{2}(1-s), \qquad h = \frac{1}{3}h_2(1-r)^{1-\frac{1}{\alpha}} = \frac{1}{3}\delta_2(1-r),$$

the hypotheses of Theorems 3.3.3, 3.3.5 and 3.3.4 are satisfied with $\delta = \frac{1}{3}\delta_2 < \frac{1}{6}$. Therefore, by (3.3.10), (3.3.17) and (3.3.12), there exists $r_o < 1$ such that

$$n_g(r,h) \leq n_G(s,h_2)$$

$$\leq \frac{1+2\delta_2}{1-s}\{\frac{1}{2\pi}\int_0^{2\pi} V^+(e^{i\Psi})\,d\Psi\}$$

$$\leq \frac{(1+2\delta_2)}{(1-s)}\{\frac{(1+\varepsilon)}{\alpha\pi}(\cos\frac{\pi\alpha}{2})^{-\frac{1}{\alpha}}\int_0^1 M(x,v^+)(1-x)\,dx + M(r_o,v^+)\}$$

$$\leq (1+2\delta_2)(\frac{1}{1-r})^{\frac{1}{\alpha}}\{\frac{1+\varepsilon}{\alpha\pi}(\cos\frac{\pi\alpha}{2})^{-\frac{1}{\alpha}}\int_0^1 M(x,v^+)(1-x)^{\frac{1}{\alpha}-1}\,dx + M(r_o,v^+)\}.$$

This gives inequality (3.3.27), after suitable adjustments to the bound on the right-hand side of the above inequality and an appropriate rotation of the point $z = re^{i\theta}$ to bring it to the real axis.

Next, by Theorem 3.3.7, applied with $h = \frac{1}{8}(1-r)$, we have that there exists a set $F^* \subset (r, r + h/2)$ such that the Lebesgue measure of F^* is at least $\frac{h}{4}$ and, for $|z| \in F^*$,

$$N_g(z, \frac{h}{2}) \leq n_o \log \frac{B}{h}$$

where $n_o \geq n_g(z,h)$. It follows, therefore, on using (3.3.27) with $\varepsilon = \frac{1}{8}$, that, for $|z| \in F^*$,

$$N_g(z, \frac{h}{2}) \leq (\frac{1}{1-r})^{\frac{1}{\alpha}} \log \frac{B}{1-r} \times \tag{3.3.28}$$

$$\times \{\frac{2}{\alpha\pi} (\cos \frac{\pi\alpha}{2})^{-\frac{1}{\alpha}} \int_0^1 M(x,v^+)(1-x)^{\frac{1}{\alpha}-1} dx+2M(r_o,v^+)\}.$$

In view of (3.3.9), (3.3.26) and the above inequality, we have for $|z| \in F^*$,

$$\log |g(z)| = \omega_g(z, \frac{h}{2}) - N_g(z, \frac{h}{2})$$

$$\geq - (\frac{1}{1-r})^{\frac{1}{\alpha}} \log \frac{A(\alpha)}{1-r} \times$$

$$\times \{\frac{2}{\alpha\pi} (\cos \frac{\pi\alpha}{2})^{-\frac{1}{\alpha}} \int_0^1 M(x,v^+)(1-x)^{\frac{1}{\alpha}-1} dx + 2M(r_o,v^+)\}.$$

Let $R|z| = \sigma$ where $|z| \in F^*$. Then, for $z = \sigma e^{i\theta}$, $0 \leq \theta < 2\pi$,

$$\log |f(z)| = \log |g(\frac{z}{R})| \geq - (\frac{1}{1-r})^{\frac{1}{\alpha}} \log \frac{A(\alpha)}{1-r} \times$$

$$\times \{\frac{2}{\alpha\pi} (\cos \frac{\pi\alpha}{2})^{-\frac{1}{\alpha}} \int_0^1 M(x,v^+)(1-x)^{\frac{1}{\alpha}-1} dx+2M(r_o,v^+)\}. \tag{3.3.29}$$

But $v(z) = \log |g(z)| = \log |f(Rz)| = u(Rz)$, so that

$$M(x,v^+) = M(Rx,u^+).$$

Therefore,

$$\int_0^1 M(x,v^+)(1-x)^{\frac{1}{\alpha}-1} dx = \int_0^1 M(Rx,u^+)(1-x)^{\frac{1}{\alpha}-1} dx$$

$$= R^{-\frac{1}{\alpha}} \int_0^R M(x,u^+)(R-x)^{\frac{1}{\alpha}-1} dx$$

and by (3.3.29) we finally have, for $z = \sigma e^{i\theta}$,

$$u(z) = \log|f(z)| \geq -\left(\frac{1}{1-r}\right)^{\frac{1}{\alpha}} \log \frac{A(\alpha)}{1-r} \times$$

$$\times \left\{\frac{2}{\alpha\pi} \left(R \cos \frac{\pi\alpha}{2}\right)^{-\frac{1}{\alpha}} \int_0^R M(x,u^+)(R-x)^{\frac{1}{\alpha}-1} dx + 2M(r_o,v^+)\right\}. \tag{3.3.30}$$

Since $M(r_o,v^+) \leq M(r_o,u^+)$, inequality (3.3.25) follows from the above inequality.

Now, let $F = \{\sigma : \text{(3.3.30) holds for } z = \sigma e^{i\theta}\}$. Let $\delta = \frac{1}{8}$. Since $h = \delta(1-r) = \frac{1}{8}(1-r)$ and

$$F^* \subset (r, r + \frac{h}{2}) = (r, r + \frac{1}{16}(1-r)),$$

the set F is contained in the interval $(Rr, R(r + \frac{1}{16}(1-r))$. Further, since F^* has Lebesgue measure at least $\frac{h}{4} = \frac{1}{32}(1-r)$ the set F has the Lebesgue measure at least $\frac{1}{32} R(1-r)$. □

Using Theorem 3.3.8, the behaviour of the minimum modulus $m(r)$ for functions analytic in U and having order ρ, $1 < \rho < \infty$ can now be studied.

Theorem 3.3.9. *Let* f *be analytic in* U *and have order* ρ (cf.(1.5.1)). *If* $1 < \rho < \infty$, *then there exists a constant* $K(\rho)$, *such that*

$$-K(\rho) \leq \limsup_{r\to 1} \frac{\log m(r)}{\log M(r) \log\log M(r)} \leq 0$$

where $m(r)$ *is the minimum of* $|f(z)|$ *on* $|z| = r$, $0 \leq r < 1$.

Proof. We assume without loss of generality that $f(0) = 1$. Let $\rho(r)$ be a proximate order (cf. Section 1.6) for f such that

$$\limsup_{r\to 1} \frac{\log M(r)}{(1-r)^{-\rho(r)}} = 1.$$

Then there exists $r_1 < 1$ such that

$$M(r,u^+) = \log M(r) < 2(1-r)^{-\rho(r)}$$

for all r in $r_1 < r < 1$ where $u(z) = \log|f(z)|$ and $M(r) \equiv M(r,f)$.

Define α such that

$$\frac{1}{\alpha} = \begin{cases} \frac{1}{4}(3+\rho) & \text{if } 1 < \rho \leq 5 \\ 2 & \text{if } 5 < \rho < \infty . \end{cases}$$

Then $\frac{1}{2} \leq \alpha < 1$. Further, since $1 < \rho < \infty$,

$$\frac{1}{\alpha} \leq \frac{1}{4}(3+\rho) < \tfrac{1}{2}(1+\rho).$$

Therefore, since (cf. (1.6.17)) $(1-r)^{((1+\rho)/2)) -\rho(r)}$ is an increasing function of r for $r_2 < r < 1$, for $r_o = \max(r_1, r_2)$

$$\int_{r_o}^{r} M(x,u^+)(r-x)^{\frac{1}{\alpha}-1}\,dx < 2\int_{r_o}^{r}(1-x)^{\frac{1}{\alpha}-1-\rho(x)}\,dx$$

$$< 2(1-r)^{\frac{1+\rho}{2}-\rho(r)}\int_{r_o}^{r}(1-x)^{\frac{1}{\alpha}-\frac{1}{2}(3+\rho)}\,dx$$

$$< K(\rho)\left(\frac{1}{1-r}\right)^{\rho(r)-\frac{1}{\alpha}}.$$

Further,

$$\int_0^{r_o} M(x,u^+)(r-x)^{\frac{1}{\alpha}-1}\,dx \leq M(r_o,u^+)\int_0^{r_o}(r-x)^{\frac{1}{\alpha}-1}\,dx$$

$$< M(r_o,u^+).$$

Therefore, for $0 < r_o < r < 1$,

$$\int_0^{r} M(x,u^+)(r-x)^{\frac{1}{\alpha}-1}\,dx < K(\rho)\left(\frac{1}{1-r}\right)^{\rho(r)-\frac{1}{\alpha}} + M(r_o,u^+). \tag{3.3.31}$$

Since,

$$(Rr,\ R(r+\frac{1}{16}(1-r))) \subset (2r-1,r)$$

if $R = r$ and r is sufficiently near to 1, it follows by Theorem 3.3.8 and (3.3.31) that there exists $\sigma \in (2r-1,r)$, $0 < r_o < r < 1$, such that

$$\log m(\sigma) > -K(\rho)\log\frac{A}{1-r}\left\{\left(\frac{1}{1-r}\right)^{\rho(r)} + M(r_o,u^+)\left(\frac{1}{1-r}\right)^{\alpha}\right\}. \tag{3.3.32}$$

Next, by the definition of the proximate order $\rho(r)$, there exists a sequence $r_n \to 1$ as $n \to \infty$, such that

$$\log M(r_n) > \tfrac{1}{2}(1-r_n)^{-\rho(r_n)}. \tag{3.3.33}$$

Let $s_n = (1+r_n)/2$. Then for n sufficiently large, there exists a σ_n in each interval $(2s_n-1, s_n) = (r_n, \frac{1}{2}(1+r_n))$ such that, by (3.3.32),

$$\log m(\sigma_n) > -K(\rho)(\log \frac{1}{1-s_n}) \{(\frac{1}{1-s_n})^{\rho(s_n)} \tag{3.3.34}$$

$$+ M(r_o,u^+)(\frac{1}{1-s_n})^{\frac{1}{\alpha}}\}.$$

Since

$$1-\sigma_n < 1 - r_n = 2(1-s_n)$$

for n sufficiently large, (3.3.33) and Exercise 1.6.2 give that

$$\log M(\sigma_n) \geq \log M(r_n) > \tfrac{1}{2}(\frac{1}{1-r_n})^{\rho(r_n)}$$

$$\sim 2^{-\rho-1} (\frac{1}{1-s_n})^{\rho(s_n)}$$

so that, for n sufficiently large,

$$\log \log M(\sigma_n) > K(\rho) \log \frac{1}{1-s_n} .$$

The above inequality and (3.3.34), in view of the relation $1/\alpha < \frac{1}{2}(1+\beta)$, give that, for n sufficiently large,

$$\frac{\log m(\sigma_n)}{\log M(\sigma_n)} > -K(\rho) (\log \frac{1}{1-s_n}) (1+o(1))$$

$$> -K(\rho) (\log \log M(\sigma_n)) (1+o(1))$$

and so

$$\limsup_{r \to 1} \frac{\log m(r)}{\log M(r) \log \log M(r)} \geq -K(\rho).$$

Further, since $\rho > 1$, $M(r) \to \infty$ as $r \to 1$. Thus

$$\frac{\log m(r)}{\log M(r) \log \log M(r)} < \frac{\log M(r)}{\log M(r) \log \log M(r)}$$

which implies that

$$\limsup_{r \to 1} \frac{\log m(r)}{\log M(r) \log \log M(r)} \leq 0. \qquad \square$$

EXERCISES 3.3

1. Show that for every β, $1 \leq \beta < \infty$, there exists a function f_o, analytic in U and having order β, such that

$$\limsup_{r \to 1} \frac{\log m(r,f_o)}{\log M(r,f_o) \log\log M(r,f_o)} < 0 .$$

(Linden [1])

2. Let f be analytic in U and have infinite order. Then prove that

$$\limsup_{r \to 1} \frac{\log m(r,f)}{\log M(r,f) \log\log M(r,f)} = 0.$$

(Linden [2])

3. Let f be analytic in U and have order not less than σ, $0 < \sigma < \infty$. Then, prove that

$$\limsup_{r \to 1} \frac{\log m(r,f)}{\log M(r,f) \log\log M(r,f)} \geq -c\ (1 + \frac{1}{\sigma}).$$

(Linden [2])

4. Let $f_1(z) = \exp\ (-\frac{1}{1-z})\ (\log^{[q]} \frac{K}{1-z})^{\beta}$, where q is a positive integer, $\beta > -2$ and K is sufficiently large. Show that

(i) f_1 is of order 1;

(ii) $\log m(r,f_1) = -\frac{1}{1-r}\ (\log^{[q]} \frac{K}{1-r})^{\beta}$;

(iii) $\log M(r,f_1) < \frac{K_1(q,\beta)}{1-r} \Pi_{k=1}^{q}\ (\log^{[k]} \frac{1}{1-r})^{-2} (\log^{[q]} \frac{1}{1-r})^{\beta}$;

(iv) $\limsup_{r \to 1} \frac{\log m(r,f_1)}{\log M(r,f_1) \log\log M(r,f_1)} = -\infty.$

(Linden [1])

5. Let f be a nonvanishing analytic function in U satisfying f(0) = 0 and

$$\log M(r,f) < \frac{A}{1-r} \Pi_{k=1}^{q}\ (\log^{[k]} \frac{1}{1-r})^{-2}\ \Pi_{k=q}^{n}\ (\log^{[k]} \frac{1}{1-r})^{\alpha_k}$$

for $r_o < r < 1$, where α_k is real and $\alpha_q > 0$. Prove that there exists a constant $L \equiv L(q,n,r_o,\alpha_q,\dots,\alpha_n)$ such that, for $r_o < r < 1$,

$$\log m(r,f) > -\frac{LA}{1-r} \Pi_{k=q}^{n}\ (\log^{[k]} \frac{1}{1-r})^{\alpha_k}.$$

(Linden [1])

6. Prove that, if f is analytic in U and the order of f is 1, then for every $\varepsilon > 0$ there exists a constant K and a sequence $\{r_n\}$ such that $r_n \to 1$ as $n \to \infty$ and

$$\log m(r_n,f) > -K\ (1-r_n)^{1-\varepsilon}.$$

7. Prove that the number $K(\rho)$ in Theorem 3.3.9 is finite if $1 < \rho < \infty$ and $K(\rho) \to \infty$ as $\rho \to 1$.

3.4 MINIMUM MODULUS OF FUNCTIONS HAVING ORDER LESS THAN ONE

In this section we focus our attention on the behaviour of the minimum modulus of those analytic functions that have order (cf. (1.5.1)) less than one. In such cases, instead of transformation (3.3.1) used for investigating the similar problem for analytic functions having order greater than one, we employ the transformation

$$\omega = S^*(z) = z - D(1-z)^{\beta} \tag{3.4.1}$$

where $0 < D \leq 1/12$ and $1 < \beta < 3$. We choose that branch of $(1-z)^{\beta}$ in (3.4.1) which is positive for $0 < z < 1$. As in the case of the transformation (3.3.1), it is easily seen that S^* is also analytic in $\mathcal{U}$ and maps $\mathcal{U}$ onto a region that contains $|\omega| \leq 1$ except for the point $\omega = 1$.

For a function f, analytic in $\mathcal{U}$, let f^* be defined by

$$f^*(\omega) = f(S^{*-1}(\omega)). \tag{3.4.2}$$

Then f^* is analytic in $|\omega| \leq 1$ except for the point $\omega = 1$.

For clarity in the description of the essential techniques, we shall assume throughout in this section that f is such that f^* is in the Nevanlinna class N (cf. Section 1.2), although it can be proved that if f is of order less than 1, then $f^* \in N$ (cf. Exercise 3.4.6). Set

$$W_f(z,h) = \log |f(z)| + N_f(z,h)$$

$$W_{f^*}(\omega,h) = \log |f^*(\omega)| + N_{f^*}(\omega,h).$$

where, for a function f_o, analytic in $\mathcal{U}$,

$$N_{f_o}(z,h) = \int_0^h \frac{n_{f_o}(z,s)}{s}\, ds, \tag{3.4.3}$$

$n_{f_o}(z,s)$ being the number of zeros of f_o in $\{|\xi - z| < s\} \subset \mathcal{U}$. A comparison of the general behaviour of f and f^* is as follows.

Theorem 3.4.1. *Let* f *be analytic in* U *and* f* *be as defined in* (3.4.2). *Then there exists* $r_o \equiv r_o(D,\beta)$ *such that, for* $0 < r_o < r < 1$, $0 < h < 1-r$ *and* $s = S^*(r)$,

$$n_f^*(s, \frac{h}{4}) \le n_f(r, \frac{h}{2}) \le n_{f^*}(s,h) \tag{3.4.4}$$

and

$$W_{f^*}(s, \frac{h}{4}) \le \omega_f(r, \frac{h}{2}) \le W_{f^*}(s,h). \tag{3.4.5}$$

Proof. From (3.4.1) it is clear that

$$\omega - s = z-r + D\{(1-r)^{\beta} - (1-z)^{\beta}\}. \tag{3.4.6}$$

If $|z-r| = \frac{\tau}{2} < \frac{1}{2}(1-r)$, then

$$|(1-r)^{\beta}-(1-z)^{\beta}|=|\int_0^1 \beta\{(1-r) + \sigma(r-z)\}^{\beta-1}(z-r)d\sigma|$$

$$\le \beta|z-r|(\frac{3}{2}(1-r))^{\beta-1}$$

and therefore, it follows from (3.4.6) that

$$|\omega-s| = |z-r| \{1+H(r,z)\}$$

where $|H(r,z)| \le \beta D(\frac{3}{2}(1-r)^{\beta-1}) \to 0$ as $r \to 1$. This implies that there exists a constant $r_o \equiv r_o(\beta,D)$ in $(0,1)$ such that

$$\frac{\tau}{4} < |\omega-s| < \tau \tag{3.4.7}$$

whenever $|z-r| = \tau/2$, $r_o < r < 1$ and $0 < \tau < 1-r$. Now (3.4.7) gives that

$$n_{f^*}(s, \frac{\tau}{4}) \le n_f(r, \frac{\tau}{2}) \le n_{f^*}(s,\tau) \tag{3.4.8}$$

and (3.4.4) follows from the above inequalities by replacing τ by h.

Dividing the terms of (3.4.8) by τ and integrating, we have

$$N_{f^*}(s, \frac{h}{4}) \le N_f(s, \frac{h}{2}) \le N_{f^*}(s,h)$$

and inequalities (3.4.5) follow from the above inequalities and (3.4.3) together with the fact that $f(r) = F(s)$. □

Before proving the main result of this section, we prove a lemma which is needed in the sequel.

Lemma 3.4.1. *Let* f *be analytic in* U, $f(0) = 1$, *and, for* $0 < \alpha < 1$ *and* $z \in U$,

$$\log |f(z)| < \frac{A}{(1-|z|)^{\alpha}} \, . \tag{3.4.9}$$

Let $r_o \in (0,1)$ *and* $a_1, a_2, \ldots, a_m$ *denote the zeros of* f *in* $\{z: |z| \leq (1+r_o)/2\}$. *Then* f_1 *defined by*

$$f(z) = f_1(z) \, \Pi_{n=1}^{m} \, (1 - \frac{z}{a_n}) \tag{3.4.10}$$

satisfies

$$\log |f_1(z)| < \frac{AK_1(r_o,\alpha)}{(1-|z|)^{\alpha}} \qquad \text{for } z \in U \tag{3.4.11}$$

$$|f_1(z)| > AK_2(r_o,\alpha) \qquad \text{for } |z| \leq r_o \tag{3.4.12}$$

and

$$\log m(r,f\ ) > AK_3(r_o,\alpha) + \log m(r,f_1) \qquad \text{for } \tfrac{1}{4}(3+r_o) \leq |z| \leq 1. \tag{3.4.13}$$

where $m(r,f)$ *and* $m(r,f_1)$ *are respectively minima of* $|f(z)|$ *and* $|f_1(z)|$ *for* $|z| = r$.

Proof. It is easily seen that f_1 is analytic in U and that $f_1(0) = 1$. From Jensen's theorem (cf. Appendix A.9), we get

$$m \leq \frac{A(\frac{1}{4}(1-r_o))^{-\alpha}}{\log\ ((3+r_o)/(2+2r_o))} \tag{3.4.14}$$

$$= A\ K_4(r_o,\alpha).$$

Thus, if $\frac{1}{4}(3+r_o) \leq |z| < 1$,

$$\Pi_{n=1}^{m} \ |1 - \frac{z}{a_n}| \geq (\tfrac{1}{4}\ (1-r_o))^{m} \tag{3.4.15}$$

and so (3.4.11) follows from (3.4.10).

Since f_1 has no zeros in $|z| < \frac{1}{2}(1+r_o)$, (3.4.12) can also be easily deduced.

To get (3.4.13), we combine (3.4.10), (3.4.14) and (3.4.15). □

Theorem 3.4.2. *Let* f *be analytic in* U *and* f(0) = 1. *Let, for* $0 \leq \alpha$, $r < 1$,

$$\log M(r) < A(1-r)^{-\alpha}.$$

Then there exist numbers $K = K(A,\alpha) > 0$, $s_o \equiv s_o(\alpha)$ *in* (0,1) *and a set* $F_s \subset (s, \frac{1}{2}(1+s))$ *for* $s \in (s_o,1)$, *such that the Lebesgue measure of* F_s *is at least* $\frac{1}{4}(1-s)$ *and, for* $|z| = r \in F_s$,

$$\log m(r) > -\frac{K}{1-r}\log\frac{1}{1-r} \tag{3.4.16}$$

where M(r) *is the maximum and* m(r) *is the minimum of* $|f(z)|$ *on* $|z| = r$.

Proof. In view of Lemma 3.4.1, we may assume without loss of generality that

$$\log|f(z)| < \frac{AK_1(r_o,\alpha)}{(1-|z|)^{\alpha}}, \qquad \text{for } |z| < 1, \tag{3.4.17}$$

and

$$|f(z)| > AK_2(r_o,\alpha), \qquad \text{for } |z| \leq r_o. \tag{3.4.18}$$

Let f* be as defined in (3.4.2). According to our assumption $f^* \in N$, so that

$$\int_0^{2\pi} \log^+ |f^*(\sigma e^{i\theta})|d\theta < K_4$$

for $0 < \sigma \leq 1$. By (3.4.18), it follows that $|f^*(0)| = |f(r_o)| > AK_2$. Thus, using Theorem 3.3.5, with $h = \frac{1}{2}(1-s)$ and $s = S^*(r)$,

$$|W_{f^*}(s,h)| \leq \frac{K_5}{1-s} \text{ and } n_{f^*}(s,h) \leq \frac{K_6}{1-s}.$$

Clearly, $1-s > 1-r$ and for any θ in $|0,2\pi)$, $f(ze^{i\theta})$ satisfies the same hypothesis as f(z). Therefore, by Theorem 3.4.1, for $k = \frac{1}{2}(1-r)$, $r_o < r < 1$,

$$\omega_f(re^{i\theta}, \tfrac{1}{2}k) \geq -\frac{K_7}{1-r} \text{ and } n_f(re^{i\theta}, \tfrac{1}{2}k) \leq \frac{K_8}{1-r}. \tag{3.4.19}$$

Since $n_f(re^{i\theta}, \frac{1}{2}k)$ denotes the number of zeros of f in $|z-re^{i\theta}| \leq \frac{1}{2}k$, by applying the second inequality in (3.4.15) at a suitable number of points we can cover $|z-re^{i\theta}| < k$ with a set of discs in each of which f has at most $3K_8/(1-r)$ zeros. The required number of such circles is independent of both r and θ, so that, for $\frac{1}{2}(1+r_o) < r < 1$,

$$n_f(re^{i\theta},k) \leq \frac{K_9}{1-r}\,. \tag{3.4.20}$$

Application of Theorem 3.3.7 now shows that, if $s \in (\frac{1}{2}(1+r_o),1)$, then, for $0 \leq \theta < 2\pi$,

$$N_f(re^{i\theta}, \frac{k}{2}) \leq \frac{K_{10}}{1-r} \log \frac{1}{1-r} \tag{3.4.21}$$

on a set of r in $(s, \frac{1}{2}(1+s))$ of measure at least $\frac{1}{4}(1-s)$. Inequality (3.4.16) with the stipulated conditions now follows by using (3.4.21) and (3.4.19) together with (3.4.3). □

The following result is a direct consequence of Theorem 3.4.2.

Theorem 3.4.3. *Let* f, *analytic and nonconstant in* U, *have order* $\rho(0 \leq \rho < 1)$. *Then there exist numbers* $K > 0$, $L \geq \frac{1}{4}$, $s_o \in (0,1)$ *and a set* $F_s \subset (s, (1+s)/2)$, *for* $s \in (s_o,1)$, *such that the Lebesgue measure of* F_s *is at least* $L(1-s)$ *and, for* $|z| = r \in F_s$,

$$\log m(r) > -\frac{K}{1-r} \log \frac{1}{1-r}$$

where $m(r)$ *is the minimum of* $|f(z)|$ *on* $|z| = r$.

EXERCISES 3.4

1. Let $f(z) = b_o + \Sigma_{n=1}^{\infty} b_n z^{\lambda_n}$ be analytic in U and have order ρ, $0 \leq \rho \leq \infty$. If

$$\liminf_{n\to\infty} \frac{\log(\lambda_{n+1} - \lambda_n)}{\log \lambda_{n+1}} > \frac{1}{2}\left(\frac{2+\rho}{1+\rho}\right)$$

then prove that

$$\limsup_{r\to 1} \frac{m(r,f)}{M(r,f)} = 1. \qquad \text{(Wiman [1])}$$

2. Prove that for each ρ, $0 < \rho < \infty$, there exists a function

$f(z) = b_o + \Sigma_{n=1}^{\infty} b_n z^{\lambda_n}$ such that

(i) f has order ρ;

(ii) $\liminf_{n\to\infty} \dfrac{\log(\lambda_{n+1} - \lambda_n)}{\log \lambda_{n+1}} = \frac{1}{2}\left(\frac{2+\rho}{1+\rho}\right)$;

(iii) $\limsup_{r \to 1} \frac{m(r,f)}{M(r,f)} \leq \frac{e^{4/\rho}}{e^{4/\rho}+2}$. (Nicholls and Sons [1])

*3. If $f(z) = b_o + \Sigma_{n=1}^{\infty} b_n z^{\lambda_n}$ analytic in U and having order $\rho(0 \leq \rho \leq \infty)$, satisfies

$$\liminf_{n \to \infty} \frac{\log^{[2]}(\lambda_{n+1} - \lambda_n)}{\log^{[2]} \lambda_{n+1}} = 1,$$

then does

$$\limsup_{r \to 1} \frac{\log m(r,f)}{\log M(r,f)} = 1$$

hold?

4. Let $f(z) = \Sigma_{n=0}^{\infty} b_n z^{\lambda_n}$ be analytic in U and its maximum term (cf. Section 1.4) $\mu(r,f)$ on $|z| = r$ satisfy $\mu(r,f) \to \infty$ as $r \to 1$. If f has finitely many zeros in U, show that

$$\lim_{r \to 1} \frac{\mu(r,f)}{m(r,f)} = \infty .$$

If f has infinitely many zeros in U, then show that

$$\limsup_{r \to 1} \frac{\mu(r,f)}{m(r,f)} = \infty. \quad \text{(Clunie and Hayman [1])}$$

*5. Does the result analogous to Theorem 3.4.3 hold when the condition on the order of the functions in this theorem is replaced by a similar condition on the lower order of the functions?

6. Let f, analytic in U, have order $\rho(0 \leq \rho < 1)$ and $f(0) = 1$. If f^*, given by (3.4.2), is the associated function of f, where $\rho\beta < 1$, then $f^* \in N$ (cf. Section 1.2).

(Linden [4])

REFERENCES

Clunie and Hayman [1],
Dinghas [1],
Hayman [1],
Heins [1],
Linden [1], [2], [4], [6] ,
Nicholls and Sons [1],
Wiman [1].

4 Applications – approximation and interpolation of analytic functions

4.1. INTRODUCTION

The theory of approximation of functions originated in the classical approximation theorem of Weierstrass. Chebyshev's concept of best approximation and the converse theorem of Bernstein on the existence of a function with a given sequence of best approximants gave it a broad base. Runge's celebrated theorem that if E is a connected region in the complex plane then an analytic function can be approximated, on arbitrary compact subsets of E, by rational functions aroused interest in the approximation of analytic functions by polynomials and rational functions on subsets of the complex plane. The most significant contribution in this direction is due to Mergelyan who showed that a necessary and sufficient condition for every function f continuous on a compact set E and analytic in the interior points of E to be uniformly approximable by polynomials is that E should not separate the plane.

It has long been observed that the order of magnitude of the minimal error in approximating a function f by polynomials of degree n is related to the differentiability properties of f. The more differentiable the function is, the better suited it is to approximation. Thus, analytic functions happen to occupy a privileged position as regards approximation by polynomials. The degree of approximation is seen to depend not only on the region of analyticity of the function in question but also on its growth.

In the present chapter we consider the representation of analytic functions in the complex domain by sequence of polynomials, the sequences being defined either by interpolation or by best approximation in some sense. Using the results of earlier chapters, we also investigate the interrelations that exist between the degree of approximation and the growth parameters of the function. Thus, in Section 4.2, we first introduce the concept of Faber polynomials and obtain a representation theorem for the expansion of an analytic function in a Faber series. The coefficients of the Faber series are then related to the growth parameters of the function. In Section 4.3 we consider the Fourier series expansions in the Hilbert space $H_2(D)$ of square-integrable analytic functions and, after characterizing the Bergman kernel function, determine the growth estimates involving Fourier coefficients. In Section 4.4 the results developed for Fourier series expansions of analytic functions are used to investigate the decay of their minimum error in L^2-norm with respect to an orthonormal system. These results are also extended to cover the case of the minimum error in L^p-norm, $1 < p < \infty$, and Chebyshev norm. Finally, in Section 4.5, the approximation of analytic functions by Lagrange's and Newton's interpolatory polynomials is investigated. Intimate connection of lemniscates with Jacobi interpolatory series has been exploited to obtain growth relations for a related class of analytic functions in this section.

4.2. FABER POLYNOMIALS AND FABER SERIES

Let E be a closed and bounded set containing at least two points such that its complement E' with respect to the extended complex plane is a simply connected domain containing the point at infinity. According to the Riemann mapping theorem, there exists a one-to-one analytic function $z = \Psi(w)$ which maps $\{w : |w| > 1\}$ conformally onto E' such that $\Psi(\infty) = \infty$ and $\Psi'(\infty) > 0$. Thus, in a neighbourhood of infinity, the function has the representation

$$z = \Psi(w) = \rho\left[w + d_o + \frac{d_{-1}}{w} + \ldots\right]$$

where the number $\rho(\, > 0)$ is called the conformal radius or *transfinite diameter* of E. If we define $\eta(w) = \Psi(w/\rho)$, then η maps $\{w : |w| > \rho\}$ onto E' in a one-one conformal manner. If $w = \Omega(z)$ is the inverse function of η, then $\Omega(\infty) = \infty$, $\lim_{z\to\infty} (\Omega(z)/z) = 1$ and, in a neighbourhood of infinity, the function $\Omega(z)$ has a Laurent expansion of the form

$$\Omega(z) = z + b_o + \frac{b_{-1}}{z} + \ldots . \tag{4.2.1}$$

Thus, for each positive integer n and for sufficiently large $|z|$, one has an expansion of the form

$$[\Omega(z)]^n = z^n + b_{n-1,n} z^{n-1} + b_{n-2,n} z^{n-2} + \ldots + b_{1,n} z + b_{o,n} + \frac{b_{-1,n}}{z} + \ldots . \tag{4.2.2}$$

The polynomials

$$P_n(z) = z^n + b_{n-1,n} z^{n-1} + \ldots + b_{1,n} z + b_{o,n}, \quad n = 0,1,2,\ldots, \tag{4.2.3}$$

which comprise nonnegative powers of z in the Laurent series development of $[\Omega(z)]^n$ about infinity, are called the *Faber polynomials* for E. In other words, Faber polynomial $P_n(z)$ is the 'principal part' of the Laurent series expansion of $[\Omega(z)]^n$ in a neighbourhood of the point at infinity. Thus

$$P_n(z) = [\Omega(z)]^n + \chi_n(z)$$

where $\chi_n(z)$ is a function regular in the domain E' and $\chi_n(\infty) = 0$.

It is easy to check that if E is the closed disc $|z-z_o| \leq \rho$, then

$$w = \Omega(z) = z - z_o$$

and so the Faber polynomials for the closed disc are given by

$$P_n(z) = (z-z_o)^n, \quad n = 0,1,2,\ldots .$$

Thus, the Taylor polynomials $(z-z_o)^n$ are a special case of Faber polynomials. In fact, it will be seen shortly that the Taylor series and the concepts related to it stand generalized in the wider perspective of Faber series etc.

A few more examples of Faber polynomials are given in the exercises at the end of this section.

We now obtain a useful generating formula for Faber polynomials.

Let $L_r = \{z: z = \eta(w), |w| = r > \rho\}$. Since $\Omega(z)$ is analytic and univalent, L_r's are analytic, Jordan curves. If E_r denotes the domain bounded by L_r, then $E \subset E_r$ for each $r > \rho$. Further, $L_r \subset E_{r_1}$ for $r < r_1$. For $z \in E_r$ and $w = \Omega(\zeta)$, one has

$$\frac{1}{2\pi i} \int_{L_r} \frac{[\Omega(\zeta)]^n}{\zeta - z} d\zeta = \frac{1}{2\pi i} \int_{|w|=r} \frac{w^n \eta'(w)}{\eta(w) - z} dw. \tag{4.2.4}$$

Since the integrand $[\Omega(\zeta)]^n/(\zeta-z)$ is analytic outside L_r, the path of integration in the left integral can be taken as a circle C of sufficiently large radius that L_r lies inside C and the Laurent series (4.2.2) of $[\Omega(\zeta)]^n$ and that of

$$\frac{1}{\zeta-z} = \frac{1}{\zeta} + \frac{z}{\zeta^2} + \ldots + \frac{z^n}{\zeta^{n+1}} + \ldots$$

converge uniformly on this circle. Multiplying these series, integrating term by term and making use of the relations

$$\int_C \zeta^n \, d\zeta = 0 \quad \text{for} \quad n \neq -1, \quad \int_C \zeta^{-1} \, d\zeta = 2\pi i,$$

we find that the contribution due to all terms with negative exponents is zero and those with nonnegative exponents yield $P_n(z)$. Thus

$$P_n(z) = \frac{1}{2\pi i} \int_{|w|=r} \frac{w^n \eta'(w)}{\eta(w) - z} \, dw, \qquad n = 0,1,\ldots, \quad z \in E_r \tag{4.2.5}$$

Relation (4.2.5) shows that $P_n(z)$ is the coefficient of w^{-n-1} ($n = 0,1,2,\ldots$) in the Laurent series expansion of the function $\eta'(w)/(\eta(w)-z)$ at infinity. Since

$$\lim_{w\to\infty} \frac{\eta(w)}{w} = \lim_{z\to\infty} \frac{z}{\Omega(z)} = 1,$$

η has, in a neighbourhood of infinity, an expansion of the form

$$\eta(w) = w + \beta_o + \frac{\beta_{-1}}{w} + \ldots$$

so that

$$\lim_{w\to\infty} \frac{\eta'(w)}{\eta(w)-z} = \lim_{w\to\infty} \frac{1 - (\beta_{-1}/w^2) - \ldots}{w+\beta_o+(\beta_{-1}/w)+\ldots-z} = 0.$$

This shows that $\eta'(w)/(\eta(w)-z)$ has a zero at infinity. Hence, for large $|w|$,

$$\frac{\eta'(w)}{\eta(w)-z} = \Sigma_{n=0}^{\infty} \frac{P_n(z)}{w^{n+1}} \tag{4.2.6}$$

and the series on the right converges uniformly for $|w|$ large. This is the desired generating function for $P_n(z)$.

Now consider $P_n(z) = P_n(\eta(w)) = H_n(w)$, say. $H_n(w)$ is analytic for $|w| > \rho$. Let $\rho < r_1 < r_2$. Then, for w satisfying $r_1 < |w| < r_2$, $z = \eta(w)$ lies in the domain bounded by the curves L_{r_1} and L_{r_2}. For positive integer n, the function $\zeta^n\eta'(\zeta)/(\eta(\zeta) - \eta(w))$, as a function of ζ, is analytic for all ζ lying outside L_{r_1} except for a simple pole at $\zeta = w$ which is situated between the circles $|\zeta| = r_1$ and $|\zeta| = r_2$. The residue at the pole $\zeta = w$ is given by

$$\lim_{\zeta \to w} \frac{(\zeta - w)\, \zeta^n\, \eta'(\zeta)}{\eta(\zeta) - \eta(w)} = w^n.$$

Thus, in view of Cauchy's residue theorem and (4.2.5), we have

$$P_n(z) = H_n(w) = w^n + \frac{1}{2\pi i} \int_{|\zeta| = r_1} \frac{\zeta^n\, \eta'(\zeta)}{\eta(\zeta) - \eta(w)}\, d\zeta. \tag{4.2.7}$$

The integral on the right, as a function of w, is analytic for all w satisfying $|w| > r_1$ and, at infinity, its value is zero. Thus, in this region, it must be representable by a Laurent series of the form $\Sigma_{k=1}^{\infty} a_{nk} w^{-k}$. (4.2.7) then yields

$$H_n(w) = w^n + \Sigma_{k=1}^{\infty} a_{nk} w^{-k}. \tag{4.2.8}$$

Since $H_n(w)$ is analytic for $|w| > \rho$, relation (4.2.8) should hold for all w satisfying $|w| > \rho$ and so the series on the right must converge uniformly in $|w| \geq r > \rho$. (4.2.8) is then easily seen to lead to the relation

$$a_{nk} = -\frac{1}{4\pi^2} \int_{|w| = r} \int_{|\zeta| = r_1} \frac{w^{k-1}\, \zeta^n\, \eta'(\zeta)}{\eta(\zeta) - \eta(w)}\, d\zeta\, dw; n,k = 1,2,3,\ldots . \tag{4.2.9}$$

The coefficients a_{nk} are called the Faber coefficients for E.

Useful Cauchy estimates can be obtained from (4.2.7) and (4.2.9) as follows. For $|w| = r^*$, $r_1 < r^* < r_2$, (4.2.7) gives

$$|P_n(z)| \leqq r^{*n} + r_1^n\, \theta(r^*, r_1)$$

where $\theta(r^*, r_1) = \max\limits_{\substack{|\zeta| = r_1 \\ |w| = r^*}} \left| \frac{\eta'(\zeta)}{\eta(\zeta) - \eta(w)} \right|$.

The above relation can be stated as

$$|P_n(z)| \leqq M r^{*n} \tag{4.2.10}$$

where M is a constant depending on r^* and r_1 but independent of n. Using this estimate in (4.2.9) one has

$$|a_{nk}| \leqq M\, r^{*n+k+1}, \qquad n,k = 1,2,\ldots \;. \tag{4.2.11}$$

Our main concern now is to obtain the representation of a regular function in terms of a series of Faber polynomials. We first need an important lemma.

Lemma 4.2.1. *On every compact subset of* E', *one has the uniform limit*

$$\lim_{n\to\infty} |P_n(z)|^{1/n} = |\Omega(z)|. \tag{4.2.12}$$

Proof. Since Ω maps E' onto $\{w : |w| > \rho\}$, given a compact subset $K \subset E'$ there exists $r > \rho$ such that, for $w \in \Omega(K)$, one has $|w| > r$. For such w, we have, by (4.2.11),

$$\begin{aligned} |w^{-n} \Sigma_{k=1}^{\infty} a_{nk}\, w^{-k}| &\leqq \Sigma_{k=1}^{\infty} |a_{nk}|\; |w|^{-(k+n)} \\ &\leqq M\, \Sigma_{k=1}^{\infty}\, r^{n+k+1}\; |w|^{-(k+n)} \\ &= M\, r\, \frac{(r|w|^{-1})^n}{1-r|w|^{-1}} \to 0 \quad \text{as} \quad n \to \infty. \end{aligned}$$

Thus, by (4.2.8), for $z \in K$, we have

$$\begin{aligned} \lim_{n\to\infty} \frac{P_{n+1}(z)}{P_n(z)} &= \lim_{n\to\infty} \frac{H_{n+1}(w)}{H_n(w)} \\ &= \lim_{n\to\infty} \frac{w\left[1 + w^{-n-1}\, \Sigma_{k=1}^{\infty}\, a_{n+1,k}\, w^{-k}\right]}{1 + w^{-n}\, \Sigma_{k=1}^{\infty}\, a_{n,k}\, w^{-k}} \\ &= w = \Omega(z) \end{aligned}$$

so that

$$\lim_{n\to\infty} |P_n(z)|^{1/n} = |\Omega(z)|. \quad \square$$

Let $H(E;R)$ denote the class of all functions f that are regular in E_R with a singularity on L_R $(\rho < R < \infty)$. Our next theorem gives a representation for functions in $H(E;R)$ in a series of Faber polynomials and is thus a generalization of the classical Cauchy-Hadamard theorem and

Taylor theorem.

Theorem 4.2.1. *Every* $f \in H(E;R)$ *can be represented by the Faber series*

$$f(z) = \Sigma_{n=0}^{\infty} a_n P_n(z) \quad , \quad z \in E_R \tag{4.2.13}$$

with

$$a_n = \frac{1}{2\pi i} \int_{|\zeta|=r} f(\eta(\zeta))\, \zeta^{-n-1}\, d\zeta \quad , \quad \rho < r < R \tag{4.2.14}$$

if and only if

$$\limsup_{n \to \infty} |a_n|^{1/n} = \frac{1}{R}. \tag{4.2.15}$$

The series in (4.2.13) *converges absolutely and uniformly on every compact subset of* E_R *and diverges outside* L_R.

Proof. Let $f \in H(E;R)$. If $z \in E_R$, then $z \in E_r$ for some r satisfying $\rho < r < R$. By Cauchy's integral formula,

$$f(z) = \frac{1}{2\pi i} \int_{L_r} \frac{f(t)dt}{t-z}$$

$$= \frac{1}{2\pi i} \int_{|\zeta|=r} \frac{f(\eta(\zeta))\, \eta'(\zeta)}{\eta(\zeta)-z}\, d\zeta$$

$$= \frac{1}{2\pi i} \int_{|\zeta|=r} \left(\Sigma_{n=0}^{\infty} \frac{f(\eta(\zeta))}{\zeta^{n+1}} P_n(z)\right) d\zeta, \qquad \text{by (4.2.6).}$$

Since the series under the integral sign converges uniformly on $|\zeta| = r$, it can be integrated term by term. Thus we have

$$f(z) = \Sigma_{n=0}^{\infty} a_n P_n(z), \qquad z \in E_R$$

where a_n is given by (4.2.14). If

$$\overline{M}(r,f) = \max_{|\zeta|=r} |f(\eta(\zeta))|, \tag{4.2.16}$$

then (4.2.14) gives

$$|a_n| \leqq \frac{\overline{M}(r,f)}{r^n}, \qquad n = 0,1,2,\ldots, \tag{4.2.17}$$

which are analogues of Cauchy's inequality for Taylor series.

From (4.2.17), we immediately have

$$\limsup_{n\to\infty} |a_n|^{1/n} \leq \frac{1}{r}.$$

Since this holds for every r satisfying $\rho < r < R$, we have

$$\limsup_{n\to\infty} |a_n|^{1/n} \leq \frac{1}{R}.$$

We now show that inequality is not permissible in the above relation. If

$$\limsup_{n\to\infty} |a_n|^{1/n} = \frac{1}{R_o} < \frac{1}{R},$$

let r satisfy $\rho < r < R_o$; then for every ε such that $0 < \varepsilon < R_o - r$, one has

$$|a_n| < \frac{1}{(R_o-\varepsilon/2)^n} < \frac{1}{(r+\frac{\varepsilon}{2})^n} \qquad \text{for } n \geq n_o.$$

On the other hand, by (4.2.12), for every $z \in L_r$, we have

$$|P_n(z)| < (r+\frac{\varepsilon}{4})^n \qquad \text{for } n \geq n_1.$$

Thus, for all $z \in L_r$,

$$|a_nP_n(z)| < (\frac{r+\varepsilon/4}{r+\varepsilon/2})^n < 1 \quad \text{for } n \geq \max(n_o, n_1).$$

The above inequality shows that the series $\Sigma_{n=0}^{\infty} a_nP_n(z)$ converges uniformly on L_r and hence on E_r. Since this is true for every r satisfying $\rho < r < R_o$, it follows that the series $\Sigma_{n=0}^{\infty} a_nP_n(z)$ converges uniformly on every compact subset of E_{R_o} to a function, F(z), say. The function F(z) must be regular in E_{R_o} since each term of the series is a regular function. However, on $E_R \subset E_{R_o}$, the series converges to f(z) so that F(z) is the analytic continuation of f(z) to E_{R_o}. Since this contradicts the hypothesis that f has a singular point on $L_R(R < R_o)$, we must have (4.2.15).

To show that the series (4.2.13) diverges outside L_R, let z_o lie outside L_R. Then we have $|\Omega(z_o)| > R$. In view of (4.2.12) and (4.2.15), one has

$$\limsup_{n\to\infty} |a_n P_n(z_o)|^{1/n} > 1,$$

showing that the series $\Sigma_{n=0}^{\infty} a_n P_n(z_o)$ diverges.

Conversely, if (4.2.15) holds, then, as above, we can show that the series $\Sigma_{n=0}^{\infty} a_n P_n(z)$ converges uniformly on compact subsets of E_R to a regular function, f(z), say. If f(z) had no singular point on L_R then it would be possible to extend f(z) analytically to a bigger domain E_{R_o}, say. But then the first part of the theorem would give that $\limsup_{n\to\infty} |a_n|^{1/n} \leqq 1/R_o < 1/R$, a contradiction. □

We now introduce the concept of measurement of growth for functions in $H(E;R)$. Thus, growth parameters, analogous to those introduced for functions regular in U, may be defined for functions regular in E_R as follows.

We say that $f \in H(E,R)$, $\rho < R < \infty$ is of *E-order* α in E_R if

$$\alpha = \limsup_{r\to R} \frac{\log^+\log^+ \overline{M}(r)}{\log(Rr/(R-r)} \tag{4.2.18}$$

where $\overline{M}(r) = \max_{z\in L_r} |f(z)|$, $\rho < r < R$. The *lower E-order* β is similarly defined as

$$\beta = \liminf_{r\to R} \frac{\log^+\log^+ \overline{M}(r)}{\log(Rr/(R-r))} . \tag{4.2.19}$$

To compare the growth of functions in $H(E;R)$ that have the same nonzero finite E-order, the concepts of E-type and lower E-type may be introduced. Then $f \in H(E;R)$ having E-order α, $(0 < \alpha < \infty)$, is said to be of *E-type* τ and *lower E-type* ν, $0 \leq \nu, \tau \leq \infty$, if

$$\tau = \limsup_{r\to R} \frac{\log \overline{M}(r)}{\{Rr/(R-r)\}^{\alpha}} \tag{4.2.20}$$

and

$$\nu = \liminf_{r\to R} \frac{\log \overline{M}(r)}{\{Rr/(R-r)\}^{\alpha}} . \tag{4.2.21}$$

The concepts of *regular E-growth*, *perfectly regular E-growth* etc. can also be introduced in the same way as for functions analytic in U(cf.Section 1.5).

It may be easily checked that when E is the disc $\{z : |z| \le \rho < 1\}$, then $H(E;1)$ is the class of functions analytic in U and having a singularity on $|z| = 1$ and the above definitions reduce to those of order, lower order etc. introduced in Section 1.5.

We now obtain characterization of the growth parameters introduced above in terms of the coefficients of the Faber series (4.2.13). To obtain the results in a general setting we shall assume that $f \in H(E;R)$ is represented by the gap Faber series

$$f(z) = \Sigma_{n=0}^{\infty} a_n P_{\lambda_n}(z) \quad , \qquad z \in E_R \tag{4.2.22}$$

where $\{\lambda_n\}_{n=0}^{\infty}$ is a strictly increasing sequence of positive integers and $a_n \neq 0$ for all n.

First we have

Theorem 4.2.2. *Let* $f \in H(E;R)$ *be represented by the Faber series* (4.2.22) *and be of* E-*order* α $(0 \le \alpha \le \infty)$. *Then*

$$\frac{\alpha}{1+\alpha} = \limsup_{n \to \infty} \frac{\log^+ \log^+ (|a_n| R^{\lambda_n})}{\log \lambda_n} \tag{4.2.23}$$

where the left-hand side in (4.2.23) *is interpreted to be* 1 *when* $\alpha = \infty$.

Proof. Let $\alpha < \infty$. By (4.2.18), given $\varepsilon > 0$ there exists $r_o = r_o(\varepsilon)$ such that

$$\log \bar{M}(r) < \left(\frac{Rr}{R-r}\right)^{\alpha+\varepsilon} \quad \text{for } r_o < r < R. \tag{4.2.24}$$

By (4.2.17) and (4.2.24), we therefore have, for all r satisfying $r_o < r < R$ and $n = 1,2,\ldots$,

$$\log(|a_n| R^{\lambda_n}) < \left(\frac{Rr}{R-r}\right)^{\alpha+\varepsilon} - \lambda_n \log \frac{r}{R}$$

$$< \left(\frac{Rr}{R-r}\right)^{\alpha+\varepsilon} + \lambda_n \frac{R-r}{r} .$$

Choosing r such that $\frac{rR}{R-r} = \{\frac{\lambda_n R}{\alpha+\varepsilon}\}^{1/(\alpha+\varepsilon+1)}$, we get, for $n > n_o$,

$$\log^+(|a_n| R^{\lambda_n}) < \{\frac{\lambda_n R}{\alpha+\varepsilon}\}^{(\alpha+\varepsilon)/(\alpha+\varepsilon+1)} \times (\alpha+\varepsilon+1)$$

which gives

$$\limsup_{n \to \infty} \frac{\log^+\log^+(|a_n|R^{\lambda_n})}{\log \lambda_n} \leqq \frac{\alpha}{\alpha+1}. \tag{4.2.25}$$

This obviously holds if $\alpha = \infty$.

To prove the inequality that is the converse of (4.2.25), we note that, by (4.2.12), there exists n_o such that, for $0 < \lambda < 1$ and $z \in L_r$, one has

$$|P_{\lambda_n}(z)| < \{\lambda r + (1-\lambda)R\}^{\lambda_n} \quad \text{for} \quad n > n_o.$$

Thus, by (4.2.22),

$$\overline{M}(r) \leqq \Sigma_{n=0}^{\infty} |a_n| \, |P_{\lambda_n}(z)|_{z\in L_r}$$

$$\leqq \Sigma_{n=0}^{n_o} |a_n| \, |P_{\lambda_n}(z)|_{z\in L_r} + \Sigma_{n=n_o+1}^{\infty} |a_n| R^{\lambda_n} \left\{\frac{\lambda r+(1-\lambda)R}{R}\right\}^{\lambda_n}$$

$$< K + M\left(\frac{\lambda r+(1-\lambda)R}{R}, F\right) \tag{4.2.26}$$

where K is a constant, $F(s) = \Sigma_{n=0}^{\infty} (|a_n|R^{\lambda_n})s^{\lambda_n}$ and $M(t,F) = \max_{|s|=t} |F(s)|$. Since the Faber series for f(z) converges uniformly on compact subsets of E_R, it follows that F(s) is analytic for $|s| < 1$. If the order of F(s), as defined by (1.5.1), be ρ_o, then (4.2.26) gives

$$\alpha = \limsup_{r \to R} \frac{\log^+\log^+\overline{M}(r)}{\log (Rr/(R-r))}$$

$$\leq \limsup_{r \to R} \frac{\log^+\log^+ M(\{(\lambda r+(1-\lambda)R)/R\}, F)}{\log (Rr/(R-r))}$$

$$= \limsup_{r \to R} \frac{\log^+\log^+ M(\{(\lambda r+(1-\lambda)R)/R\}, F)}{-\log(1-\{(\lambda r+(1-\lambda)R)/R\})+\log (\lambda r)}$$

$$= \rho_o$$

and so Theorem 2.2.1, applied to F(s), yields

$$\frac{\alpha}{1+\alpha} \leqq \limsup_{n \to \infty} \frac{\log^+\log^+(|a_n|R^{\lambda_n})}{\log \lambda_n} \tag{4.2.27}$$

and (4.2.25) and (4.2.27) lead to (4.2.23). □

Proceeding in an analogous manner we may obtain the following theorem that gives a characterization for E-type in terms of Faber coefficients.

Theorem 4.2.3. Let $f \in H(E;R)$ *be represented by the Faber series* (4.2.22) *and, for* $0 < \alpha < \infty$, *set*

$$J \equiv \limsup_{n \to \infty} \frac{[\log^+ (|a_n| R^{\lambda_n})]^{\alpha+1}}{\lambda_n^{\alpha}} . \tag{4.2.28}$$

If $0 < J < \infty$, *the function* f *is of* E-*order* α *and* E-*type* τ, *if and only if*

$$J = \frac{(\alpha+1)^{\alpha+1}}{\alpha^{\alpha}} R^{\alpha} \tau . \tag{4.2.29}$$

If $J = 0$ *or* ∞, *the function* f *is respectively of* E-*growth* $(\alpha,0)$ *or of* E-*growth not less than* (α,∞) *and conversely.*

The results for the lower E-growth parameters, corresponding to those for functions analytic in U, also hold good. For example, analogues of Theorems 2.3.1, 2.3.2 and 2.3.8 for functions represented by Faber series (4.2.22) read as follows:

Theorem 4.2.4. Let $f \in H(E;R)$, *represented by the Faber series* (4.2.22), *be of lower* E-*order* $\beta (0 \leq \beta \leq \infty)$. *Then for any increasing sequence* $\{n_k\}$ *of natural numbers*

$$1 + \beta \geq \liminf_{k \to \infty} \frac{\log \lambda_{n_{k-1}}}{\log \lambda_{n_k} - \log^+ \log^+ (|a_{n_k}| R^{\lambda_{n_k}})} . \tag{4.2.30}$$

Further, if f *has* E-*order* $\alpha (0 < \alpha < \infty)$ *and lower* E-*type* ν $(0 \leq \nu \leq \infty)$, *then*

$$\frac{(\alpha+1)^{\alpha+1}}{\alpha^{\alpha}} R^{\alpha} \nu \geq \liminf_{k \to \infty} \{\lambda_{n_{k-1}} \left(\frac{\log^+ (|a_{n_k}| R^{\lambda_{n_k}})}{\lambda_{n_k}} \right)^{\alpha+1} \} . \tag{4.2.31}$$

Theorem 4.2.5. Let $f \in H(E;R)$, *represented by the Faber series* (4.2.22), *be of* E-*order* $\alpha (0 < \alpha \leq \infty)$ *and lower* E-*order* β $(0 \leq \beta \leq \infty)$. *Further, let*

$\Psi(n) = |a_{n-1}/a_n|^{1/(\lambda_n - \lambda_{n-1})}$ *form a nondecreasing function of* n *for* $n > n_o$; *then*

$$1+\beta \leq \liminf_{n\to\infty} \left\{ \frac{\log \lambda_n}{\log \lambda_n - \log^+ \log^+ (|a_n| R^{\lambda_n})} \right\} . \tag{4.2.32}$$

Theorem 4.2.6. *Let* $f \in H(E;R)$, *represented by the Faber series* (4.2.22), *be of* E-*order* $\alpha (0 < \alpha < \infty)$ *and lower* E-*type* ν $(0 \leq \nu \leq \infty)$. *If* $\Psi(n) = |a_{n-1}/a_n|^{1/(\lambda_n - \lambda_{n-1})}$ *forms a nondecreasing function of* n *for* $n > n_o$, *then*

$$\frac{(\alpha+1)^{\alpha+1}}{\alpha^\alpha} R^\alpha \nu \leq \liminf_{n\to\infty} \left\{ \lambda_n \left(\frac{\log^+ (|a_n| R^{\lambda_n})}{\lambda_n} \right)^{\alpha+1} \right\} . \tag{4.2.33}$$

The above theorems can be proved either by proceeding along the same lines as in the case of corresponding theorems for functions analytic in U or by making use of Cauchy inequality (4.2.17) and the auxiliary function F(s) as in the proof of Theorem 4.2.2. We omit the details of the proofs of these theorems.

For other results for functions in $H(E;R)$ that correspond to those of Chapter 2, see the exercises at the end of this section.

Faber polynomials can be effectively used in the study of approximation by polynomials of functions analytic on a closed and bounded set which complements a simply connected domain in the extended complex plane. Before turning to this, we prove a lemma which is of fundamental importance.

Lemma 4.2.2. *Let* E *be a closed and bounded set with transfinite diameter* $\rho (> 0)$ *such that its complement* E' *with respect to the extended complex plane is simply connected. Let* $Q_n(z)$ *be a polynomial of degree* $\leq$ n *such that* $|Q_n(z)| \leq M$ *for* $z \in E$. *Then for* $z \in L_r$, $r > \rho$, *we have*

$$|Q_n(z)| \leq M \left(\frac{r}{\rho}\right)^n . \tag{4.2.34}$$

Proof. Let $w = \Omega(z)$ be the univalent analytic function mapping E' onto $\{w : |w| > \rho\}$ such that $\Omega(\infty) = \infty$ and $\lim_{z\to\infty} (\Omega(z)/z) = 1$. Consider the function

$\chi(z) = Q_n(z)/[\Omega(z)]^n$. If we define the value of $\chi(z)$ at infinity to be the leading coefficient of $Q_n(z)$, then $\chi(z)$ is seen to be analytic in E' including at infinity. Hence, by the maximum modulus principle, for z lying outside $L_{r'}$, $\rho < r' < r$, one has

$$|\chi(z)| \leqq \max_{\zeta \in L_{r'}} |\chi(\zeta)| = \max_{\zeta \in L_{r'}} \frac{|Q_n(\zeta)|}{r'^n}.$$

In particular, for $z \in L_r$, this gives

$$|Q_n(z)| \leqq \max_{\zeta \in L_{r'}} |Q_n(\zeta)| \left(\frac{r}{r'}\right)^n < M(r';Q_n) \left(\frac{r}{\rho}\right)^n$$

where $M(r';Q_n) = \max_{\zeta \in L_{r'}} |Q_n(\zeta)|$. Taking limits as $r' \to \rho+$, we get, for $z \in L_r$,

$$|Q_n(z)| \leqq \lim_{r' \to \rho+} M(r';Q_n) \left(\frac{r}{\rho}\right)^n$$

$$\leqq M\left(\frac{r}{\rho}\right)^n,$$

where existence of $\lim_{r' \to \rho+} M(r';Q_n)$ is ensured by the fact that $M(r';Q_n)$ is nonincreasing as $r' \to \rho+$. □

We now have

Theorem 4.2.7. *Let* E *be as in Lemma* 4.2.2 *and let* f *be analytic on* E_R $(R > \rho)$. *Then for every* $n \geq 0$ *there exists a polynomial* $Q_n(z)$ *of degree* n *such that, for all* $z \in E$, *one has*

$$|f(z) - Q_n(z)| < M\left(\frac{\rho}{R}\right)^n \tag{4.2.35}$$

where M *is a constant independent of* z *and* n.

Further, if $\rho < r'' < r' < r < R$, *then for all* $n > n_0(r',r'')$ *one has, for* $z \in L_{r''}$

$$|f(z) - Q_n(z)| < \overline{M}(r) \left(\frac{r}{r-r'}\right) \left(\frac{r'}{r}\right)^n \tag{4.2.36}$$

where $\overline{M}(r) = \max_{z \in L_r} |f(z)|$.

Proof. Let $\{P_n(z)\}$ be the sequence of Faber polynomials for E. By Theorem 4.2.1, the series $\Sigma_{n=0}^{\infty} a_n P_n(z)$ converges uniformly on compact subsets of E_R to f(z). If we set $Q_n(z) = \Sigma_{k=0}^{n} a_k P_k(z)$, then $Q_n(z)$ is a polynomial of degree n. Let r, r', r" be such that $\rho < r'' < r' < r < R$. If $z \in L_{r''}$ then, by (4.2.12), there exists an integer $n_o = n_o(r', r'')$ such that

$$|P_n(z)| < r'^n \qquad \text{for} \quad n > n_o. \tag{4.2.37}$$

Hence, if $n > n_o$ and $z \in L_{r''}$, we have

$$\begin{aligned} |f(z) - Q_n(z)| &\leq \Sigma_{k=n+1}^{\infty} |a_k| \, |P_k(z)| \\ &< \overline{M}(r) \, \Sigma_{k=n+1}^{\infty} \left(\frac{r'}{r}\right)^k \\ &= \overline{M}(r) \left(\frac{r'}{r-r'}\right) \left(\frac{r'}{r}\right)^n \end{aligned}$$

where we have used (4.2.37) and the Cauchy's inequalities (4.2.17). This gives (4.2.36). To obtain (4.2.35) we note that by replacing $\overline{M}(r)r'/(r-r')$ by a larger number M, if necessary, we can assume that (4.2.36) holds for all n. Thus, for $n \geq 0$ and $z \in L_{r''}$, we have

$$|f(z) - Q_n(z)| < M\left(\frac{r'}{R}\right)^n .$$

By the maximum modulus principle, the above inequality holds for $z \in E_{r''}$ also and in particular for $z \in E \subset E_{r''}$. Since this holds for every $r' > \rho$, making $r' \to \rho$ we have (4.2.35). □

The converse of the above theorem also holds in the sense of the following:

Theorem 4.2.8. *Let* E *be as in Lemma* 4.2.2. *For a function* f *defined on* E *suppose that for every integer* $n \geq 0$ *there is a polynomial* $Q_n(z)$ *of degree not exceeding* n *such that, for all* $z \in E$,

$$|f(z) - Q_n(z)| < M\left(\frac{\rho}{R}\right)^n , \qquad \rho < R \tag{4.2.38}$$

where M *is a constant independent of* z *and* n; *then the function* f *can be extended as an analytic function on* E_R.

Proof. We first show that the polynomial sequence $\{Q_n(z)\}$ converges uniformly on compact subsets of E_R. We note that, for $z \in E$,

$$|Q_n(z) - Q_{n-1}(z)| = |(f(z) - Q_{n-1}(z)) - (f(z) - Q_n(z))|$$

$$\leq M\left(\frac{\rho}{R}\right)^{n-1} + M\left(\frac{\rho}{R}\right)^{n} \leq 2M\left(\frac{\rho}{R}\right)^{n-1} .$$

Application of Lemma 4.2.2 gives, for $z \in L_r \cup E_r$, $r > \rho$,

$$|Q_n(z) - Q_{n-1}(z)| \leq 2M\left(\frac{\rho}{R}\right)^{n-1} \left(\frac{r}{\rho}\right)^{n} = 2M \frac{R}{\rho} \left(\frac{r}{R}\right)^{n} .$$

Thus it follows that for any r, $\rho < r < R$, the series $Q_1(z) + (Q_2(z) - Q_1(z)) + \ldots + (Q_n(z) - Q_{n-1}(z)) + \ldots$ converges uniformly in $L_r \cup E_r$. Since any compact subset of E_R is contained in $L_r \cup E_r$ for some r, $\rho < r < R$, it follows that the sequence $\{Q_n(z)\}$ converges uniformly on compact subsets of E_R to some regular function, $\tilde{f}(z)$, say. However, (4.2.38) gives that $\{Q_n(z)\}$ converges uniformly on E to $f(z)$. It follows that $f(z) = \tilde{f}(z)$ for $z \in E$ and so f can be continued analytically to E_R. □

EXERCISES 4.2

1. Let E be the segment $\{x : -1 \leq x \leq 1\}$ of the real axis. Show that Faber polynomials for E are given by

$$P_n(z) = \frac{1}{2^{n-1}} \cos (n \arccos z)$$

$$= \frac{1}{2^n} \left[(z + \sqrt{z^2-1})^n + (z - \sqrt{z^2-1})^n\right], \quad n = 1,2,\ldots .$$

2. Let E be the ellipse $\{z : (\mathrm{Re}\, z/a)^2 + (\mathrm{Im}\, z/b)^2 \leq 1\}$. Show that Faber polynomials for E are given by

$$P_n(z) = 2 \sum_{k=0}^{[n/2]} \binom{n}{2k} z^{n-2k} (z^2-4c^2)^k \text{ where } c = \frac{(a^2-b^2)^{1/2}}{2},$$

$$n = 1,2,\ldots .$$

3. Find the first few Faber polynomials for the lemniscate $|z^2-1| = 1$.

In the following problems E is a closed and bounded set of transfinite diameter $\rho > 0$ containing at least two points such that its complement E' with respect to the extended complex plane is a simply connected domain containing the point at infinity.

4. Show that the Faber polynomials for E are the same as the Faber polynomials for $\overline{E}_r$, $\rho < r < \infty$.

5. Show that a_{nk}, as defined by (4.2.9), satisfy the Grunsky law of symmetry, i.e., $ka_{nk} = na_{kn}$, $k,n = 1,2,\ldots$.

(Curtiss [1])

6. Show that if $P_n(z)$ $(n = 1,2,\ldots)$ are Faber polynomials for E there exist absolute constants A and $\alpha < \frac{1}{2}$ such that

$$\max_{z\in E} |P_n(z)| \leqq A\, n^{\alpha}.$$

(Kovari and Pommerenke [1])

7. If $\mathcal{P}_n$ denotes the set of all polynomials of degree not exceeding n, for f continuous on E and analytic in the interior of E, set

$$\Delta_n^{(\infty)}(f;E) = \inf_{g\in\mathcal{P}_n} \max_{z\in E} |f(z) - g(z)|.$$

Show that if $S_n(z) = \Sigma_{k=0}^{n} a_k P_k(z)$ denotes the nth partial sum of the Faber series for f then, for $z \in E$,

$$|f(z) - S_n(z)| \leqq A\, n^{\alpha}\, \Delta_n^{(\infty)}(f;E)$$

where A and $\alpha < \frac{1}{2}$ are absolute constants.

(Kovari and Pommerenke [1])

8. Show that, if E is convex, the zeros of Faber polynomials $P_n(z)$ for E lie in E.

(Kovari and Pommerenke [1])

9. If H(E) denotes the convex hull of E, show that the zeros of the derivatives of Faber polynomials $\{P_n'(z)\}$ lie in H(E).

(Ullman [1])

10. Show by an example that the zeros of Faber polynomials of E need not lie in H(E) (cf. Exercise 9).

(Goodman [1])

11. Let $F(w)=\Sigma_{n=0}^{\infty} c_n w^{-n}$, $c_o \neq 0$ be a function regular in $|w| > \rho$. If $\eta'(w)/(\eta(w)-z)$ is the generating function (4.2.6) for Faber polynomials for E, define

$$H(z,w) = \frac{\eta'(w)}{\eta(w)-z} F(w) = \Sigma_{n=0}^{\infty} \frac{G_n(z)}{w^{n+1}}$$

where $G_n(z) = c_o P_n(z) + c_1 P_{n-1}(z)+\ldots+ c_n P_o(z)$. The polynomials $G_n(z)$ are called generalized Faber polynomials for E. Extend Theorem 4.2.1 and other results of this section to the case of generalized Faber polynomials.

12. Show that for $f \in H(E;R)$, represented by Faber series (4.2.22), if q-E-order $\alpha(q)$ is defined as

$$\alpha(q) = \limsup_{r \to R} \frac{\log^{[q]} \overline{M}(r)}{\log (Rr/(R-r))}$$

where $\overline{M}(r) = \max_{z \in L_r} |f(z)|$, then

$$\alpha(q) + A(q) = \limsup_{n \to \infty} \frac{\log^{[q-1]} \lambda_n}{\log \lambda_n - \log^+ \log^+ (|b_n| R^{\lambda_n})}$$

where $A(q) = 1$ if $q = 2$ and $A(q) = 0$ if $q = 3,4,\ldots$.

Define analogously lower q-E-order, q-E-type and lower q-E-type and obtain their characterizations in terms of Faber coefficients.

Generalize other aspects discussed in Chapter 2, e.g., decomposition theorems, coefficient estimates involving proximate order etc. for function having q-E-order $\alpha(q)$.

(Juneja [3])

13. For $f \in H(E;R)$, represented by Faber series (4.2.22), define its logarithmic E-order α_o as

$$\alpha_o = \limsup_{r \to R} \frac{\log^+ \log^+ \overline{M}(r)}{\log \log (Rr/(R-r))}$$

where $\overline{M}(r) = \max_{z \in L_r} |f(z)|$. Show that α_o satisfies

$$\sigma_o \le \alpha_o \le \max(1, \sigma_o)$$

$$\text{where } \sigma_o = \limsup_{n \to \infty} \frac{\log^+ \log^+ (|b_n| R^{\lambda_n})}{\log \log \lambda_n} .$$

Define by analogy lower logarithmic E-order, logarithmic E-type etc. and obtain the analogues of corresponding results of Chapter 2 in this case.

(Juneja [3])

14. For $f \in H(E;R)$, introduce the concepts of (α,β) E-order $\rho_o(\alpha,\beta)$ and generalized E-order $\zeta_o(h)$ as

$$\rho_o(\alpha,\beta) = \limsup_{r \to R} \frac{\alpha(\log \overline{M}(r))}{\beta(Rr/(R-r))}$$

$$\zeta_o(h) = \limsup_{r \to R} \frac{h(\log \overline{M}(r))}{\log(Rr/(R-r))}$$

where $\alpha(x)$, $\beta(x)$ and $h(x)$ satisfy the conditions stated in Exercises 1.5. Obtain the results involving the above growth parameters and the coefficients of Faber series. Also introduce, as in Exercises 1.5, other growth parameters and obtain the results corresponding to those of Chapter 2 in this case.

(Juneja [3])

4.3. FOURIER SERIES EXPANSION OF ANALYTIC FUNCTIONS

Let D be a bounded domain in the complex plane and let $H_2(D)$ denote the class of all functions regular in D that satisfy

$$\iint_D |f(z)|^2 dx\, dy < \infty, \qquad z = x+iy. \tag{4.3.1}$$

Here the integral is to be taken as

$$\lim_{n\to\infty} \iint_{K_n} |f(z)|^2 dx\, dy \tag{4.3.2}$$

where $\{K_n\}$ is a compact exhaustion of D; i.e., $\{K_n\}$ is a sequence of compact subsets of D satisfying

(i) $K_n \subset K_{n+1}$ for $n = 1,2,$

(ii) $D = \cup_{n=1}^{\infty} K_n$ and

(iii) for each compact subset K of D there is a K_n such that $K \subset K_n$.

Since for $a,b \in \mathbb{C}$ and $f,g \in H_2(D)$,

$$\iint_D |af(z)+bg(z)|^2 dx\, dy \le \iint_D (|af(z)|+|bg(z)|)^2 dx\, dy$$
$$\le 2 \iint_D (|a|^2|f(z)|^2+|b|^2|g(z)|^2)\, dx\, dy < \infty,$$

it follows that $af + bg \in H_2(D)$. Thus $H_2(D)$ is a linear space over the field of complex numbers $\mathbb{C}$. For $f,g \in H_2(D)$, define

$$(f,g) = \iint_D f(z)\, \overline{g(z)}\, dx\, dy; \qquad (4.3.3)$$

since, by the Cauchy-Schwarz inequality,

$$\left|\iint_D f(z)\, \overline{g(z)}\, dx\, dy\right|^2 \le \left(\iint_D |f(z)|^2 dx\, dy\right) \left(\iint_D |g(z)|^2 dx\, dy\right) < \infty,$$

(f,g) is well-defined. It is easy to verify that (f,g), defined by (4.3.3), satisfies the axioms of an inner product, i.e., for $f,g,h \in H_2(D)$ and $a,b \in \mathbb{C}$, one has

$$(f,g) = \overline{(g,f)}$$

$$(af+bg,h) = a(f,h) + b(g,h)$$

and

$$(f,f) \ge 0$$

where equality occurs in the last relation if and only if $f(z) = 0$ for all $z \in D$.

The inner product, defined by (4.3.3), induces on $H_2(D)$ a norm $\|\cdot\|$, given by

$$\|f\| = (f,f)^{1/2} = \left(\iint_D |f(z)|^2 dx\, dy\right)^{1/2}, \qquad (4.3.4)$$

thereby making $H_2(D)$ a normed linear space. We shall now show that $H_2(D)$ is complete with respect to the metric generated by this norm. First we prove two lemmas.

Lemma 4.3.1. *Let* $K \subset D$ *be a compact set and* ρ_K *be the distance of* K *from the boundary* ∂D *of* D. *Then, for* $f \in H_2(D)$ *one has*

$$\iint_D |f(z)|^2 dx\,dy \geq \pi\rho_K^2\,|f(z_o)|^2 \quad \textit{for every} \quad z_o \in K. \tag{4.3.5}$$

Proof. Let $z_o \in K$ and ρ_{z_o} be the distance of z_o from ∂D. Since f is regular in D, by Taylor's theorem

$$f(z) = \Sigma_{n=0}^{\infty} a_n (z-z_o)^n, \qquad a_n = \frac{f^{(n)}(z_o)}{n!}$$

where the series on the right-hand side converges absolutely and uniformly on the disc $B = \{z: |z-z_o| \leq \rho < \rho_{z_o}\}$. Substituting $z = z_o + re^{i\theta}$, we have

$$\iint_B |f(z)|^2 dx\,dy = \int_0^{2\pi}\int_0^{\rho} |\Sigma_{n=0}^{\infty} a_n r^n e^{in\theta}|^2\; r\,dr\,d\theta$$

$$= \int_0^{\rho} dr \int_0^{2\pi} \Sigma_{n=0}^{\infty}\,\Sigma_{m=0}^{\infty}\, a_n \overline{a}_m r^{n+m+1} e^{i(n-m)\theta} d\theta$$

$$= 2\pi \int_0^{\rho} \Sigma_{n=0}^{\infty} |a_n|^2\, r^{2n+1} dr$$

$$= \pi\, \Sigma_{n=0}^{\infty} \frac{|a_n|^2}{n+1}\, \rho^{2n+2}.$$

Thus

$$\iint_D |f(z)|^2 dx\,dy \geq \iint_B |f(z)|^2 dx\,dy \geq \pi |a_o|^2\, \rho^2.$$

Making $\rho \to \rho_{z_o}$, we get

$$\iint_D |f(z)|^2 dx\,dy \geq \pi |f(z_o)|^2\, \rho_{z_o}^2,$$

$$\geq \pi\rho_K^2\, |f(z_o)|^2,$$

which is (4.3.5). □

Lemma 4.3.2. *If* $f \in H_2(D)$ *and if* $\{f_n\}$ *is a sequence of functions in* $H_2(D)$ *that converges in the norm to* f, *then* $\{f_n(z)\}$ *converges uniformly on compact subsets of* D *to* $f(z)$.

Proof. Since $\{f_n\}$ converges in the norm to f, given $\varepsilon > 0$, there exists an integer $n_o = n_o(\varepsilon)$ such that

$$\iint_D |f(z) - f_n(z)|^2 dx\, dy < \varepsilon^2 \quad \text{for} \quad n \geq n_o.$$

Let K be a compact subset of D and $\rho_K (> 0)$ be the distance of K from the boundary of D. If z_o is any point of K, by Lemma 4.3.1 we have

$$\pi\rho_K^2 |f(z_o) - f_n(z_o)|^2 \leq \pi\rho_{z_o}^2 \; |f(z_o) - f_n(z_o)|^2$$

$$\leq \iint_D |f(z)-f_n(z)|^2 dx\, dy < \varepsilon^2 \quad \text{for} \quad n \geq n_o.$$

Thus, for any $z_o \in K$,

$$|f(z_o) - f_n(z_o)| < \frac{\varepsilon}{\sqrt{\pi}\rho_K} \qquad \text{for} \quad n \geq n_o$$

which shows uniform convergence on K. □

We are now in a position to prove

Theorem 4.3.1. *$H_2(D)$ is a Hilbert space.*

Proof. We have already seen that $H_2(D)$ is an inner product space. We now show that it is complete with respect to the norm induced by the inner product. Let $\{f_n\}$ be a Cauchy sequence in $H_2(D)$. Then, given $\varepsilon > 0$, there exists an integer $n_o = n_o(\varepsilon)$ such that

$$\| f_m - f_n \| < \varepsilon \qquad \text{for} \quad m,n \geq n_o. \tag{4.3.6}$$

If K is any compact subset of D, by Lemma 4.3.1 we have, for any $z \in K$,

$$|f_m(z) - f_n(z)| \leq \frac{1}{\sqrt{\pi}\rho_K} \; \| f_m - f_n \|$$

$$< \frac{\varepsilon}{\sqrt{\pi}\rho_K} \quad \text{for} \quad m,n \geq n_o.$$

The above inequality shows that the sequence $\{f_n(z)\}$ converges uniformly on K to a limit function, f(z), say. Since K is any compact subset of D, the sequence $\{f_n(z)\}$ converges uniformly to f(z) on every compact subset of D. Hence by a classical theorem of Weierstrass (cf. Appendix A.4), f(z) must be regular in D. Now applying Fatou's lemma (cf. Appendix A.3) to (4.3.6), we get

$$\iint_D |f(z) - f_n(z)|^2 dx\,dy \leq \varepsilon^2 \quad \text{for } n \geq n_o. \tag{4.3.7}$$

This shows that $f - f_n \in H_2(D)$. However, $H_2(D)$ being a linear space, we must have $f \in H_2(D)$ and then (4.3.7) gives that $\{f_n\}$ converges in the norm to f. □

Our main concern in the deliberations that follow is to study how 'best' one can approximate a function $f \in H_2(D)$ by a sequence of polynomials, the 'closeness' of approximation being measured in terms of the norm of the space $H_2(D)$. Since this study is closely related to that of orthonormal sets in space, we introduce this concept first.

A sequence $\chi_o(z), \chi_1(z), \ldots, \chi_n(z), \ldots,$ of functions in $H_2(D)$ is said to be an *orthogonal sequence* on D if

$$\iint_D \chi_n(z)\, \overline{\chi_m(z)}\, dx\,dy = 0 \qquad \text{for } n \neq m; \tag{4.3.8}$$

if, in addition,

$$\iint_D |\chi_n(z)|^2 dx\,dy = 1 \quad \text{for} \quad n = 0,1,2,\ldots, \tag{4.3.9}$$

the sequence $\{\chi_n(z)\}_{n=0}^{\infty}$ is said to be *orthonormal* on D. For $f \in H_2(D)$,

$$a_n = \iint_D f(z)\, \overline{\chi_n(z)}\, dx\,dy \tag{4.3.10}$$

is said to be the nth Fourier coefficient of f with respect to the orthonormal system $\{\chi_n(z)\}$ and $\Sigma_{n=0}^{\infty} a_n\chi_n(z)$ is called the *Fourier series* of f with respect to this system. We now show that, among all linear combinations of the form $\Sigma_{k=0}^{n} c_k\chi_k(z)$, $\Sigma_{k=0}^{n} a_k\chi_k(z)$ gives the best approximation to f in the norm of $H_2(D)$. Precisely, we have

Theorem 4.3.2. *Let* $f \in H_2(D)$ *and its Fourier coefficients* a_n *be given by* (4.3.10). *Then the smallest value of the integral*

$$\iint_D |f(z) - \Sigma_{k=0}^{n} c_k\chi_k(z)|^2 dx\,dy$$

is attained for $c_k = a_k, \quad 0 \leq k \leq n,$
and equals

$$\iint_D |f(z)|^2 dx\, dy - \Sigma_{k=0}^n |a_k|^2 \geq 0. \tag{4.3.11}$$

Proof. We have

$$\iint_D |f(z) - \Sigma_{k=0}^n c_k\chi_k(z)|^2 dx\, dy = \iint_D \{f(z) - \Sigma_{k=0}^n c_k\chi_k(z)\}$$

$$\times \{\overline{f(z)} - \Sigma_{k=0}^n \bar{c}_k \overline{\chi_k(z)}\}\, dx\, dy$$

$$= \iint_D |f(z)|^2 dx\, dy - \Sigma_{k=0}^n c_k \iint_D \overline{f(z)}\, \chi_k(z) dx\, dy - \Sigma_{k=0}^n \bar{c}_k \iint_D f(z)\overline{\chi_k(z)} dx\, dy$$

$$+ \Sigma_{k=0}^n c_k\bar{c}_k \iint_D |\chi_k(z)|^2 dx\, dy$$

$$= \iint_D |f(z)|^2 dx\, dy - \Sigma_{k=0}^n c_k\bar{a}_k - \Sigma_{k=0}^n \bar{c}_k a_k + \Sigma_{k=0}^n |c_k|^2,$$

(by (4.3.8), (4.3.9) and (4.3.10))

$$= \iint_D |f(z)|^2 dx\, dy - \Sigma_{k=0}^n |a_k|^2 + \Sigma_{k=0}^n |c_k - a_k|^2 \geq 0.$$

Thus it follows that the integral in question has the least value for $c_k = a_k$, $k = 0,1,\ldots,n$ and this value equals the one given by (4.3.11). □

Making $n \to \infty$ in (4.3.11), we get the following

COROLLARY *(Bessel's inequality). For* $f \in H_2(D)$, *we have*

$$\iint_D |f(z)|^2 dx\, dy \geq \Sigma_{k=0}^\infty |a_k|^2 \tag{4.3.12}$$

where a_k *are the Fourier coefficients of* f.

Remark. If equality holds in (4.3.12), i.e., if

$$\iint_D |f(z)|^2 dx\, dy = \Sigma_{k=0}^\infty |a_k|^2 \tag{4.3.13}$$

the result is known as *Parseval's formula.* Theorem 4.3.2 therefore shows that *if* f *satisfies Parseval's formula, then*

$$\lim_{n\to\infty} \iint_D |f(z) - \Sigma_{k=0}^n a_k\chi_k(z)|^2 dx\, dy = 0,$$

i.e., *the Fourier series of* f *converges to* f *in the norm of* $H_2(D)$. Our next theorem shows that the converse of this result also holds. In fact we have the following stronger result.

Theorem 4.3.3. *If* $\{c_k\}$ *is a sequence of complex numbers such that* $\Sigma_{k=0}^{\infty} |c_k|^2$ *converges, then the series* $\Sigma_{k=0}^{\infty} c_k\chi_k(z)$ *converges in the norm of* $H_2(D)$ *to some function* $f \in H_2(D)$, *the Fourier coefficients of* f *coincide with the numbers* c_k, *and Parseval's formula is satisfied for* f.

Proof. Set $s_n(z) = \Sigma_{k=0}^{n} c_k\chi_k(z)$; then $s_n \in H_2(D)$ for each n and it is easy to check that

$$\iint_D |s_{n+p}(z) - s_n(z)|^2 dx\,dy = \Sigma_{k=n+1}^{n+p} |c_k|^2. \tag{4.3.14}$$

Since $\Sigma_{k=0}^{\infty} |c_k|^2$ converges, it follows from (4.3.14) that $\{s_n(z)\}$ is a Cauchy sequence in $H_2(D)$. Since $H_2(D)$ is complete (cf. Theorem 4.3.1), there exists a function $f \in H_2(D)$ such that $\{s_n(z)\}$ converges in the norm of $H_2(D)$ to f. If a_k are the Fourier coefficients of f, then, as in the proof of Theorem 4.3.2, we have

$$\begin{aligned}\iint_D |f(z)-\Sigma_{k=0}^{n} c_k\chi_k(z)|^2 dx\,dy &= \iint_D |f(z)|^2 dx\,dy - \Sigma_{k=0}^{n}|a_k|^2+\Sigma_{k=0}^{n}|c_k-a_k|^2 \\ &= \iint_D |f(z)-\Sigma_{k=0}^{n} a_k\chi_k(z)|^2 dx\,dy +\Sigma_{k=0}^{n}|c_k-a_k|^2 \\ &\geq \Sigma_{k=0}^{n} |c_k-a_k|^2 \end{aligned} \tag{4.3.15}$$

Since the left-hand side tends to zero as $n \to \infty$ and each term on the right-hand side is nonnegative, it follows that $c_k = a_k$ for all k and then relation (4.3.15) gives that Parseval's formula holds for f. □

It is clear from the above theorem and the remark following the preceding theorem that a function $f \in H_2(D)$ satisfies Parseval's formula if and only if it is limit in the norm of its Fourier series. Lemma 4.3.2 then gives that this Fourier series converges uniformly on compact subsets of D to f(z). An orthonormal system $\{\chi_n(z)\}_{n=0}^{\infty}$ is said to be *closed* in $H_2(D)$ if Parseval's formula holds for *every* function in $H_2(D)$. Closed systems have interesting properties. If a closed orthonormal system $\{\chi_n(z)\}$ is given, then the minimum of the integral of Theorem 4.3.2 tends to zero as $n \to \infty$ and so Bessel's inequality is replaced by the Parseval formula.

Further, any $f \in H_2(D)$ can be approximated in the norm of $H_2(D)$ by linear combinations of the functions χ_n. If this latter property is taken to be the definition of a 'closed' system then this definition can be extended to include the case of any set of linearly independent functions which are not necessarily orthonormal. It may be recalled that a set of functions $\phi_1(z),\ldots,\phi_n(z)$ is linearly independent if there do not exist constants $a_1,\ldots,a_n$, not all zero such that

$$a_1\phi_1(z)+\ldots+a_n\phi_n(z) \equiv 0. \tag{4.3.16}$$

An infinite set of functions $\{\phi_n(z)\}_{n=0}^{\infty}$ is linearly independent if each of its finite subsets is linearly independent. It is easy to check that if $\{\chi_n(z)\}_{n=0}^{\infty}$ forms an orthogonal sequence then it must be a linearly independent set. A concept closely related to 'closedness' is that of 'completeness'. An orthonormal system $\{\chi_n(z)\}$ is said to be *complete* in $H_2(D)$ if there is no function $f \in H_2(D)$ such that $f \not\equiv 0$ and f is orthogonal to all the functions $\chi_n(z)$.

It is significant that above two concepts are equivalent in $H_2(D)$.

Theorem 4.3.4. *An orthonormal system* $\{\chi_n(z)\}_{n=0}^{\infty}$ *is complete in* $H_2(D)$ *if and only if it is closed.*

Proof. Suppose that the orthonormal system $\{\chi_n(z)\}$ is closed and let $\phi \in H_2(D)$ be a function such that ϕ is orthogonal to all the functions $\chi_n(z)$, i.e.

$$\iint_D \phi(z)\,\overline{\chi_n(z)}\,dx\,dy = 0 \qquad \text{for } n = 0,1,2,\ldots .$$

It follows from (4.3.10) that all the Fourier coefficients of $\phi(z)$ are zero. Since the system is closed, Parseval's formula holds for $\phi(z)$ and so (4.3.13) gives

$$\iint_D |\phi(z)|^2 dx\,dy = 0.$$

Continuity of $\phi(z)$ now gives that $\phi(z) \equiv 0$ in D. Hence the system is complete.

Now suppose that the orthonormal system $\{\chi_n(z)\}$ is complete. If a_n are the Fourier coefficients of a function $f \in H_2(D)$, then by Bessel's inequality (4.3.12), the series $\Sigma_{n=0}^{\infty} |a_n|^2$ converges. Thus, by Theorem 4.3.3, the series $\Sigma_{n=0}^{\infty} a_n\chi_n(z)$ converges in the norm to some function $g \in H_2(D)$ which has a_n as its Fourier coefficients and for which Parseval's formula holds. Hence it follows that

$$\iint_D [f(z) - g(z)] \overline{\chi_n(z)}\, dx\, dy = 0 \quad \text{for} \quad n = 0,1,2,\ldots .$$

Since the system $\{\chi_n(z)\}$ is complete, the above relations give that $f(z) \equiv g(z)$ in D. Thus Parseval's formula holds for any $f \in H_2(D)$ and so the system is closed. □

There always exist, in $H_2(D)$, closed orthonormal systems. This is demonstrated in

Theorem 4.3.5. *For every bounded domain* D *there exists a closed (and hence complete) system of orthonormal functions in* $H_2(D)$.

Proof. Let A denote the class of all functions $h \in H_2(D)$ that satisfy, at a fixed but arbitrary point $z_o \in D$, the conditions

$$h(z_o) = h'(z_o) = \ldots = h^{(n-1)}(z_o) = 0,\ h^{(n)}(z_o) = 1. \tag{4.3.17}$$

Note that the class A is non-empty since the function $s(z) = (z-z_o)^n/n!$ is in A. We consider the problem of minimizing the integral

$$I(h) = \iint_D |h(z)|^2 dx\, dy$$

for $h \in A$. We shall show that this problem has a solution in the class A. If

$$\lambda = \underset{h \in A}{\text{g.}\ell\text{.b.}}\ I(h) \tag{4.3.18}$$

then there exists a sequence $\{h_j\}$ of functions in A such that $\lim_{j\to\infty} I(h_j) = \lambda$. Since every convergent sequence is bounded, it follows that there exists a constant $M > 0$ such that $I(h_j) \leq M$ for all j. Consider the family $F = \{h_j;\ j = 1,2,\ldots\}$. If K is any compact subset of D, by Lemma 4.3.1,

$$|h_j(z)|^2 \leq \frac{I(h_j)}{\pi\rho_K^2} \leq \frac{M}{\pi\rho_K^2} \quad \text{for every } z \in K \text{ and } j = 1,2,\ldots .$$

Thus the family F is uniformly bounded on compact subsets of D and so is a normal family (cf. Appendix A.7). It therefore follows that there exists a subsequence $\{h_{j_\ell}(z)\}$ that converges uniformly on compact subsets of D to a regular function $f_n(z)$, say. Thus, for every compact subset K of D,

$$\iint_K |f_n(z)|^2 dx\, dy = \lim_{\ell\to\infty} \iint_K |h_{j_\ell}(z)|^2 dx\, dy \leq \lambda$$

and so, by (4.3.2),

$$I(f_n) \leq \lambda. \tag{4.3.19}$$

Since $\lim_{\ell\to\infty} h_{j_\ell}(z) = f_n(z)$ and $h_{j_\ell} \in A$ for $\ell = 1,2,\ldots$, it follows that $f_n(z)$ satisfies conditions (4.3.17) and so $f_n \in A$. (4.3.18) and (4.3.19) then give that $I(f_n) = \lambda$ and so our problem of minimizing the integral $I(h)$ over A has a solution in A.

We claim that this solution is unique. First we show that if $\Psi \in H_2(D)$ is such that $\Psi^{(j)}(z_o) = 0$ for $j = 0,1,\ldots,n$, then

$$(\Psi,f_n) \equiv \iint_D \Psi(z)\, \overline{f_n(z)}\, dx\, dy = 0. \tag{4.3.20}$$

It is a simple matter to verify that for every constant c the function $f_n+c\Psi \in A$. If the integral in (4.3.20) were nonzero, then for $c = -(f_n,\Psi)/\|\Psi\|^2$ we would have

$$I(f_n+c\Psi) = \iint_D |f_n(z)|^2 dx\, dy + c(\Psi,f_n) + \bar{c}(f_n,\Psi) + c\bar{c}\|\Psi\|^2$$

$$= \lambda - |(f_n,\Psi)|^2/\|\Psi\|^2 < \lambda,$$

which would contradict (4.3.18). Thus (4.3.20) holds true. If $g_n(z)$ is another solution of our problem, then for $\Psi = f_n-g_n$ we have $\Psi^{(j)}(z_o) = 0$ for $j = 0,1,\ldots,n$ and so (4.3.20) gives $(f_n-g_n,f_n) = 0$ and $(f_n-g_n,g_n) = 0$ which yields $(f_n-g_n, f_n-g_n) = 0$, i.e., $f_n = g_n$.

Now we consider the sequence $\{f_n(z)\}_{n=0}^\infty$. If we set

$$\chi_n(z) = \frac{f_{n-1}(z)}{\|f_{n-1}\|}, \qquad n = 1,2,\ldots \tag{4.3.21}$$

then it follows, in view of (4.3.20), that the system of functions $\{\chi_n(z)\}$ is orthonormal. We show that this system is closed, i.e., every $f \in H_2(D)$ satisfies Parseval's formula with respect to this system. Let $f \in H_2(D)$ and a_n be its Fourier coefficients, defined by (4.3.10), with respect to the system (4.3.21). By Theorem 4.3.3, the series $\Sigma_{n=1}^{\infty} a_n\chi_n(z)$ converges in the norm of $H_2(D)$ to some function $h \in H_2(D)$ whose Fourier coefficients are a_n and for which Parseval's formula holds. If we show that $f(z) \equiv h(z)$, we are through. Since, for any positive integer k, the system $\{\chi_1(z),\ldots,\chi_k(z)\}$ is linearly independent (being an orthonormal system), using relation (4.3.16) it follows that

$$\begin{vmatrix} \chi_1(z_o) & \chi_2(z_o) & \ldots & \chi_k(z_o) \\ \chi_1^{(1)}(z_o) & \chi_2^{(1)}(z_o) & \ldots & \chi_k^{(1)}(z_o) \\ - - - & - - - & - - - & - - - \\ \chi_1^{(k-1)}(z_o) & \chi_2^{(k-1)}(z_o) & \ldots & \chi_k^{(k-1)}(z_o) \end{vmatrix} \neq 0.$$

Thus, there exist uniquely determined constants $b_{1k},\ldots,b_{kk}$ satisfying the equations

$$\begin{aligned} b_{1k}\,\chi_1(z_o) &+ \ldots + b_{kk}\,\chi_k(z_o) = f(z_o) \\ b_{1k}\,\chi_1^{(1)}(z_o) &+ \ldots + b_{kk}\,\chi_k^{(1)}(z_o) = f^{(1)}(z_o) \\ &- - - - - - - - - - - - \\ b_{1k}\,\chi_1^{(k-1)}(z_o) &+ \ldots + b_{kk}\,\chi_k^{(k-1)}(z_o) = f^{(k-1)}(z_o). \end{aligned} \qquad (4.3.22)$$

We assert that $b_{jk} = a_j$ for $j = 1,2,\ldots,k$. If we set $\phi_k(z) = \Sigma_{j=1}^{k} b_{jk}\,\chi_j(z)$, equations (4.3.22) may be written as

$$(\phi_k - f)^{(\nu)}(z_o) = 0 \quad \text{for} \quad \nu = 0,1,\ldots,k-1. \qquad (4.3.23)$$

The relation (4.3.20), applied to $\Psi = \phi_k - f$, gives

$$\iint_D (\phi_k - f)(z)\,\overline{\chi_j(z)}\,dx\,dy = 0 \quad \text{for} \quad j = 1,2,\ldots,k$$

$$\iint_D \phi_k(z)\,\overline{\chi_j(z)}\,dx\,dy = \iint_D f(z)\,\overline{\chi_j(z)}\,dx\,dy$$

i.e.,

$$b_{jk} = a_j \quad \text{for} \quad j = 1,2,\ldots,k.$$

Since $f(z) - h(z) = \lim_{k\to\infty} (f(z) - \Sigma_{j=1}^{k} a_j\chi_j(z))$, we have, for $\nu \geq 0$,

$$f^{(\nu)}(z_o) - h^{(\nu)}(z_o) = \lim_{k\to\infty} (f^{(\nu)}(z_o) - \Sigma_{j=1}^{k} a_j\chi_j^{(\nu)}(z_o))$$

$$= \lim_{k\to\infty} (f-\phi_k)^{(\nu)}(z_o)$$

$$= 0, \quad \text{by } (4.3.23).$$

Thus f and h coincide in a neighbourhood of z_o and so, by analytic continuation, $h(z) \equiv f(z)$ in D. □

Our next theorem introduces the important notion of kernel function for the domain D which may be seen to play a crucial role in further development of the theory.

Theorem 4.3.6. *Given an orthonormal system* $\{\chi_n(z)\}_{n=0}^{\infty}$ *in* $H_2(D)$, *for each fixed* $s \in D$, *the series* $\Sigma_{n=0}^{\infty} \overline{\chi_n(s)}\, \chi_n(z)$ *converges uniformly (in z) on compact subsets of* D *to a regular function* $K(z,s)$ *in* $H_2(D)$,

$$K(z,s) = \Sigma_{n=0}^{\infty} \overline{\chi_n(s)}\, \chi_n(z), \quad s \in D,\ z \in D. \tag{4.3.24}$$

$\overline{\chi_n(s)}$ *are the Fourier coefficients of the function* $K(z,s)$ *and Parseval's formula*

$$\iint_D |K(z,s)|^2 dx\, dy = \Sigma_{n=0}^{\infty} |\chi_n(s)|^2 = K(s,s) \tag{4.3.25}$$

holds for the function $K(z,s)$.

Proof. Given the orthonormal system $\{\chi_n(z)\}_{n=0}^{\infty}$, we consider the sum

$$K_n(z,s) = \Sigma_{k=0}^{n} \overline{\chi_k(s)}\, \chi_k(z), \qquad s \in D,\ z \in D. \tag{4.3.26}$$

Using the orthonormality of the system, it follows easily that

$$\iint_D |K_n(z,s)|^2 dx\, dy = \Sigma_{k=0}^{n} |\chi_k(s)|^2 = K_n(s,s) \geq 0.$$

An application of Lemma 4.3.1 gives, for any compact subset F of D,

$$\iint_D |K_n(z,s)|^2 dx\, dy = K_n(s,s) \geq \pi\rho_F^2 \{K_n(s,s)\}^2, \quad s \in F$$

i.e.,

$$K_n(s,s) = \Sigma_{k=0}^{n} |\chi_k(s)|^2 \leq \frac{1}{\pi\rho_F^2}, \qquad s \in F.$$

Since this inequality holds for every n, we must have

$$\Sigma_{k=0}^{\infty} |\chi_k(s)|^2 \le \frac{1}{\pi\rho_F^2} < \infty, \qquad s \in F.$$

Applying Theorem 4.3.3, with $c_k = \overline{\chi_k(s)}$ we get the result stated in the theorem. □

The function K(z,s), defined by (4.3.24), is known as the Bergman kernel function. It acts as a 'reproducing kernel' for functions in $H_2(D)$, as shown in the following:

Theorem 4.3.7. *For* $f \in H_2(D)$, $\{\chi_n(z)\}_{n=0}^{\infty}$ *an arbitrary, closed orthonormal system in* $H_2(D)$, *and for* K(z,s) *defined by* (4.3.24), *one has*

$$f(s) = \iint_D K(z,s)\, f(z)dx\, dy, \qquad s \in D. \tag{4.3.27}$$

Proof. The integral in (4.3.27) exists since both f(z) and K(z,s) are in $H_2(D)$ and the integral is the inner product (f,K). If a_n is the nth Fourier coefficient of f(z) with respect to the orthonormal system $\{\chi_n(z)\}_{n=0}^{\infty}$, then

$$\iint_D \overline{K(z,s)} f(z)dx\, dy = \iint_D \overline{K_n(z,s)} f(z)dx\, dy + \iint_D \{\overline{K(z,s)-K_n(z,s)}\} f(z)dx\, dy$$

$$= \Sigma_{k=0}^{n} a_k\chi_k(s) + \iint_D \{\Sigma_{k=n+1}^{\infty} \chi_k(s)\, \overline{\chi_k(z)}\} f(z)dx\, dy,$$

(by (4.3.26)).

If we show that the integral on the right tends to zero as $n \to \infty$, the theorem is proved. To see this, we note that, by the Schwarz inequality,

$$\left|\iint_D \{\Sigma_{k=n+1}^{\infty} \chi_k(s)\overline{\chi_k(z)}\} f(z)dx\, dy\right|^2 \le \left(\iint_D \left|\Sigma_{k=n+1}^{\infty} \chi_k(s)\overline{\chi_k(z)}\right|^2 dx\, dy\right)$$

$$\times \left(\iint_D |f(z)|^2 dx\, dy\right) = \left(\Sigma_{k=n+1}^{\infty} |\chi_k(s)|^2\right) \|f\|^2$$

$$\to 0 \quad \text{as} \quad n \to \infty$$

since the series $\Sigma_{k=0}^{\infty} |\chi_k(s)|^2$ converges. □

Corollary. *An orthonormal set* $\{\chi_n(z)\}_{n=0}^{\infty}$ *forms a closed system in* $H_2(D)$ *if and only if the kernel function* K(z,s), *defined by* (4.3.24), *satisfies the reproducing property* (4.3.27) *for every* $f \in H_2(D)$.

It may be shown that the kernel function for a bounded domain D is unique (see Exercise 4.3.11). This would imply that the kernel function for a bounded domain D is independent of the choice of the closed orthonormal system defining it. There is an intimate relation between the kernel of a bounded, simply connected domain D and the Riemann mapping function mapping D onto $\mathcal{U}$. This is demonstrated in the next theorem. Incidentally, this theorem also shows that the kernel function is independent of the choice of the orthonormal system defining it.

Theorem 4.3.8. *Let* D *be a bounded, simply connected domain and let* $w = f(z)$ *be a regular function mapping* D *univalently onto* $\mathcal{U} = \{w: |w| < 1\}$ *such that* $f(z_o) = 0$, $f'(z_o) > 0$, $z_o \in D$. *If* $K(z,z_o)$ *is the kernel function of* D, *defined by* (4.3.24) *with respect to a closed orthonormal system* $\{\chi_n(z)\}$, *then for any* $z \in D$,

$$f'(z) = \frac{f'(z_o)}{K(z_o,z_o)} K(z,z_o) = \frac{\sqrt{\pi}}{\sqrt{\Sigma_{n=0}^{\infty} |\chi_n(z_o)|^2}} \Sigma_{n=0}^{\infty} \overline{\chi_n(z_o)}\, \chi_n(z). \quad (4.3.28)$$

Proof. Since

$$\iint_D |f'(z)|^2 dx\, dy = \iint_{|w|<1} du\, dv = \pi,$$

it follows that $f' \in H_2(D)$. If a_n is the nth Fourier coefficient of f' with respect to the orthonormal system $\{\chi_n(z)\}_{n=0}^{\infty}$, then

$$\overline{a}_n = \iint_D \overline{f'(z)}\, \chi_n(z) dx\, dy.$$

If $z = g(w)$ denotes the function inverse to $w = f(z)$ and D_r, $0 < r < 1$, is the image of the disc $|w| < r$ under g, then

$$\overline{a}_n = \lim_{r\to 1} \iint_{D_r} \frac{\chi_n(z)}{f'(z)} |f'(z)|^2 dx\, dy$$

$$= \lim_{r\to 1} \iint_{|w|<r} \frac{\chi_n(g(w))}{f'(g(w))} du\, dv, \qquad w = u+iv.$$

Since $\chi_n(g(w))/f'(g(w))$ is a regular function of w for $|w| < 1$, writing $\chi_n(g(w))/f'(g(w)) = \Sigma_{k=0}^{\infty} c_k^{(n)} w^k$, we get

$$\bar{a}_n = \lim_{r\to 1} \iint_{|w|<r} (\Sigma_{k=0}^{\infty} c_k^{(n)} w^k) du\, dv$$

$$= \lim_{r\to 1} \int_o^r \int_o^{2\pi} (\Sigma_{k=0}^{\infty} c_k^{(n)} \rho^k e^{ik\theta}) \rho\, d\rho\, d\theta$$

$$= \pi\, c_o^{(n)}$$

$$= \pi \left[\frac{\chi_n(g(w))}{f'(g(w))}\right]_{w=0}$$

$$= \pi\, \chi_n(z_o)/f'(z_o).$$

Therefore,

$$f'(z) = \Sigma_{n=0}^{\infty} a_n \chi_n(z)$$

$$= \frac{\pi}{f'(z_o)} \Sigma_{n=0}^{\infty} \overline{\chi_n(z_o)}\, \chi_n(z)$$

$$= \frac{\pi}{f'(z_o)} K(z,z_o)$$

which gives (4.3.28) in view of the obvious relation $K(z_o,z_o) = \frac{1}{\pi}[f'(z_o)]^2$. □

As stated earlier, we wish to consider those domains D such that every $f \in H_2(D)$ can be approximated by a sequence of polynomials in the norm of $H_2(D)$. This is so if the polynomials form a closed set in $H_2(D)$. Although for every bounded domain D, as shown earlier, there exists a closed orthonormal set of functions in $H_2(D)$, this set need not consist of polynomials. In fact, there exist bounded simply-connected domains D for which the set of polynomials is not closed in $H_2(D)$. For example, let D be the open unit disc slit along the positive real axis. The system $\{\sqrt{((n+1)/\pi)}\, z^n\}_{n=0}^{\infty}$ which is orthonormal in the unit disc U is also orthonormal in D since the slit does not influence the values of the area integrals. If this system were closed with respect to D, every function $f \in H_2(D)$, regular in D, would be represented by a series of functions belonging to the system $\{\sqrt{((n+1)/\pi)}\, z^n\}_{n=0}^{\infty}$, i.e., by a power series, thus leading to the absurd conclusion that every function which is regular in D and belongs to $H_2(D)$ is regular in the unit disc U (consider, for example, an analytic branch of $\sqrt{z}$ in D). We shall show that a special type of bounded, simply connected domains D, namely, those whose complement is the closure of a domain, has the property that polynomials form a closed system in $H_2(D)$. In particular, if D is a domain bounded by a closed Jordan curve, it satisfies this property. Thus, we have the following theorem:

Theorem 4.3.9. *Let* D *be a bounded, simply-connected domain whose complement is the closure of a domain. Then every* $f \in H_2(D)$ *can be approximated in the norm of* $H_2(D)$ *by polynomials, or, in other words, polynomials form a closed set in* $H_2(D)$.

Proof. We first prove, by induction, that there exists a unique sequence of polynomials $\{P_n(z)\}_{n=0}^{\infty}$, orthonormal on D, such that $P_n(z)$, of degree n, has, as the coefficient of z^n, a real positive number. If A is the area of D, set $P_o(z) = 1/\sqrt{A}$. Now suppose $P_o(z),\ldots,P_m(z)$ have already been obtained. Then $P_{m+1}(z)$ can be expressed in the form

$$P_{m+1}(z) = \lambda_m z^{m+1} + d_m^{(m)} P_m(z)+\ldots+d_o^{(m)} P_o(z) \tag{4.3.29}$$

where $\lambda_m > 0$. Since $\{P_n(z)\}$ is to form an orthonormal sequence, we must have

$$\iint_D P_{m+1}(z) \overline{P_n(z)} dx\, dy = 0 \quad \text{for} \quad n = 0,1,\ldots,m \tag{4.3.30}$$

and

$$\iint_D |P_{m+1}(z)|^2 dx\, dy = 1. \tag{4.3.31}$$

(4.3.29) and (4.3.30) give $d_n^{(m)} = -\lambda_m c_n^{(m)}$, $n = 0,1,\ldots,m$, where

$$c_n^{(m)} = \iint_D z^{m+1} \overline{P_n(z)}\, dx\, dy.$$

Thus

$$P_{m+1}(z) = \lambda_m\left[z^{m+1} - c_m^{(m)} P_m(z) - \ldots - c_o^{(m)} P_o(z)\right] = \lambda_m q_m(z), \text{ say.}$$

(4.3.31) then gives $\lambda_m = 1/\| q_m \|$. This determines the sequence $\{P_n(z)\}_{n=0}^{\infty}$ uniquely. Set

$$H(z,s) = \Sigma_{k=0}^{\infty} P_k(z) \overline{P_k(s)}, \qquad z,s \in D. \tag{4.3.32}$$

By Theorem 4.3.6, for every fixed $s \in D$, the series on the right converges absolutely and uniformly on every compact subset of D and the sum function

$H(z,s)$ is an analytic function of z in D. We shall show that $H(z,s)$ and the kernel function $K(z,s)$ associated with the domain D are identical. This would, in turn, imply, by corollary to Theorem 4.3.7 that $\{P_n(z)\}_{n=0}^{\infty}$ form a closed system in $H_2(D)$.

We note that, if $z^n = a_o + a_1P_1(z)+\ldots+a_nP_n(z)$, then

$$\iint_D \overline{H(z,s)}\; z^n dx\, dy = \iint_D (\Sigma_{k=0}^{\infty} \overline{P_k(z)}\; P_k(s))(\Sigma_{k=0}^{n} a_k P_k(z))dx\, dy$$

$$= \Sigma_{k=0}^{n} a_k P_k(s) \iint_D \overline{P_k(z)} P_k(z) dx\, dy$$

$$= \Sigma_{k=0}^{n} a_k P_k(s) = s^n. \tag{4.3.33}$$

It therefore follows from (4.3.33) that $H(z,s)$ exhibits the same reproducing property (4.3.27) as possessed by $K(z,s)$ for such functions $f(z)$ which are either polynomials or can be approximated by polynomials in $H_2(D)$.

Since D is bounded there exists an R such that $|z| < R$ for all $z \in D$. Let $\xi \in \mathbb{C}$ be such that $|\xi| > R$. Then the expansion

$$\frac{1}{\xi-z} = \frac{1}{\xi} + \frac{z}{\xi^2} + \frac{z^2}{\xi^3} + \ldots$$

converges uniformly for all $z \in D$. Hence

$$\iint_D \overline{H(z,s)}\; (\frac{1}{\xi-z})dx\, dy = \Sigma_{n=0}^{\infty} \xi^{-n-1} \iint_D \overline{H(z,s)}\; z^n dx\, dy$$

$$= \Sigma_{n=0}^{\infty} \xi^{-n-1} s^n = \frac{1}{\xi-s}. \tag{4.3.34}$$

Since both sides of (4.3.34) are analytic functions of ξ for $\xi \notin D$ and the relation (4.3.34) holds for all $|\xi| > R$, it should continue to hold for all $\xi \notin D$ by analytic continuation.

Now let $f(z)$ be a function which is regular in the closure of D. It then follows that $f(z)$ must be regular in a domain D' which includes both D and its boundary ∂D. Let Γ be a closed contour that is contained in D' but which has no points in common with either D or ∂D. For $s \in D$, we then have

$$f(s) = \frac{1}{2\pi i} \int_\Gamma \frac{f(t)dt}{t-s}$$

$$= \frac{1}{2\pi i} \int_\Gamma f(t) \left[\iint_D \overline{H(z,s)}\; (\frac{1}{t-z})dx\, dy\right] dt \quad \text{(by (4.3.34))}.$$

A change of order of integration gives

$$f(s) = \iint_D \overline{H(z,s)} \left[\frac{1}{2\pi i} \int_\Gamma \frac{f(t)}{t-z}\, dt\right] dx\, dy$$

$$= \iint_D \overline{H(z,s)}\, f(z)\, dx\, dy.$$

The above relation shows that H(z,s) has the reproducing property (4.3.27) for the functions f(z) that are analytic in the closure of D and therefore also for those functions in $H_2(D)$ that can be approximated by functions regular in the closure of D. If it can be shown that any function of $H_2(D)$ can be approximated by functions that are regular in the closure of D, it will then follow that H(z,s) has the reproducing property (4.3.27) for all the functions of $H_2(D)$. Since kernel function for a domain is unique, it then follows that H(z,s) and K(z,s) are identical and so, as remarked earlier, this implies that $\{P_n(z)\}_{n=0}^{\infty}$ form a closed system in $H_2(D)$.

To conclude, we therefore show that any function f regular in D can be approximated by functions regular in the closure of D if D is given in the theorem. It then follows that any function in $H_2(D)$ can be so approximated. Since the complement of D is a closed domain by hypothesis, it should be possible to find a decreasing sequence $\{D_n\}_{n=1}^{\infty}$ of simply connected, bounded domains such that $D_n \supset D$ for every n, $\lim_{n\to\infty} D_n = D$ and the boundary of D has no point in common with the boundaries of the domains D_n. Here $\lim_{n\to\infty} D_n = D$ is to be understood in the sense that D is the kernel of every infinite subsequence of the sequence $\{D_n\}$. By the Riemann mapping theorem (cf. Appendix A.12) there exists a one-one, analytic function $w = \phi_n(z)$ mapping D_n onto D such that $\phi_n(s) = s$, $\phi_n'(s) > 0$, $s \in D$. Since $\lim_{n\to\infty} D_n = D$, it follows by Caratheodory's mapping theorem that $\lim_{n\to\infty} \phi_n(z) = z$. Since ∂D has no point in common with the boundary of D_n, $\phi_n(z)$ is regular in the closure $D \cup \partial D$ of D and maps $D \cup \partial D$ onto a closed domain entirely contained in D. Thus, the function $\Psi_n(z) = f(\phi_n(z))$ is regular in $D \cup \partial D$. Since as $n \to \infty$, $\phi_n(z) \to z$, it follows that $\Psi_n(z) \to f(z)$. This shows that f(z) can be approximated by functions regular in $D \cup \partial D$. □

For a bounded and simply connected domain D whose complement K is a closed domain it follows that the interior K^o of K is a simply connected domain in the extended complex plane. If ρ is the transfinite diameter of $\overline{D}$ it follows, as in Section 4.2, that there exists a univalent analytic function $w = \Omega(z)$ which maps K^o onto $|w| > \rho$ so that $\Omega(\infty) = \infty$ and $\lim_{z\to\infty} \Omega(z)/z = 1$. If $z = \eta(w)$ is the inverse function of Ω, then $L_r = \{z:z = \eta(w), |w| = r > \rho\}$ is an analytic Jordan curve for each $r > \rho$. If D_r denotes the domain bounded by L_r then $D \subset D_r$ for each $r > \rho$.

Before proceeding to our next theorem, we need a lemma.

Lemma 4.3.3. *Let* D *be a bounded, simply connected domain, with* $\overline{D}$ *of transfinite diameter* $\rho > 0$, *having complement* K *as a closed domain and let* $\{\chi_n(z)\}_{n=0}^{\infty}$ *be a closed orthonormal system of polynomials in* $H_2(D)$. *Then for any* $r^* > \rho$ *there exists a constant* M' *such that*

$$|\chi_n(z)| \le M' \left(\frac{r^*}{\rho}\right)^n, \qquad n = 0,1,\ldots, \qquad z \in \overline{D}. \tag{4.3.35}$$

Proof. Let $w = \phi(z)$ be a univalent analytic function mapping D onto $|w| < \rho$ so that $\phi(z_o) = 0$ for some $z_o \in D$. Define $D^{(t)} = \{z \in D: |\phi(z)| < t\rho, 0 < t < 1\}$. Let $L_r^{(t)}$ $(r > \rho)$ be the Jordan analytic curves for $D^{(t)}$ which are defined in the same way as for D so that $D_r^{(t)}$ denotes the domain bounded by $L_r^{(t)}$. By Caratheodory's mapping theorem (cf. Appendix A.14) the domains $D^{(t)}$ converge to their kernel D as t approaches 1 whereas, for fixed r, the domains $D_r^{(t)}$ approach D_r (bounded by L_r) as $t \to 1$. Since $\overline{D} \subset D_r$ for each $r > \rho$, so for t sufficiently near 1, $\overline{D} \subset D_r^{(t)}$. By (4.3.25), for $n = 0,1,2,\ldots$ and $z \in \overline{D}^{(t)}$,

$$|\chi_n(z)| \le M_t \qquad \text{where} \qquad M_t = \max_{z\in\overline{D}^{(t)}} |K(z,z)|^{1/2}.$$

Thus, by Lemma 4.2.2,

$$|\chi_n(z)| \le M_t\left(\frac{r^*}{\rho}\right)^n \qquad \text{for} \qquad z \in \overline{D}^{(t)}_{r^*}.$$

Since $\overline{D} \subset \overline{D}^{(t)}_{r^*}$, we get (4.3.35). □

Our next theorem is a representation theorem for functions in $H(\overline{D};R)$ in terms of their Fourier series. Recall that $H(\overline{D};R)$ is the class of all functions that are regular in D_R with a singularity on L_R $(\rho < R < \infty)$. Since $\overline{D} \subset D_R$ for $R > \rho$, it follows that every $f \in H(\overline{D};R)$ is analytic in $\overline{D}$ and so belongs to $H_2(D)$. Thus $H(\overline{D};R) \subset H_2(D)$. We now have

Theorem 4.3.10. *Let* D *be a bounded simply connected domain, with* $\overline{D}$ *of transfinite diameter* $\rho > 0$, *having complement* K *as a closed domain and let* $\{\chi_n(z)\}_{n=0}^{\infty}$ *be a closed orthonormal system of polynomials in* $H_2(D)$. *Then every* $f \in H(\overline{D};R)$ *can be represented by the Fourier series*

$$f(z) = \Sigma_{n=0}^{\infty} a_n \chi_n(z) \tag{4.3.36}$$

with

$$a_n = \iint_D f(z)\, \overline{\chi_n(z)}\, dx\, dy \tag{4.3.37}$$

if and only if

$$\limsup_{n\to\infty} |a_n|^{1/n} = \rho/R. \tag{4.3.38}$$

The series in (4.3.36) *converges absolutely and uniformly on every compact subset of* D_R *and diverges outside* L_R.

Proof. Since the orthonormal system of polynomials $\{\chi_n(z)\}_{n=0}^{\infty}$ is closed in $H_2(D)$, by Lemma 4.3.2, the expansion (4.3.36) is valid uniformly on compact subsets of D.

By Theorem 4.2.7, for every $n \geq 0$ there exists a polynomial $Q_n(z)$ of degree not exceeding n such that

$$|f(z) - Q_n(z)| < M \left(\frac{\rho}{R}\right)^n, \qquad z \in \overline{D}, \tag{4.3.39}$$

where M is a constant independent of z and n. Now (4.3.37) can be written as

$$a_n = \iint_D (f(z) - Q_{n-1}(z))\overline{\chi_n(z)}\, dx\, dy + \iint_D Q_{n-1}(z)\overline{\chi_n(z)}\, dx\, dy.$$

Since $\chi_n(z)$ is orthogonal to any polynomial of degree less than n, the last integral on the right is zero. Schwarz inequality and (4.3.39) then give, for $\rho < r < R$,

$$|a_n|^2 \leq \left(\iint_D |f(z) - Q_{n-1}(z)|^2 dx\, dy\right)\left(\iint_D |\chi_n(z)|^2 dx\, dy\right)$$

$$\leq AM^2\left(\frac{\rho}{R}\right)^{2n}, \quad \text{A being the area of D.} \tag{4.3.40}$$

Thus

$$\limsup_{n\to\infty} |a_n|^{1/n} \le \frac{\rho}{R}. \tag{4.3.41}$$

If

$$\limsup_{n\to\infty} |a_n|^{1/n} = \frac{\rho}{R_o} < \frac{\rho}{R},$$

let R_1 be such that $R < R_1 < R_o$. By Lemmas 4.2.2 and 4.3.3, we then have, for $z \in D_{R_1}$ and r such that $\rho < r < R_1$,

$$\limsup_{n\to\infty} |a_n\chi_n(z)|^{1/n} \le \frac{\rho}{R_o} \cdot \frac{r}{\rho} \cdot \frac{R_1}{\rho} = \frac{rR_1}{\rho R_o}.$$

Since $R_1 < R_o$, $r > \rho$ can be chosen so that $rR_1 < \rho R_o$. It then follows that the series $\Sigma_{n=0}^{\infty} a_n\chi_n(z)$ converges uniformly on compact subsets of D_{R_1} to a regular function. Since $D \subset D_{R_1}$ and the series (4.3.36) converges in D to f(z) it follows that f(z) is regular in D_{R_1}, a contradiction to the hypothesis that f(z) has a singularity on L_R. Thus equality must hold in (4.3.41). This proves (4.3.38).

Conversely, if (4.3.38) holds then $|a_n| < ((\rho+\varepsilon)/R)^n < 1$ for $n > n_o$ and so $\Sigma_{n=0}^{\infty} |a_n|^2$ converges. Theorem 4.3.3 then gives that the series $\Sigma_{n=0}^{\infty} a_n\chi_n(z)$ is a Fourier series for some $f \in H_2(D)$ and so converges to f uniformly on compact subsets of D. Proceeding as above it follows that the series converges to f uniformly on compact subsets of D_R so that f is regular in D_R. The function f must have a singularity on L_R for otherwise f would be regular in D_{R_1} for some $R_1 > R$ and so the first part of the theorem would give

$$\limsup_{n\to\infty} |a_n|^{1/n} = \frac{\rho}{R_1} < \frac{\rho}{R},$$

a contradiction. Thus $f \in H(\overline{D};R)$. □

Corollary. *If* $\overline{M}(r) = \max_{z\in L_r} |f(z)|$, $f \in H(\overline{D};R)$ *and* a_n *is given by* (4.3.37) *then for* $\rho < r' < r < R$, *we have*

$$|a_n| \le \sqrt{A}\, \overline{M}(r) \frac{r}{r-r'} \left(\frac{r'}{r}\right)^n, \qquad n > n_o = n_o(r'). \tag{4.3.42}$$

The corollary follows from (4.3.40) and Theorem 4.2.7.

We shall now determine, for functions in $H(\bar{D};R)$, the growth parameters as defined by (4.2.18) to (4.2.21) in terms of the Fourier coefficients. Since the methods of proof of these results are similar to those adopted while characterizing them in terms of Faber coefficients, we shall prove only the first theorem in this series and leave others to the reader.

To obtain the results in a general setting we shall assume that $f \in H(\bar{D};R)$ is represented by the gap Fourier series

$$f(z) = \Sigma_{n=0}^{\infty} a_n \chi_{\lambda_n}(z), \qquad z \in D_R, \tag{4.3.43}$$

where $\{\lambda_n\}_{n=0}^{\infty}$ is a strictly increasing sequence of positive integers and $a_n \neq 0$ for all n.

Our first theorem relates Fourier coefficients to the $\bar{D}$-order of f.

Theorem 4.3.11. *Let* $f \in H(\bar{D};R)$ *be represented by the Fourier series* (4.3.43) *with respect to the closed orthonormal system* $\{\chi_n(z)\}_{n=0}^{\infty}$ *of polynomials in* $H_2(D)$ *and let it be of* $\bar{D}$*-order* α $(0 \leq \alpha < \infty)$. *Then*

$$\frac{\alpha}{\alpha+1} = \limsup_{n \to \infty} \frac{\log^+\log^+(|a_n|(R/\rho)^{\lambda_n})}{\log \lambda_n} \tag{4.3.44}$$

where the left-hand side in (4.3.44) *is to be interpreted as* 1 *when* $\alpha = \infty$.

Proof. Let $\alpha < \infty$. By (4.2.18), given $\varepsilon > 0$ there exists $r_o = r_o(\varepsilon)$ such that

$$\log \bar{M}(r) < \left(\frac{Rr}{R-r}\right)^{\alpha+\varepsilon} \qquad \text{for} \quad r_o < r < R. \tag{4.3.45}$$

By (4.3.42) and (4.3.45) we therefore have, for $n > n_o$ and $\rho < r' < r < R$, $r_o < r$,

$$\log(|a_n|\left(\frac{R}{\rho}\right)^{\lambda_n}) < \left(\frac{Rr}{R-r}\right)^{\alpha+\varepsilon} + \log\frac{r}{r-r'} + \lambda_n \log\frac{r'}{r} + \lambda_n \log\frac{R}{\rho} + \log\sqrt{A}.$$

Choosing $r' = \lambda\rho + ((1-\lambda)R\rho/r)$, $0 < \lambda < 1$, this gives

$$\log(|a_n|\left(\frac{R}{\rho}\right)^{\lambda_n}) < \left(\frac{Rr}{R-r}\right)^{\alpha+\varepsilon} + \lambda_n(2-\lambda)\frac{R-r}{r} + \log\frac{r^2}{r^2-\lambda\rho r+(1-\lambda)R\rho} + \log\sqrt{A}.$$

Choosing r such that $rR/(R-r) = (\lambda_n R/(\alpha+\varepsilon))^{1/(\alpha+\varepsilon+1)}$, we get, for $n > n_o'$,

$$\log^+(|a_n|\left(\frac{R}{\rho}\right)^{\lambda_n}) < \left(\frac{\lambda_n R}{\alpha+\varepsilon}\right)^{(\alpha+\varepsilon)/(\alpha+\varepsilon+1)} \{1+(2-\lambda)(\alpha+\varepsilon)\}$$

which gives

$$\limsup_{n\to\infty} \frac{\log^+\log^+(|a_n|(R/\rho)^{\lambda_n})}{\log \lambda_n} \le \frac{\alpha}{\alpha+1} \tag{4.3.46}$$

which obviously holds if $\alpha = \infty$.

To prove the inequality that is the converse of (4.3.46) we note that, by Lemmas 4.3.3 and 4.2.2, one has for $z \in L_r$, $R > r^* > \rho$, $R > r$ and $n = 0,1,2,...$

$$|\chi_{\lambda_n}(z)| \le M'\left(\frac{r^*}{\rho}\right)^{\lambda_n}\left(\frac{r}{\rho}\right)^{\lambda_n}$$

where M' is a constant independent of n.

Thus, by (4.3.43), for $\rho < r < R$,

$$\begin{aligned}\overline{M}(r) &\le \Sigma_{n=0}^{\infty} |a_n|\ |\chi_{\lambda_n}(z)|_{z\in L_r} \\ &\le M'\ \Sigma_{n=0}^{\infty} |a_n| \left(\frac{r^*}{\rho}\right)^{\lambda_n}\left(\frac{r}{\rho}\right)^{\lambda_n} \\ &= M'\ \Sigma_{n=0}^{\infty} |a_n| \left(\frac{R}{\rho}\right)^{\lambda_n}\left(\frac{rr^*}{\rho R}\right)^{\lambda_n}.\end{aligned} \tag{4.3.47}$$

Taking $r^* = \lambda\rho + (1-\lambda)(\rho R/r)$, $0 < \lambda < 1$, (4.3.47) can be written as

$$\overline{M}(r) \le M'M\left(\frac{\lambda r + (1-\lambda)R}{R}, F\right)$$

where $F(s) = \Sigma_{n=0}^{\infty} |a_n|\ (R/\rho)^{\lambda_n} s^{\lambda_n}$ and $M(t,F) = \max_{|s|=t} |F(s)|$. Since the Fourier series for f(z) converges uniformly on compact subsets of D_R, it follows that F(s) is analytic for $|s| < 1$. If the order of F(s), as defined by (1.5.1), is ρ_o, then inequality derived from (4.3.47) gives

$$\alpha = \limsup_{r \to R} \frac{\log^+\log^+ \overline{M}(r)}{\log (Rr/(R-r))}$$

$$\leqq \limsup_{r \to R} \frac{\log^+\log^+ M(\{(\lambda r+(1-\lambda)R)/R\}, F)}{\log (Rr/(R-r))}$$

$$= \limsup_{r \to R} \frac{\log^+\log^+ M(\{(\lambda r+(1-\lambda)R)/R\}, F)}{-\log(1-\{(\lambda r+(1-\lambda)R)/R\})+\log(\lambda r)}$$

$$= \rho_o,$$

and so Theorem 2.2.1, applied to F(s), yields

$$\frac{\alpha}{\alpha+1} \leq \limsup_{n \to \infty} \frac{\log^+\log^+(|a_n|(R/\rho)^{\lambda_n})}{\log \lambda_n} \tag{4.3.48}$$

and (4.3.46) and (4.3.48) lead to (4.3.44). □

Proceeding in a similar manner, the following theorem, giving a characterization for $\overline{D}$-type in terms of Fourier coefficients, may be obtained.

Theorem 4.3.12. *Let* $f \in H(\overline{D};R)$ *be represented by the Fourier series* (4.3.43) *with respect to the closed orthonormal system* $\{\chi_n(z)\}_{n=0}^{\infty}$ *of polynomials in* $H_2(D)$ *and, for* $0 < \alpha < \infty$, *set*

$$H \equiv \limsup_{n \to \infty} \frac{\left[\log^+(|a_n|(R/\rho)^{\lambda_n})\right]^{\alpha+1}}{\lambda_n^{\alpha}} . \tag{4.3.49}$$

If $0 < H < \infty$, *the function* f *is of* $\overline{D}$-*order* α *and* $\overline{D}$-*type* τ *if and only if*

$$H = \frac{(\alpha+1)^{\alpha+1}}{\alpha^{\alpha}} \left(\frac{R}{\rho}\right)^{\alpha} \tau . \tag{4.3.50}$$

If $H = 0$ *or* ∞, *the function* f *is respectively of* $\overline{D}$-*growth* $(\alpha,0)$ *or of* $\overline{D}$-*growth not exceeding* (α,∞) *and conversely.*

Analogues of Theorems 2.3.1, 2.3.2 and 2.3.8 for functions represented by Fourier series (4.3.43) run as follows:

Theorem 4.3.13. *Let* $f \in H(\overline{D};R)$ *be represented by the Fourier series* (4.3.43) *with respect to the closed orthonormal system* $\{\chi_n(z)\}_{n=0}^{\infty}$ *of polynomials in*

$H_2(D)$ *and let it be of lower* $\bar{D}$*-order* $\beta (0 \leq \beta \leq \infty)$. *Then for any increasing sequence* $\{n_k\}$ *of natural numbers*

$$1+\beta \geq \liminf_{k \to \infty} \left\{ \frac{\log \lambda_{n_{k-1}}}{\log \lambda_{n_k} - \log^+ \log^+ (|a_{n_k}| (R/\rho)^{\lambda_{n_k}})} \right\}. \tag{4.3.51}$$

Further, if f *has* $\bar{D}$*-order* α $(0 < \alpha < \infty)$ *and lower* $\bar{D}$*-type* ν $(0 \leq \nu \leq \infty)$, *then*

$$\frac{(\alpha+1)^{\alpha+1}}{\alpha^\alpha} \left(\frac{R}{\rho}\right)^\alpha \nu \geq \liminf_{k \to \infty} \left\{ \lambda_{n_{k-1}} \left(\frac{\log^+ (|a_{n_k}| \left(\frac{R}{\rho}\right)^{\lambda_{n_k}})}{\lambda_{n_k}} \right)^{\alpha+1} \right\}. \tag{4.3.52}$$

Theorem 4.3.14. *Let* $f \in H(\bar{D};R)$ *be represented by the Fourier series* (4.3.43) *with respect to the closed orthonormal system* $\{\chi_n(z)\}_{n=0}^\infty$ *of polynomials in* $H_2(D)$ *and let it be of* $\bar{D}$*-order* α $(0 < \alpha \leq \infty)$ *and lower* $\bar{D}$*-order* β $(0 \leq \beta \leq \infty)$. *Further, let* $\Psi(n+1) = |a_n/a_{n+1}|^{1/(\lambda_{n+1}-\lambda_n)}$ *form a nondecreasing sequence for* $n > n_o$; *then*

$$1+\beta \leq \liminf_{n \to \infty} \left\{ \frac{\log \lambda_n}{\log \lambda_n - \log^+ \log^+ (|a_n| (R/\rho)^{\lambda_n})} \right\}. \tag{4.3.53}$$

Theorem 4.3.15. *Let* $f \in H(\bar{D};R)$ *be represented by the Fourier series* (4.3.43) *with respect to the closed orthonormal system* $\{\chi_n(z)\}_{n=0}^\infty$ *of polynomials in* $H_2(D)$ *and let it be of* $\bar{D}$*-order* α $(0 < \alpha < \infty)$ *and lower type* ν $(0 \leq \nu \leq \infty)$. *If* $\Psi(n) = |a_{n-1}/a_n|^{1/(\lambda_n - \lambda_{n-1})}$ *forms a nondecreasing function of* n *for* $n > n_o$, *then*

$$\frac{(\alpha+1)^{\alpha+1}}{\alpha^\alpha} \left(\frac{R}{\rho}\right)^\alpha \nu \leq \liminf_{n \to \infty} \{ \lambda_n^{-\alpha} (\log^+ (|a_n| \left(\frac{R}{\rho}\right)^{\lambda_n}))^{\alpha+1} \}. \tag{4.3.54}$$

As remarked earlier, the proofs of these results and formulation and proofs of other results corresponding to those of Chapter 2 are left as exercises for the reader.

EXERCISES 4.3

1. Show, by an example, that the converse of Lemma 4.3.2 need not be true, i.e., there exists a sequence $\{f_n\} \subset H_2(D)$ converging uniformly on compact subsets of D to a regular function f but $\{f_n\}$ need not converge to f in the norm. Even f need not belong to $H_2(D)$.

2. Show that limit in (4.3.2) is independent of the particular compact exhaustion of D.

3. Let $C_2(D)$ denote the class of all functions f continuous in a bounded domain D and satisfying (4.3.1). Show that if $\{f_n\}$ is a Cauchy sequence in $C_2(D)$, then it need not converge pointwise in D nor need there exist a function $f \in C_2(D)$ such that $\lim_{n\to\infty} \| f_n - f \| = 0$.

4. Show that the space $C_2(D)$ (cf. Exercise 3) is not complete.

5. Show that if $U = \{z : |z| < 1\}$, then the system of functions $\chi_n(z) = \sqrt{(n+1)/\pi}\, z^n$, $n = 0,1,2,\ldots$, forms a complete orthonormal system in $H_2(U)$.

6. Show that if D denotes either of the two domains bounded by the curve $|z^2-1| = a$, $a > 0$, then the system of polynomials $\chi_n(z) = (2\sqrt{n+1}/(\sqrt{\pi}\, a^{n+1}))z(z^2-1)^n$, $n = 0,1,2,\ldots$, forms a complete orthonormal system in $H_2(D)$.

7. Show that for the elliptic region $D = \{z : (\mathrm{Re}\, z/a)^2 + (\mathrm{Im}\, z/b)^2 < 1\}$, the set of polynomials

$$\chi_n(z) = 2\sqrt{\left(\frac{n+1}{\pi}\right)}(d^{n+1} - d^{-n-1})^{-1/2}\, U_n(z), \quad n = 0,1,2,\ldots, (a+b)^2 = d,$$

where $U_n(z)$ is the Chebyshev polynomial of second kind, i.e., $U_n(z) = (1-z^2)^{-1/2} \sin\left[(n+1)\cos^{-1} z\right]$, forms a closed set in $H_2(D)$.

8. Show that if D is a simply connected domain with at least two boundary points and w maps D onto $|w| < 1$, then the set of functions $\chi_n(z) = \sqrt{(n+1)/\pi}\, [w(z)]^n\, w'(z)$, $n = 0,1,2,\ldots$, forms a complete orthonormal system in $H_2(D)$.

9. Show that, in general, not every $f \in H_2(D)$ is the limit in the norm of its Fourier series.

10. Show that the Bergman kernel for the unit disc U is given by

$$K(z,s) = \frac{1}{\pi(1-\bar{s}z)^2}.$$

Determine the Bergman kernel for the disc $|z-z_o| < R$.

11. Prove that kernel function for a given bounded domain D is unique.

12. If D is a bounded domain, then show that for $f \in H_2(D)$

$$|f(z)|^2 \leqq K(z,z) \quad \|f\|^2$$

$$|K(z,s)|^2 \leqq K(z,z)\ K(s,s). \qquad z,s \in D.$$

13. Let D^* be a subdomain of D and let $K^*(z,s)$ and $K(z,s)$ denote, respectively, the kernel functions of D^* and D. Show that

$$K(s,s) \leqq K^*(s,s), \qquad s \in D^*.$$

14. Show that if, for a simply connected domain D, the kernel function $K(z,s)$ is bounded for some $s \in D$, it is bounded for every choice of $s \in D$.

15. Consider the problem of determining a function f, regular in a domain D, with $f(s) = 1$ at a fixed point $s \in D$ such that the integral $\iint_D |f(z)|^2\, dx\, dy$ is minimum. Show that the solution to this problem is given by the function $f(z) = K(z,s)/K(s,s)$ and that the minimum value of this integral is $1/K(s,s)$.

16. Let D_1 and D_2 be bounded simply connected domains in the complex plane, each containing the origin, and let D be the component of $D_1 \cap D_2$ which contains the origin. Let $\{W_n\}_{n=1}^{\infty}$ and $\{V_n\}_{n=1}^{\infty}$ be complete orthonormal sets in $H_2(D_1)$ and $H_2(D_2)$ respectively. Show that the set $\{W_n : n = 1,2,\ldots\} \cup \{V_n : n = 1,2,\ldots\}$ spans $H_2(D)$.

(Suffridge [1])

Let D be a bounded simply connected domain and let D_∞ be the component of the complement of $\bar{D}$ containing the point at infinity. D is said to be a *Caratheodory domain* if D and D_∞ have the same boundary.

17. Show that a bounded simply connected domain D is a Caratheodory domain if and only if there exists a decreasing sequence of bounded simply connected domains $\{D_n\}$ converging to D as its kernel.

18. Show that Theorems 4.3.9 and 4.3.10 continue to hold if the domain D under consideration is a Caratheodory domain.

What other results of this section continue to hold for Caratheodory domains?

19. For $f \in H(\overline{D};R)$, represented by the Fourier series (4.3.43) with respect to the closed orthonormal system $\{\chi_n(z)\}_{n=0}^{\infty}$ of polynomials in $H_2(D)$, define q-$\overline{D}$-order $\alpha(q)$ as in Exercise 4.2.12. Show that

$$\alpha(q)+A(q) = \limsup_{n \to \infty} \left\{ \frac{\log^{[q-1]} \lambda_n}{\log \lambda_n - \log^+ \log^+ (|a_n| (R/\rho)^{\lambda_n})} \right\}$$

where $A(q) = 1$ if $q = 2$ and $A(q) = 0$ if $q = 3,4,\ldots$.

Obtain by analogy characterizations for lower q-$\overline{D}$-order, q-$\overline{D}$-type and lower q-$\overline{D}$-type in terms of Fourier coefficients of f.

Also obtain decomposition theorems, estimates involving proximate order etc. in terms of Fourier coefficients for functions having q-$\overline{D}$-order $\alpha(q)$.

(Juneja [4])

20. Find interrelations of the following growth parameters of a function $f \in H(\overline{D};R)$ and its Fourier coefficients (cf. Exercises 1.5):

(i) (α,β)-$\overline{D}$-order, lower (α,β)-$\overline{D}$-order, (α,β)-$\overline{D}$-type and lower (α,β)-$\overline{D}$ type;

(ii) logarithmic $\overline{D}$-order, lower logarithmic $\overline{D}$-order, logarithmic $\overline{D}$-type and lower logarithmic $\overline{D}$-type;

(iii) generalized $\overline{D}$-order, lower generalized $\overline{D}$-order, generalized $\overline{D}$-type and lower generalized $\overline{D}$-type.

(Juneja [4])

4.4. APPROXIMATION OF ANALYTIC FUNCTIONS IN L^p-NORM AND CHEBYSHEV NORM

Let D denote a bounded, simply connected domain such that $\overline{D}$ has transfinite diameter $\rho > 0$ and such that the complement of D is the closure of a domain. We have seen in the last section that for such domains the set of polynomials is 'closed' in $H_2(D)$. Let $\{\chi_n(z)\}_{n=0}^{\infty}$ be a closed orthonormal system of polynomials in $H_2(D)$. We now introduce the concept of minimum error or order of approximation of a function $f \in H_2(D)$ with respect to the closed orthonormal system $\{\chi_n(z)\}_{n=0}^{\infty}$. If we set, for $f \in H_2(D)$,

$$\Delta_n^{(2)}(f) = \inf_{\{c_o,\ldots,c_{n-1}\}} \left[\iint_D |f(z) - \Sigma_{k=0}^{n-1} c_k\chi_k(z)|^2 dx\, dy\right]^{\frac{1}{2}} \tag{4.4.1}$$

where infimum is taken over all $(c_o,\ldots,c_{n-1}) \in \mathbb{C}^n$ (here $\mathbb{C}^n$ denotes the Cartesian product of n copies of the complex plane $\mathbb{C}$), then $\Delta_n^{(2)}(f)$ is called the minimum error of degree n of f with respect to the given closed orthonormal system. If P_{n-1} denotes the set of all polynomials of degree not exceeding n-1, then $\Delta_n^{(2)}(f)$ may also be expressed as

$$\Delta_n^{(2)}(f) = \inf_{g \in P_{n-1}} \| f-g\| \tag{4.4.2}$$

where norm $\|\cdot\|$ is defined by (4.3.3). Since polynomials are dense in $H_2(D)$, it is clear that for every $f \in H_2(D)$, $\Delta_n^{(2)}(f)$ is a nonincreasing sequence tending to zero as $n \to \infty$. The following lemma gives the exact value of $\Delta_n^{(2)}(f)$ in terms of Fourier coefficients of f.

Lemma 4.4.1. *Let* $f \in H_2(D)$ *be represented by the Fourier series*

$$f(z) = \Sigma_{n=0}^{\infty} a_n\chi_n(z) \tag{4.4.3}$$

then

$$\Delta_n^{(2)}(f) = (\Sigma_{k=n}^{\infty} |a_k|^2)^{1/2}, \qquad n = 0,1,2,\ldots . \tag{4.4.4}$$

Proof. By Theorem 4.3.2, the smallest value of the integral

$$\iint_D |f(z) - \Sigma_{k=0}^{n} c_k\chi_k(z)|^2\, dx\, dy$$

is attained for $c_k = a_k$, $k = 0,1,\ldots,n-1$, and equals

$$\iint_D |f(z)|^2\, dx\, dy - \Sigma_{k=0}^{n-1} |a_k|^2 . \tag{4.4.5}$$

Since the orthonormal system $\{\chi_n(z)\}_{n=0}^{\infty}$ is given as closed, Parseval's formula

$$\iint_D |f(z)|^2\, dx\, dy = \Sigma_{k=0}^{\infty} |a_k|^2$$

holds. Using this relation in (4.4.5) we get (4.4.4). □

We shall now obtain relations that show how the growth of a function $f \in H(\overline{D};R)$ affects its minimum error with respect to a closed orthonormal system and *vice versa*. It may be recalled that $f \in H(\overline{D};R)$ if and only if it is analytic in D_R with a singularity on L_R. First we have

Lemma 4.4.2. *If* $f \in H(\overline{D};R)$ *then for* $\rho < r' < r < R$ *and* $n > n_o(r')$, *one has*

$$\Delta_n^{(2)}(f) \leqq B\, \overline{M}(r) \left(\frac{r}{r-r'}\right)^{3/2} \left(\frac{r'}{r}\right)^n \tag{4.4.6}$$

where $\overline{M}(r) = \max_{z \in L_r} |f(z)|$ *and* B *is a constant independent of* r, r' *and* n.

The lemma is an easy consequence of (4.3.42) and (4.4.4), so we omit its proof.

Our next theorem gives a necessary and sufficient condition, in terms of $\Delta_n^{(2)}(f)$, for a function f to be in $H(\overline{D};R)$.

Theorem 4.4.1. *A function* $f \in H_2(D)$ *is in* $H(\overline{D};R)$, *if and only if,*

$$\limsup_{n \to \infty} \left[\Delta_n^{(2)}(f)\right]^{1/n} = \frac{\rho}{R} . \tag{4.4.7}$$

Proof. If $f \in H(\overline{D};R)$, by (4.4.6), we have, for $\rho < r' < r < R$

$$\limsup_{n \to \infty} \left[\Delta_n^{(2)}(f)\right]^{1/n} \leqq \frac{r'}{r}$$

which gives

$$\limsup_{n \to \infty} \left[\Delta_n^{(2)}(f)\right]^{1/n} \leqq \frac{\rho}{R} . \tag{4.4.8}$$

On the other hand, by (4.4.4),

$$\Delta_n^{(2)}(f) \geqq |a_n|. \tag{4.4.9}$$

This yields, in view of (4.3.38),

$$\limsup_{n \to \infty} \left[\Delta_n^{(2)}(f)\right]^{1/n} \geqq \frac{\rho}{R} \tag{4.4.10}$$

and (4.4.8) and (4.4.10) give (4.4.7). Conversely, if (4.4.7) holds for an $f \in H_2(D)$, then, by (4.4.9), we must have

$$\limsup_{n \to \infty} |a_n|^{1/n} \leqq \frac{\rho}{R}. \tag{4.4.11}$$

If

$$\limsup_{n \to \infty} |a_n|^{1/n} = \frac{\rho}{R_o} < \frac{\rho}{R}$$

then, by Theorem 4.3.10, $f \in H(\overline{D};R_o)$ and so (4.4.8) gives

$$\limsup_{n \to \infty} \left[\Delta_n^{(2)}(f)\right]^{1/n} \leqq \frac{\rho}{R_o}$$

which would contradict (4.4.7). Thus equality holds in (4.4.11) and so $f \in H(\overline{D};R)$. □

Theorems 4.4.1 and 4.3.10 show that the sequences $\{a_n\}$ and $\{\Delta_n^{(2)}(f)\}$ are almost of the same order of magnitude. How 'close' they can come together is demonstrated by the following:

Theorem 4.4.2. *For any* $f \in H(\overline{D};R)$ *there exists an increasing sequence* $\{n_k\}$ *of positive integers such that*

$$1 \leqq \limsup_{k \to \infty} \frac{\Delta_{n_k}^{(2)}(f)}{|a_{n_k}|} \leqq \frac{1}{\sqrt{1 - (\rho/R)^2}}. \tag{4.4.12}$$

Proof. By (4.3.38), we have

$$\limsup_{n \to \infty} |a_n|^{1/n} = \frac{\rho}{R}.$$

Let $\{\varepsilon_k\}$ be an arbitrary decreasing sequence of positive numbers each less than 1 such that $\varepsilon_k \to 0$ as $k \to \infty$. For each $\varepsilon_k > 0$ there exists an integer $N(k)$ such that

$$|a_n|^{1/n} \leqq \frac{1}{(R'/\rho) - \varepsilon_k} \qquad \text{for} \quad n \geq N(k), \tag{4.4.13}$$

where R' is any fixed number such that $\rho < R' < R$. Let $\{n_k\}$ be the sequence such that

$$\max_{n \geq N(k)} |a_n|^{1/n} = |a_{n_k}|^{1/n_k} \tag{4.4.14}$$

then (4.4.13) gives

$$|a_{n_k}|^{1/n_k} \leqq \frac{1}{(R'/\rho) - \varepsilon_k} \qquad \text{for} \quad k = 1,2,\ldots . \tag{4.4.15}$$

Moreover, (4.4.14) and (4.4.15) give

$$|a_{n_k+j}| \leqq |a_{n_k}|^{(n_k+j)/n_k} < |a_{n_k}| \frac{1}{((R'/\rho) - \varepsilon_k)^j} \quad \text{for } j = 0,1,2,\ldots . \tag{4.4.16}$$

Using (4.4.4), (4.4.9) and (4.4.16), we get

$$1 \leqq \limsup_{k \to \infty} \frac{\Delta_{n_k}^{(2)}(f)}{|a_{n_k}|} \leqq \frac{1}{\sqrt{1 - (\rho/R')^2}} .$$

Since the above relation holds for every R' satisfying $\rho < R' < R$, we must have (4.4.12). □

Both the inequalities in (4.4.12) are sharp (cf. Exercise 4.4.1) in the sense that there exist functions in $H(\overline{D};R)$ for which equality in (4.4.12) holds.

We shall now characterize, in terms of $\Delta_n^{(2)}(f)$, the various growth parameters for functions in $H(\overline{D};R)$. To consider the results in their general setting we shall, as usual, assume that the analytic function $f \in H(\overline{D};R)$ is represented by the 'gap' Fourier series

$$f(z) = \Sigma_{n=0}^{\infty} a_n \chi_{\lambda_n}(z), \qquad z \in D_R, \tag{4.4.17}$$

where $\{\lambda_n\}_0^\infty$ is an increasing sequence of positive integers and a_n ($\neq 0$) are the Fourier coefficients of f given by

$$a_n = \iint_D f(z)\, \overline{\chi_{\lambda_n}(z)}\, dx\, dy, \qquad n = 0,1,2,\ldots \,. \tag{4.4.18}$$

We now have

Theorem 4.4.3. *If* $f \in H(\overline{D};R)$ *is of* $\overline{D}$*-order* α $(0 \le \alpha \le \infty)$ *and is represented by the Fourier series* (4.4.17) *in* D_R, *then*

$$\frac{\alpha}{\alpha+1} = \limsup_{n\to\infty} \left\{ \frac{\log^+\log^+ \left(\Delta_n^{(2)}(f)\,(R/\rho)^{\lambda_n}\right)}{\log \lambda_n} \right\} \tag{4.4.19}$$

where the left side is to be interpreted as 1 *for* $\alpha = \infty$.

Proof. Since $f \in H(\overline{D};R)$ is of $\overline{D}$-order α, we have

$$\alpha = \limsup_{r\to R} \frac{\log^+\log^+ \overline{M}(r)}{\log\,(Rr/(R-r))}\,.$$

Thus, given $\varepsilon > 0$, there exists $r_o(\varepsilon)$ such that for $r_o < r < R$, one has

$$\log^+ \overline{M}(r) < \left(\frac{Rr}{R-r}\right)^{\alpha+\varepsilon}.$$

(4.4.6) then gives, for $r_o < r < R$ and $n > n_o$,

$$\log \Delta_n^{(2)}(f) < \left(\frac{Rr}{R-r}\right)^{\alpha+\varepsilon} + \lambda_n \log \frac{r'}{r} + \frac{3}{2} \log \frac{r}{r-r'} + \log B.$$

Now, proceeding as in the first half of the proof of Theorem 4.2.2, we have

$$\limsup_{n\to\infty} \frac{\log^+\log^+ \left(\Delta_n^{(2)}(f)\,(R/\rho)^{\lambda_n}\right)}{\log \lambda_n} \leqq \frac{\alpha}{\alpha+1}\,. \tag{4.4.20}$$

However, (4.4.9) and (4.3.44) give

$$\frac{\alpha}{\alpha+1} \leqq \limsup_{n\to\infty} \frac{\log^+\log^+ \left(\Delta_n^{(2)}(f)\,(R/\rho)^{\lambda_n}\right)}{\log \lambda_n} \tag{4.4.21}$$

and (4.4.20) and (4.4.21) give (4.4.19). □

The following theorems can also be proved in the same way that the corresponding theorems involving Fourier coefficients were obtained. We omit their proofs.

Theorem 4.4.4. *Let* $f \in H(\overline{D};R)$ *be of* $\overline{D}$*-order* α $(0 < \alpha < \infty)$ *and* $\overline{D}$*-type* τ $(0 \leq \tau \leq \infty)$. *Set*

$$H' = \limsup_{n \to \infty} \left\{ \frac{\left(\log^{+}(\Delta_n^{(2)}(f)\,(R/\rho)^{\lambda_n})\right)^{\alpha+1}}{\lambda_n^{\alpha}} \right\}. \tag{4.4.22}$$

If $0 < H' < \infty$, *then* f *is of* $\overline{D}$*-type* τ *if and only if*

$$H' = \frac{(\alpha+1)^{\alpha+1}}{\alpha^{\alpha}} \left(\frac{R}{\rho}\right)^{\alpha} \tau. \tag{4.4.23}$$

If $H' = 0$ *or* ∞ *the function* f *is respectively of* $\overline{D}$*-growth* $(\alpha,0)$ *or of growth not exceeding* (α,∞) *and conversely.*

Theorem 4.4.5. *Let* $f \in H(\overline{D};R)$ *be of* $\overline{D}$*-order* α $(0 < \alpha < \infty)$, *lower* $\overline{D}$*-order* β *and lower* $\overline{D}$*-type* ν, *then for any increasing sequence* $\{n_k\}$ *of natural numbers,*

$$1+\beta \geqq \liminf_{k \to \infty} \left\{ \frac{\log \lambda_{n_{k-1}}}{\log \lambda_{n_k} - \log^{+}\log^{+}(\Delta_{n_k}^{(2)}(f)\,(R/\rho))^{\lambda_{n_k}}} \right\} \tag{4.4.24}$$

$$\left(\frac{R}{\rho}\right)^{\alpha} \frac{(\alpha+1)^{\alpha+1}}{\alpha^{\alpha}}\, \nu \geqq \liminf_{k \to \infty} \{\lambda_{n_{k-1}} (\log^{+}(\Delta_{n_k}^{(2)}(f)(R/\rho)^{\lambda_{n_k}})^{1/\lambda_{n_k}})^{\alpha+1}\}. \tag{4.4.25}$$

Theorem 4.4.6. *Let* $f \in H(\overline{D};R)$ *be of* $\overline{D}$*-order* α $(0 < \alpha \leq \infty)$ *and lower* $\overline{D}$*-order* β $(0 \leq \beta \leq \infty)$ *such that* $\Psi(n) \equiv |a_{n-1}/a_n|^{1/(\lambda_n-\lambda_{n-1})}$ *forms a nondecreasing sequence for* $n > n_o$, *then*

$$1+\beta \leqq \liminf_{n \to \infty} \left\{ \frac{\log \lambda_n}{\log \lambda_n - \log^{+}\log^{+}(\Delta_n^{(2)}(f)(R/\rho)^{\lambda_n})} \right\}. \tag{4.4.26}$$

Theorem 4.4.7. *Let* $f \in H(\overline{D};R)$ *be of* $\overline{D}$*-order* α $(0 < \alpha < \infty)$ *and lower* $\overline{D}$*-type* ν $(0 \leq \nu \leq \infty)$ *such that* $\Psi(n) \equiv |a_{n-1}/a_n|^{1/(\lambda_n-\lambda_{n-1})}$ *forms a nondecreasing sequence for* $n > n_o$, *then*

$$\frac{(\alpha+1)^{\alpha+1}}{\alpha^{\alpha}}\left(\frac{R}{\rho}\right)^{\alpha} \nu \leqq \liminf_{n\to\infty} \left\{\frac{(\log^{+}(\Delta_n^{(2)}(f)(R/\rho)^{\lambda_n}))^{\alpha+1}}{\lambda_n^{\alpha}}\right\}. \tag{4.4.27}$$

The above results for the minimum error in the norm of $H_2(D)$ can also be extended to the case of L^p-norm ($1 \leq p < \infty$). For this purpose we introduce the class of pth-integrable functions. Thus, for a bounded domain D, let $H_p(D)$ ($1 \leq p < \infty$) denote the space of analytic functions f in D such that

$$\iint_D |f(z)|^p \, dx \, dy < \infty. \tag{4.4.28}$$

The integral (4.4.28) is to be understood as

$$\lim_{n\to\infty} \iint_{K_n} |f(z)|^p \, dx \, dy \tag{4.4.29}$$

where $\{K_n\}$ is a compact exhaustion of D. Since for $a,b \in \mathbb{C}$ and $f,g \in H_p(D)$, one has, by Minkowski's inequality,

$$\left(\iint_D |af(z)+bg(z)|^p \, dx \, dy\right)^{1/p} \leqq |a| \left(\iint_D |f(z)|^p \, dx \, dy\right)^{1/p}$$

$$+ |b| \left(\iint_D |g(z)|^p \, dx \, dy\right)^{1/p} < \infty,$$

it follows that $H_p(D)$ is a normed linear space where norm $\|\cdot\|_p$ is defined as

$$\|f\|_p = \left(\iint_D |f(z)|^p \, dx \, dy\right)^{1/p}, \qquad 1 \leq p < \infty. \tag{4.4.30}$$

We shall now show that $H_p(D)$ is a Banach space. We need two lemmas

Lemma 4.4.3. *Let* $K \subset D$ *be a compact set and* ρ_K *the distance of* K *from the boundary* ∂D *of* D. *Then, for* $f \in H_p(D)$ *one has*

$$\left(\iint_D |f(z)|^p \, dx \, dy\right)^{1/p} \geqq (\pi \rho_K^2)^{1/p} |f(z_o)| \tag{4.4.31}$$

for every $z_o \in K$.

Proof. Let $z_o \in K$ and ρ_{z_o}, be the distance of z_o from ∂D. Since f is regular in D, by Cauchy's integral formula, we have

$$|f(z_o)| \leqq \frac{1}{\pi\rho^2} \int_o^{2\pi} \int_o^{\rho} |f(z_o+re^{i\theta})| \; r \, dr \, d\theta \qquad \text{for} \quad \rho < \rho_{z_o}.$$

Using Holder's inequality, this leads to

$$|f(z_o)| \; (\pi\rho^2) \leqq \left(\int_o^{2\pi} \int_o^{\rho} |f(z_o+re^{i\theta})|^p \; r \, dr \, d\theta \right)^{1/p} (\pi\rho^2)^{1-\frac{1}{p}}$$

$$= \left(\iint_{|z-z_o| \leq \rho} |f(z)|^p \, dx \, dy \right)^{1/p} (\pi\rho^2)^{1-\frac{1}{p}}$$

$$\leqq \left(\iint_D |f(z)|^p \, dx \, dy \right)^{1/p} (\pi\rho^2)^{1-\frac{1}{p}}.$$

Since this holds for every $\rho < \rho_{z_o}$, we have

$$\left(\iint_D |f(z)|^p \, dx \, dy \right)^{1/p} \geqq |f(z_o)| \; (\pi \, \rho_{z_o}^2)^{1/p}$$

$$\geqq |f(z_o)| \; (\pi \, \rho_K^2)^{1/p}. \qquad \square$$

Corollary. If f *is analytic in a bounded domain* D *and satisfies the inequality*

$$\iint_D |f(z)|^p \, dx \, dy \leqq M^p, \tag{4.4.32}$$

then for every compact set $K \subset D$, *we have, for every* $z \in K$,

$$|f(z)| \leqq M \, M', \tag{4.4.33}$$

where M' *is a constant depending on* K *but neither on* f *nor on* M.

Lemma 4.4.4. *If* $f \in H_p(D)$ *and if* $\{f_n\}$ *is a sequence of functions in* $H_p(D)$ *that converges in the norm to* f, *then* $\{f_n(z)\}$ *converges uniformly on compact subsets of* D *to* f(z).

Using Lemma 4.4.3 the proof follows that of Lemma 4.3.2. Hence we omit the proof.

Theorem 4.4.8. $H_p(D)$ $(1 \leq p < \infty)$ *is a Banach space.*

Proof. To show that $H_p(D)$ is complete, let $\{f_n\}$ be a Cauchy sequence in

$H_p(D)$. Then, given $\varepsilon > 0$, there exists an integer $N = N(\varepsilon)$ such that

$$\|f_m - f_n\|_p < \varepsilon \qquad \text{for} \quad m,n \geq N. \tag{4.4.34}$$

If $K \subset D$ is a compact set, by Lemma 4.4.3, we have, for any $z \in K$,

$$|f_m(z) - f_n(z)| \leq \frac{1}{(\pi \rho_K^2)^{1/p}} \|f_m - f_n\|_p$$

$$< \frac{\varepsilon}{(\pi \rho_K^2)^{1/p}} \qquad \text{for} \quad m,n \geq N.$$

The above relation shows that the sequence $\{f_n(z)\}$ converges uniformly on compact subsets of D to a limit function, $f(z)$, say. By the Weierstrass theorem, $f(z)$ must be regular in D. Applying Fatou's lemma to (4.4.34), we get

$$\iint_D |f(z) - f_n(z)|^p \, dx\, dy \leq \varepsilon^p \qquad \text{for } n \geq N.$$

This shows that $f - f_n \in H_p(D)$. Since $H_p(D)$ is a linear space we have $f \in H_p(D)$. Thus $H_p(D)$ is complete. □

To extend the results of L^2-norm to L^p-norm $(1 \leq p < \infty)$ for functions in $H(\overline{D};R)$, we now assume that D is a bounded, simply connected domain whose closure $\overline{D}$ has transfinite diameter $\rho > 0$ and whose complement is the closure of a domain. Since any $f \in H(\overline{D};R)$ is analytic in $\overline{D}$ it follows that $\iint_D |f(z)|^p \, dx\, dy < \infty$ and so $f \in H_p(D)$. Since, by Theorem 4.2.1, there exists, for $f \in H(\overline{D};R)$, a sequence of polynomials converging uniformly on compact subsets of D_R to $f(z)$, it follows that this sequence converges in the norm of $H_p(D)$ also to f. If P_{n-1} denotes the collection of all polynomials of degree not exceeding n-1 and we set

$$\Delta_n^{(p)}(f) = \inf_{g \in P_{n-1}} \|f - g\|_p \quad , \qquad n = 1,2,\ldots, \tag{4.4.35}$$

then it is clear that $\Delta_n^{(p)}(f)$ is a nonincreasing sequence tending to zero as $n \to \infty$. Our next theorem extends Theorem 4.4.1 to the case $1 \leq p < \infty$.

Theorem 4.4.9. *If* $f \in H(\overline{D};R)$, *then*

$$\limsup_{n \to \infty} \left[\Delta_n^{(p)}(f)\right]^{1/n} = \frac{\rho}{R}. \tag{4.4.36}$$

Proof. If $f \in H(\overline{D};R)$ we have, by (4.4.35),

$$\Delta_n^{(p)}(f) = \inf_{g \in P_{n-1}} \left(\iint_D |f(z)-g(z)|^p \, dx\, dy\right)^{1/p}$$

$$\leqq \left(\iint_D |f(z)-Q_{n-1}(z)|^p \, dx\, dy\right)^{1/p}$$

$$\leqq A^{1/p} \max_{z \in \overline{D}} |f(z) - Q_{n-1}(z)|,$$

where $Q_{n-1}(z)$ is the polynomial of degree not exceeding n-1 as in Theorem 4.2.7 and A is the area of domain D. Using (4.2.36), we now get for $\rho < r' < r < R$ and $n > n_o$

$$\Delta_n^{(p)}(f) \leqq A^{1/p} \overline{M}(r) \left(\frac{r'}{r-r'}\right)\left(\frac{r'}{r}\right)^n \tag{4.4.37}$$

where $\overline{M}(r) = \max_{z \in L_r} |f(z)|$. This leads to

$$\limsup_{n \to \infty} \left[\Delta_n^{(p)}(f)\right]^{1/n} \leqq \frac{r'}{r}.$$

Since the above relation holds for all r', r satisfying $\rho < r' < r < R$, we must have

$$\limsup_{n \to \infty} \left[\Delta_n^{(p)}(f)\right]^{1/n} \leqq \frac{\rho}{R}. \tag{4.4.38}$$

To obtain the inequality that is the converse of (4.4.38), we note that, since every $f \in H(\overline{D};R)$ is in $H_2(D)$, there exists a closed orthonormal system $\{\chi_n(z)\}_{n=0}^{\infty}$ of polynomials in $H_2(D)$ such that f can be represented by its Fourier series with respect to the system $\{\chi_n(z)\}_{n=0}^{\infty}$ that converges uniformly on compact subsets of D_R to f. Thus

$$f(z) = \Sigma_{n=0}^{\infty} a_n \chi_n(z), \qquad z \in D_R, \tag{4.4.39}$$

where

$$a_n = \iint_D f(z)\, \overline{\chi_n(z)}\, dx\, dy.$$

If $g \in P_{n-1}$, then

$$|a_n| = |\iint_D (f(z)-g(z)) \overline{\chi_n(z)}\, dx\, dy|$$

$$\leqq (\iint_D |f(z)-g(z)|^p\, dx\, dy)^{\frac{1}{p}} (\iint_D |\overline{\chi_n(z)}|^{\frac{p}{p-1}}\, dx\, dy)^{1-\frac{1}{p}} .$$

Using (4.3.35) and the fact that the above inequality holds for every $g \in \mathcal{P}_{n-1}$, we get, for $r^* > \rho$,

$$|a_n| \leqq \Delta_n^{(p)}(f).M.\left(\frac{r^*}{\rho}\right)^n A^{1-\frac{1}{p}}, \tag{4.4.40}$$

so that, by (4.3.38),

$$\limsup_{n \to \infty} [\Delta_n^{(p)}(f)]^{1/n} \geqq \frac{\rho}{r^*} \limsup_{n \to \infty} |a_n|^{1/n} = \frac{\rho^2}{r^* R} .$$

Since the above inequality is valid for every $r^* > \rho$, we must have

$$\limsup_{n \to \infty} [\Delta_n^{(p)}(f)]^{1/n} \geqq \frac{\rho}{R} \tag{4.4.41}$$

and (4.4.38) and (4.4.41) give (4.4.36). □

We now obtain relations that indicate how the growth of an $f \in H(\overline{D};R)$ depends on $\Delta_n^{(p)}(f)$ and *vice versa*. The first relation is a formula for $\overline{D}$-order involving $\Delta_n^{(p)}(f)$. Precisely, we have

Theorem 4.4.10. *If* $f \in H(\overline{D};R)$ *is of* $\overline{D}$*-order* α $(0 \leq \alpha \leq \infty)$ *and is represented by the Fourier series* (4.4.17) *in* D_R, *then*

$$\frac{\alpha}{\alpha+1} = \limsup_{n \to \infty} \left\{ \frac{\log^+\log^+ (\Delta_{\lambda_n}^{(p)}(f)\, (R/\rho)^{\lambda_n})}{\log \lambda_n} \right\}, \tag{4.4.42}$$

where the left side is to be interpreted as 1 *for* $\alpha = \infty$.

Proof. If $f \in H(\overline{D};R)$ is of order $\alpha < \infty$, given $\varepsilon > 0$, there exists $r_o(\varepsilon)$ such that, for $r_o < r < R$, we have

$$\log^+ \overline{M}(r) < \left(\frac{Rr}{R-r}\right)^{\alpha+\varepsilon}.$$

Inequality (4.4.37) now gives, for $n > n_o$ and $\rho < r' < r < R$, $r_o < r'$,

$$\log(\Delta_{\lambda_n}^{(p)}(f)\left(\frac{R}{\rho}\right)^{\lambda_n}) < \frac{1}{p}\log A + \left(\frac{Rr}{R-r}\right)^{\alpha+\varepsilon} + \log\frac{r'}{r-r'} + \lambda_n \log\frac{r'}{\rho} + \lambda_n \log\frac{R}{r}$$

$$< \frac{1}{p}\log A + \left(\frac{Rr}{R-r}\right)^{\alpha+\varepsilon} + \log\frac{r'}{r-r'} + \lambda_n \frac{r'-\rho}{\rho} + \lambda_n \frac{R-r}{r}.$$

Choose r such that $Rr/(R-r) = (\lambda_n R/(\alpha+\varepsilon))^{1/(\alpha+\varepsilon+1)}$ and $r' = \lambda\rho + (1-\lambda)(R\rho/r)$, $0 < \lambda < 1$, then the above inequality gives, for $n > n_1$,

$$\log^+ (\Delta_{\lambda_n}^{(p)}(f)\left(\frac{R}{\rho}\right)^{\lambda_n}) < M(\lambda_n R)^{(\alpha+\varepsilon)/(\alpha+\varepsilon+1)}$$

where M is a constant. This leads to

$$\limsup_{n\to\infty}\left\{\frac{\log^+\log^+ (\Delta_{\lambda_n}^{(p)}(f)\ (R/\rho)^{\lambda_n})}{\log \lambda_n}\right\} \leqq \frac{\alpha}{\alpha+1}. \tag{4.4.43}$$

To prove the reverse inequality, we note that by (4.4.17) and (4.4.40) we have, for $z \in \overline{D}_r$, $\rho < r^* < r < R$,

$$|f(z)| \leqq \Sigma_{n=0}^{\infty} |a_n|\ |\chi_{\lambda_n}(z)|$$

$$\leqq M A^{1-\frac{1}{p}} \Sigma_{n=0}^{\infty} \Delta_n^{(p)}(f)\ (\tfrac{r^*}{\rho})^n |\chi_{\lambda_n}(z)|.$$

Making use of Lemmas 4.3.3 and 4.2.2 for $\chi_{\lambda_n}(z)$, we get for $\rho < r^* < r < R$,

$$\overline{M}(r) \leqq M^2 A^{1-\frac{1}{p}} M'\ \Sigma_{n=0}^{\infty} \Delta_n^{(p)}(f)\left(\frac{R}{\rho}\right)^{\lambda_n}\left(\frac{r^{*2} r}{\rho^2 R}\right)^{\lambda_n}.$$

Choosing $r^* = \rho\sqrt{\lambda+(1-\lambda)(R/r)}$, $0 < \lambda < 1$, the above inequality may be written as

$$\overline{M}(r) \leqq B\ M\left(\frac{\lambda r + (1-\lambda)R}{R}, G\right) \tag{4.4.44}$$

where B is a constant, $G(s) = \Sigma_{n=0}^{\infty} \Delta_n^{(p)}(f)\left(\frac{R}{\rho}\right)^{\lambda_n} s^{\lambda_n}$ and $M(t,G) = \max_{|s|=t} |G(s)|$. It is easy to check that $G(s)$ is analytic in $|s| < 1$. If the order of $G(s)$ in U is ρ_o then (4.4.44) gives

$$\alpha = \limsup_{r \to R} \frac{\log^+\log^+ \bar{M}(r)}{\log(Rr/(R-r))} \leq \limsup_{r \to R} \frac{\log^+\log^+ M(\{(\lambda r+(1-\lambda)R)/R\}, G)}{\log (Rr/(R-r))}$$

$$= \rho_o.$$

So, appealing to Theorem 2.2.1 for ρ_o, we get

$$\frac{\alpha}{\alpha+1} \leq \limsup_{n \to \infty} \frac{\log^+\log^+ (\Delta_{\lambda_n}^{(p)}(f)(R/\rho)^{\lambda_n})}{\log \lambda_n} \tag{4.4.45}$$

and (4.4.43) and (4.4.45) give (4.4.42). □

The following theorems involving $\Delta_n^{(p)}(f)$ and other growth parameters for $f \in H(\bar{D};R)$ follow in an analogous manner.

Theorem 4.4.11. *Let* $f \in H(\bar{D};R)$ *be of* $\bar{D}$*-order* α $(0 < \alpha < \infty)$, $\bar{D}$*-type* τ $(0 \leq \tau \leq \infty)$ *and be represented by the Fourier series* (4.4.17) *in* D_R. *Set*

$$H'' = \limsup_{n \to \infty} \left\{ \frac{(\log^+ (\Delta_{\lambda_n}^{(p)}(f)\ (R/\rho)^{\lambda_n}))^{\alpha+1}}{\lambda_n^{\alpha}} \right\}. \tag{4.4.46}$$

If $0 < H'' < \infty$, *then* f *is of* $\bar{D}$*-type* τ *if and only if*

$$H'' = \frac{(\alpha+1)^{\alpha+1}}{\alpha^{\alpha}} \left(\frac{R}{\rho}\right)^{\alpha} \tau. \tag{4.4.47}$$

If $H'' = 0$ *or* ∞ *the function* f *is respectively of* $\bar{D}$*-growth* $(\alpha,0)$ *or of growth not exceeding* (α,∞) *and conversely.*

Theorem 4.4.12. *Let* $f \in H(\bar{D};R)$ *be of* $\bar{D}$*-order* α $(0 < \alpha \leq \infty)$ *and lower* $\bar{D}$*-order* β $(0 \leq \beta \leq \infty)$; *then, for any increasing sequence* $\{n_k\}$ *of natural numbers,*

$$1+\beta \geq \liminf_{k \to \infty} \frac{\log \lambda_{n_{k-1}}}{\log \lambda_{n_k} - \log^+\log^+(\Delta_{\lambda_{n_k}}^{(p)}(f)(R/\rho)^{\lambda_{n_k}})}. \tag{4.4.48}$$

Further, if $0 < \alpha < \infty$ *and* f *is of lower* $\bar{D}$*-type* ν *then*

$$\left(\frac{R}{\rho}\right)^{\alpha} \frac{(\alpha+1)^{\alpha+1}}{\alpha^{\alpha}} \nu \geq \liminf_{k \to \infty} \{\lambda_{n_{k-1}} (\log^{+}(\Delta^{(p)}_{\lambda_{n_k}}(f)(R/\rho)^{\lambda_{n_k}})^{1/\lambda_{n_k}})^{\alpha+1}\} . \tag{4.4.49}$$

Theorem 4.4.13. *Let* $f \in H(\overline{D};R)$ *be of* $\overline{D}$*-order* α $(0 < \alpha < \infty)$ *and lower* $\overline{D}$*-order* β $(0 \leq \beta \leq \infty)$ *such that* $\Psi(n) \equiv |a_{n-1}/a_n|^{1/(\lambda_n-\lambda_{n-1})}$ *forms a nondecreasing sequence for* $n > n_o$; *then*

$$1+\beta \leq \liminf_{n \to \infty} \frac{\log \lambda_n}{\log \lambda_n - \log^{+}\log^{+}(\Delta^{(p)}_{\lambda_n}(f)(R/\rho)^{\lambda_n})} . \tag{4.4.50}$$

Theorem 4.4.14. *Let* $f \in H(\overline{D};R)$ *be of* $\overline{D}$*-order* α $(0 < \alpha < \infty)$ *and lower* $\overline{D}$*-type* ν $(0 \leq \nu \leq \infty)$ *such that* $\Psi(n) \equiv |a_{n-1}/a_n|^{1/(\lambda_n-\lambda_{n-1})}$ *forms a nondecreasing sequence for* $n > n_o$; *then*

$$\frac{(\alpha+1)^{\alpha+1}}{\alpha^{\alpha}} \left(\frac{R}{\rho}\right)^{\alpha} \nu \leq \liminf_{n \to \infty} \lambda_n \{\log^{+}(\Delta^{(p)}_{\lambda_n}(f)(R/\rho)^{\lambda_n})^{1/\lambda_n}\}^{\alpha+1} . \tag{4.4.51}$$

The results obtained above for $\Delta_n^{(p)}(f)$ $(1 \leq p < \infty)$ for $f \in H(\overline{D};R)$ can be readily extended to minimum error with respect to the Chebyshev norm which may be considered to be the limiting case of $\Delta_n^{(p)}(f)$ for $p = \infty$. More specifically, we define $H_\infty(D)$ as the space of all bounded and analytic functions in D. Define for $f \in H_\infty(D)$

$$\| f \|_\infty = \sup_{z \in D} |f(z)| . \tag{4.4.52}$$

$\| \cdot \|_\infty$ is called the Chebyshev norm or sup norm. It is a simple matter to verify that $H_\infty(D)$ is a Banach space. If one denotes by $\Delta_n^{(\infty)}(f)$ the minimum error of degree n in the Chebyshev norm then

$$\Delta_n^{(\infty)}(f) = \inf_{g \in P_{n-1}} \|f-g\|_\infty \tag{4.4.53}$$

$\Delta_n^{(\infty)}(f)$ is a nonincreasing sequence which tends to zero as $n \to \infty$. For $f \in H(\overline{D};R)$, Lemma 4.2.7 gives that

$$\limsup_{n \to \infty} \left[\Delta_n^{(\infty)}(f)\right]^{1/n} \leqq \frac{\rho}{R} \tag{4.4.54}$$

whereas (4.4.40) gives

$$\limsup_{n \to \infty} \left[\Delta_n^{(\infty)}(f)\right]^{1/n} \geqq \frac{\rho}{R}. \tag{4.4.55}$$

(4.4.54) and (4.4.55) lead to extension of Theorem 4.4.9 to the Chebyshev norm. Other results concerning $\Delta_n^{(\infty)}(f)$ and growth of $f \in H(\overline{D};R)$ follow in a similar manner. We leave them as an exercise for the reader.

EXERCISES 4.4

1. Show that inequalities in (4.4.12) (cf. Theorem 4.4.2) are sharp. (Rizvi [1])
2. Can the result of Theorem 4.4.2 be extended to $\Delta_n^{(p)}(f)$ for $1 \leq p \leq \infty$?
3. Show that polynomials are dense in $H_p(D)$ for $0 < p \leq \infty$ where D is a bounded simply connected domain whose complement is the closure of a domain.
4. Show that if a function $f(z)$ can be uniformly approximated by polynomials on each of the two sets A_1 and A_2 which are contained in two mutually exterior bounded Jordan regions then f can be uniformly approximated by polynomials on $A_1 \cup A_2$.
5. Show that polynomial of best approximation in $H_p(D)$ exists for $p > 0$ and is unique for $p > 1$.
6. Which of the results of this section can be extended to the case of Caratheodory domains?
7. State and prove results connecting $\Delta_n^{(\infty)}(f)$ and the various growth parameters, i.e., $\overline{D}$-order, $\overline{D}$-type etc. for a function $f \in H(\overline{D};R)$. (Rizvi [2])
8. For $f \in H(\overline{D};R)$ define the sequence $\{n_k\}$ as

$$\Delta_n^{(\infty)}(f) = \Delta_{n_{k-1}}^{(\infty)}(f) \qquad \text{for} \quad n_{k-1} \leq n < n_k.$$

Show that if f is of regular $\overline{D}$-growth then

$$\log n_{k-1} \sim \log n_k \quad \text{as} \quad k \to \infty.$$

Is the converse true?

9. Show that the results of this section continue to hold if error is measured with respect to a positive continuous weight function $v(z)$, i.e., define

$$\| f \|_{p,v} = (\iint_D v(z) \, |f(z)|^p \, dx \, dy)^{1/p}, \quad 1 \leq p \leq \infty$$

and the corresponding error $\Delta_{n,v}^{(p)}(f)$ and verify the results with $\Delta_n^{(p)}(f)$ replaced by $\Delta_{n,v}^{(p)}(f)$.

10. State and prove results connecting $\Delta_n^{(p)}(f)$ and the following growth parameters for a function $f \in H(\bar{D};R)$ (cf. Exercises 4.2):

(i) q-$\bar{D}$-order, lower q-$\bar{D}$-order, q-$\bar{D}$-type etc;

(ii) logarithmic $\bar{D}$-order, lower logarithmic $\bar{D}$-order, logarithmic type etc;

(iii) (α,β) $\bar{D}$-order, lower (α,β) $\bar{D}$-order etc;

(iv) generalized $\bar{D}$-order, lower generalized $\bar{D}$-order etc.

(Juneja and Kapoor [1])

4.5. INTERPOLATION OF ANALYTIC FUNCTIONS

Interpolation is a very effective way of approximating an analytic function by means of polynomials. A basic formula in this case is Lagrange's interpolation formula.

Let $z_1, z_2, \ldots, z_{n+1}$ be $(n+1)$ distinct points in the complex plane and let the values $w_1, \ldots, w_{n+1}$ be given. There exist infinitely many polynomials that take the values w_k at the points z_k $(k = 1,2,\ldots,n+1)$. However, if one is interested in a polynomial of degree not exceeding n that assumes the prescribed values w_k at the points z_k then such a polynomial is unique. For if $P_1(z)$ and $P_2(z)$ are two such polynomials then $P(z) = P_1(z) - P_2(z)$ is a polynomial of degree not exceeding n which has zeros at the $(n+1)$ distinct points z_k $(k = 1,\ldots,n+1)$. By the fundamental theorem of algebra, $P(z) \equiv 0$ and so $P_1(z) \equiv P_2(z)$. It is easy to verify that this unique polynomial is given by

$$H_n(z) = \Sigma_{k=1}^{n+1} \frac{\omega_n(z)}{(z-z_k)\ \omega_n'(z_k)} w_k \tag{4.5.1}$$

where

$$\omega_n(z) = \Pi_{k=1}^{n+1} (z-z_k) \tag{4.5.2}$$

with the interpretation that in (4.5.1) the function is defined for the exceptional points z_k by allowing the variable z to approach those values. The polynomial $H_n(z)$ is called *Lagrange's interpolation polynomial.* The aforesaid values w_k, in particular, may be given by means of the functional values of a function f(z) at the points z_k, i.e.,

$$w_k = f(z_k), \qquad k = 1,2,\ldots,n+1;$$

in this case Lagrange's interpolatory polynomial $H_n(z)$ is said to *interpolate* the function f(z) at the points z_k $(k = 1,2,\ldots,n+1)$.

Lagrange's interpolation formula takes an elegant form in the case of functions that are regular inside and on a simple closed curve. This is demonstrated in

Theorem 4.5.1 (Hermite's interpolation formula). *Let* f(z) *be analytic in a domain* G *bounded by a simple closed curve* C *and suppose that the points* $z_1,\ldots,z_{n+1}$ *all lie in* G. *Then the unique polynomial* $H_n(z)$ *of degree not exceeding* n, *which interpolates* f(z) *at the points* $z_1,\ldots,z_{n+1}$, *is given by*

$$H_n(z) = \frac{1}{2\pi i} \int_C \frac{[\omega_n(t)-\omega_n(z)]\ f(t)}{\omega_n(t)(t-z)} dt, \qquad z \in G \tag{4.5.3}$$

where $\omega_n(z)$ *is given by* (4.5.2).

Proof. Since the numerator in the integrand of (4.5.3) vanishes identically for t = z, it is divisible by t-z and so the integrand is a polynomial of degree n in z. Further,

$$H_n(z_k) = \frac{1}{2\pi i} \int_C \frac{[\omega_n(t) - \omega_n(z_k)]\ f(t)}{\omega_n(t)(t-z_k)} dt = \frac{1}{2\pi i} \int_C \frac{f(t)}{t-z_k} dt = f(z_k),$$

so that $H_n(z)$, given by (4.5.3), is the Lagrange's interpolation formula for $f(z)$. □

Remark. By Cauchy's integral formula, (4.5.3) is easily seen to be equivalent to

$$f(z)-H_n(z) = \frac{1}{2\pi i} \int_C \frac{\omega_n(z)\, f(t)dt}{\omega_n(t)(t-z)}, \qquad z \in G. \tag{4.5.4}$$

As we shall presently see, the above relation helps in obtaining an estimate for the error term $R_n(z)$ $(\equiv f(z)-H_n(z))$ when $f(z)$ is approximated by the interpolatory polynomials $H_n(z)$.

Lagrange's interpolation formula and Hermite's interpolation formula can be easily extended to cover the case of multiple interpolation. Thus, if different points $z_1,\ldots,z_k$ are given and each point z_j is associated with the quantities $w_j^{(\nu)}$, $\nu = 0,1,\ldots,m_j-1$, then the problem of constructing a polynomial $H_n(z)$ of degree not exceeding $-1 + \Sigma_{j=1}^{k} m_j$ such that $H_n^{(\nu)}(z_j) = w_j^{(\nu)}$, $\nu = 0,1,\ldots,m_j-1$, $j = 1,2,\ldots,k$, where $H_n^{(\nu)}(z)$ denotes the νth derivative of $H_n(z)$, is called the problem of *multiple interpolation*. In the case of Hermite's interpolation formula the values $w_j^{(\nu)}$ are to be replaced by $f^{(\nu)}(z_j)$ where $f^{(\nu)}(z)$ is the νth derivative of $f(z)$. Since the method of proof employed in the case of simple interpolation carries over in an analogous manner to the present situation, we leave it as an exercise for the reader.

Let E be a compact subset of the complex plane having transfinite diameter $\rho > 0$ such that its complement E' with respect to the extended complex plane is a simply connected domain containing the point at infinity. This assumption concerning E will be made throughout this section.

As seen in Section 4.2, there exists a unique analytic function $w = \Omega(z)$ mapping E' onto $|w| > \rho$ in a one-one manner such that $\Omega(\infty) = \infty$ and $\lim_{z\to\infty} \frac{\Omega(z)}{z} = 1$. Further, if $z = \eta(w)$ denotes the inverse function of $w = \Omega(z)$ then $L_r = \{z : z = \eta(w),\ |w| = r > \rho\}$ is a Jordan, analytic curve which includes E in its interior. In fact, if E_r denotes the domain bounded by L_r then $E_{r'} \subset E_{r''}$ for $\rho < r' < r''$. The following lemma concerning $\eta(w)$ will be needed in the sequel.

Lemma 4.5.1. *We have*

$$\left|\frac{\zeta\eta'(\zeta)}{\eta(\zeta)-\eta(w)}\right| \leqq \frac{|\zeta|+|w|}{|\zeta|-|w|} \qquad \textit{for} \quad \rho < |w| < |\zeta| \tag{4.5.5}$$

and

$$\left|\frac{\zeta\eta'(\zeta)}{\eta(\zeta)-z}\right| \leqq \frac{|\zeta|+\rho}{|\zeta|-\rho} \qquad \textit{for} \quad \rho < |\zeta| \quad \textit{and} \quad z \in E. \tag{4.5.6}$$

Proof. For $|w| > \rho$, we consider the function $g(s) = w/[\eta(w/s)-\eta(w)]$. Then the function $g(s)$ is regular and univalent in the disc $|s| < 1$ and satisfies $g(0) = 0$, $g'(0) = 1$. Thus, by the distortion theorem (cf. Appendix A.13) for normalized univalent functions, we get

$$\left|\frac{sg'(s)}{g(s)}\right| = \left|\frac{(w/s)\ \eta'(w/s)}{\eta(w/s) - \eta(w)}\right| \leqq \frac{1+|s|}{1-|s|}, \qquad |s| < 1.$$

Setting $s = w/\zeta$, $|\zeta| > |w|$, yields (4.5.5). The proof of (4.5.6) is similar and hence omitted. □

Let

$$z_1^{(n)}, z_2^{(n)}, \ldots, z_{n+1}^{(n)}; \qquad n = 0,1,2,\ldots \tag{4.5.7}$$

be a sequence of points in E and let $H_n(z)$ be an interpolating polynomial of degree not exceeding n that interpolates to a function $f(z)$ at the points $z_k^{(n)}$ $(k = 1,2,\ldots,n+1)$. Our next theorem gives a necessary and sufficient condition, in terms of interpolatory points (4.5.7), so that the sequence of interpolatory polynomials $\{H_n(z)\}$ converges uniformly on E to the function $f(z)$ regular in E.

Theorem 4.5.2. *A necessary and sufficient condition for the polynomial sequence* $\{H_n(z)\}$, *interpolating any regular function* $f(z)$ *in* E *at the points* (4.5.7), *to converge uniformly to* $f(z)$ *on* E *is that*

$$\lim_{n\to\infty} |\omega_n(z)|^{1/(n+1)} = |\Omega(z)| \tag{4.5.8}$$

uniformly on compact subsets of E'; *or, equivalently,*

$$\lim_{n\to\infty} M_n^{1/(n+1)} = \rho, \tag{4.5.9}$$

where

$$\omega_n(z) = \Pi_{k=1}^{n+1} (z-z_k^{(n)}) \quad \textit{and} \quad M_n = \max_{z\in E} |\omega_n(z)|. \tag{4.5.10}$$

Proof. We first prove the necessity part. Let $z_o (\neq \infty)$ be a point in E'. There exists a number $R > \rho$ such that $z_o \notin L_R$. Since the sequence $\{H_n(z)\}$ converges uniformly on E to f(z) for *any* regular function f(z) in E we consider $f(z) = 1/(z-z_o)$ which is regular in $E_R \cup L_R$. By (4.5.3),

$$H_n(z) = \frac{1}{2\pi i} \int_{L_R} \frac{[\omega_n(t) - \omega_n(z)]}{\omega_n(t)(t-z)(t-z_o)} dt$$

$$= - \lim_{t\to z_o} \frac{\omega_n(t) - \omega_n(z)}{\omega_n(t)(t-z)} = \frac{\omega_n(z_o) - \omega_n(z)}{(z-z_o)\,\omega_n(z_o)}.$$

Thus we must have

$$\lim_{n\to\infty} \left[\frac{1}{z-z_o} - \frac{\omega_n(z_o) - \omega_n(z)}{(z-z_o)\,\omega_n(z_o)}\right] = \lim_{n\to\infty} \frac{\omega_n(z)}{(z-z_o)\,\omega_n(z_o)} = 0$$

uniformly on E; i.e., for $z_o \in E'$, we have

$$\lim_{n\to\infty} \frac{M_n}{|\omega_n(z_o)|} = 0 \text{ where } M_n \text{ is given by (4.5.10).} \tag{4.5.11}$$

For $z \in E'$, set

$$\phi_n(z) = [\omega_n(z)]^{1/(n+1)}/\Omega(z), \qquad n = 0,1,2,\ldots$$

where for $[\omega_n(z)]^{1/(n+1)}$ we take that analytic branch of the function which has, in a neighbourhood of the point at infinity, an expansion of the form

$$[\omega_n(z)]^{\frac{1}{n+1}} = z + a_o^{(n)} + \frac{a_1^{(n)}}{z} + \ldots .$$

The functions $\phi_n(z)$ are single-valued, regular and nonvanishing in E'. Further, $\phi_n(\infty) = 1$. For $\rho < r$, if $M_n(r) = \max_{z\in L_r} |\omega_n(z)|$ then $M_n(r) \to M_n$ as $r \to \rho$. Also $\max_{z\in L_r} |\phi_n(z)| = \{M_n(r)\}^{1/(n+1)}/r$. By the maximum modulus principle, the quantity $\{M_n(r)\}^{1/(n+1)}/r$ is a nonincreasing function of r. Thus, for $z \in E'$, we have

$$|\phi_n(z)| \leqq \frac{M_n^{1/(n+1)}}{\rho} . \tag{4.5.12}$$

If d is the diameter of E, then $|\omega_n(z)| \leqq d^{n+1}$ for $z \in E$ and so $M_n \leqq d^{n+1}$. Thus (4.5.12), coupled with the fact that $\phi_n(\infty) = 1$, gives

$$1 \leqq \frac{M_n^{1/(n+1)}}{\rho} \leqq \frac{d}{\rho} . \tag{4.5.13}$$

On the other hand, (4.5.11) gives that for any $z \in L_r$, $r > \rho$, there exists an integer $n_o = n_o(r,z)$ such that, for $n > n_o$,

$$|\phi_n(z)| > \frac{M_n^{1/(n+1)}}{r} . \tag{4.5.14}$$

Now suppose that (4.5.9) is not valid. Then there eixsts a subsequence $\{n_k\} \to \infty$ and a number $q \neq 1$ such that, as $k \to \infty$, we have

$$M_{n_k}^{1/(n_k+1)} \to q\rho. \tag{4.5.15}$$

In view of (4.5.13), we must have $q > 1$. Since $\{\phi_n(z)\}$ is a sequence of regular functions in E' which is uniformly bounded (in view of (4.5.12) and (4.5.13)) by d/ρ, it is normal in E' (cf. Appendix A.7). Thus we can choose a subsequence of $\{\phi_{n_k}(z)\}$ which converges uniformly on compact subsets of E' to a function $\phi(z)$, regular in E'. Without loss of generality, it can be assumed that the sequence $\{\phi_{n_k}(z)\}$ itself converges to $\phi(z)$ uniformly on compact subsets of E'. Choose $R_o > \rho$ such that $1 < R_o/\rho < q$. Writing (4.5.14) for $n = n_k$ and making $k \to \infty$ we get, for $z \in L_{R_o}$, $|\phi(z)| \geqq q\rho/R_o > 1$. Since $\phi(z)$ is a nonvanishing regular function in E' such that $\phi(\infty) = \lim_{k\to\infty} \phi_{n_k}(\infty) = 1$, this contradicts the minimum modulus principle (cf. Appendix A.6). Thus (4.5.9) must hold.

To show that (4.5.8) also holds, we note that if a subsequence $\{\phi_{n_k}(z)\}$ converges uniformly on compact subsets of E' to some function $\phi(z)$ regular in E' then (4.5.9) and (4.5.12) give that $|\phi(z)| \leqq 1$ for $z \in E'$. Since $\phi(\infty) = 1$, this implies that $\phi(z) \equiv 1$ in E'. Since the sequence $\{\phi_n(z)\}$ is normal and any of its subsequences converges to 1 uniformly on compact subsets of E' it follows that the same must be true for the sequence $\{\phi_n(z)\}$. Thus (4.5.8) follows.

To prove the sufficiency part of the theorem assume that (4.5.8) holds. Let f be regular in E; then there exists $R > \rho$ such that f is regular in E_R. Let r_1 be an arbitrary number satisfying $\rho < r_1 < R$ and choose numbers r_2 and r_3 such that $\rho < r_3 < r_1 < r_2 < R$, $r_1 r_3 < \rho r_2$. Since (4.5.8) holds uniformly on compact subsets of E', in the closed domain $\overline{G}$ bounded by the curves L_{r_3} and L_{r_2} we can write

$$|\omega_n(z)| = \left[|\Omega(z)| + \xi_n(z)\right]^{n+1}, \qquad z \in \overline{G}$$

where $\xi_n(z) \to 0$ as $n \to \infty$ uniformly in $\overline{G}$. Thus, for $z \in L_{r_3}$,

$$|\omega_n(z)| \leqq (r_3+\delta_n)^{n+1}, \qquad \delta_n = \max_{z \in L_{r_3}} |\xi_n(z)|$$

whereas, for $z \in L_{r_2}$,

$$|\omega_n(z)| \geqq (r_2-\delta'_n)^{n+1}, \qquad \delta'_n = \max_{z \in L_{r_2}} |\xi_n(z)|$$

where δ_n and $\delta'_n \to 0$ as $n \to \infty$.

If $H_n(z)$ is the interpolating polynomial of degree not exceeding n for f(z), then, by (4.5.4), we have, for $z \in E$,

$$f(z) - H_n(z) = \frac{1}{2\pi i} \int_{L_{r_2}} \frac{\omega_n(z)\, f(t)}{\omega_n(t)(t-z)}\, dt$$

so that, using (4.5.6), we have

$$|f(z)-H_n(z)| \leqq \frac{(r_3+\delta_n)^{n+1}}{(r_2-\delta'_n)^{n+1}} \frac{1}{2\pi} \int_0^{2\pi} f(\eta(r_2 e^{i\theta})) \left| \frac{\eta'(r_2 e^{i\theta}) r_2 i e^{i\theta}}{\eta(r_2 e^{i\theta})-z} \right| d\theta$$

$$\leq M_o(r_2) \frac{(r_3+\delta_n)^{n+1}}{(r_2-\delta'_n)^{n+1}} \frac{r_2+\rho}{r_2-\rho} \qquad \text{where } M_o(r_2) = \sup_{0 \leq \theta \leq 2\pi} |f(\eta(r_2 e^{i\theta}))|$$

$$= \left[M_o(r_2)\left(\frac{r_2+\rho}{r_2-\rho}\right)\left(\frac{r_3+\delta_n}{r_2-\delta'_n} \cdot \frac{r_1}{\rho}\right)^{n+1}\right]\left(\frac{\rho}{r_1}\right)^{n+1} \tag{4.5.16}$$

$\to 0$ as $n \to \infty$ uniformly on E.

Thus $\{H_n(z)\}$ converges uniformly on E to f(z) if (4.5.8) is given to hold. If (4.5.9) holds then, as in the necessity part, we can show that (4.5.8) is satisfied and so by the above arguments it follows that for any function f(z) regular on E the sequence of interpolatory polynomials $H_n(z)$ converges uniformly on E to f(z). □

Corollary. If f *is regular in* E_R, $R > \rho$ *and satisfies condition* (4.5.8) *or* (4.5.9) *then for any* r_1, $\rho < r_1 < R$, *we have*

$$|f(z)-H_n(z)| \leq M'\left(\frac{\rho}{r_1}\right)^{n+1}, \qquad z \in E \tag{4.5.17}$$

where M' *is a constant independent of* n *and* z.

The corollary follows immediately from (4.5.16).

We shall now identify certain extremal points of the set E for which (4.5.8) is satisfied. These are called the Van der Monde points for the set E and are determined as follows.

For any (n+1) points $z_1^{(n)}, z_2^{(n)}, \ldots, z_{n+1}^{(n)}$ in E, form the Van der Monde determinant

$$V_n(z_1^{(n)}, \ldots, z_{n+1}^{(n)}) = \prod_{1 \leq k < j \leq n+1} (z_k^{(n)} - z_j^{(n)}). \tag{4.5.18}$$

The function $|V_n(z_1^{(n)}, \ldots, z_{n+1}^{(n)})|$ considered as a function of the (n+1) variables $z_k^{(n)}$ (k = 1,...,n+1) is a bounded continuous function on E^{n+1} and so possesses a maximum there. Let $(\zeta_1^{(n)}, \ldots, \zeta_{n+1}^{(n)})$ yield the maximum value of $|V_n|$ on E^{n+1}. It is clear that the points $\zeta_k^{(n)}$ (k = 1,...,n+1) must be all distinct.

We now consider the points

$$\zeta_k^{(n)}, \quad k = 1,2,\ldots,n+1; \qquad n = 1,2,\ldots \tag{4.5.19}$$

as a system of interpolatory points. All these points are simple and are called *Van der Monde points*. If F(z) is a regular function on the set E, the Lagrange's interpolation polynomial $H_n(z)$ of degree not exceeding n for the function F(z) with interpolatory points (4.5.19) is given by

$$H_n(z) = \Sigma_{k=1}^{n} \frac{\omega_n(z)\, F(\zeta_k^{(n)})}{(z-\zeta_k^{(n)})\, \omega_n'(\zeta_k^{(n)})} \tag{4.5.20}$$

where $\omega_n(z) = \Pi_{k=1}^{n+1} (z-\zeta_k^{(n)})$. Since, in formula (4.5.18) for $V_n(\zeta_1^{(n)},\ldots,\zeta_{n+1}^{(n)})$, the product of the factors containing $\zeta_k^{(n)}$ is precisely

$$(\zeta_1^{(n)}-\zeta_k^{(n)})\,(\zeta_2^{(n)}-\zeta_k^{(n)})\ldots(\zeta_{k-1}^{(n)}-\zeta_k^{(n)})(\zeta_k^{(n)}-\zeta_{k+1}^{(n)})\ldots(\zeta_k^{(n)}-\zeta_{n+1}^{(n)}))$$

the function

$$(z-\zeta_1^{(n)})(z-\zeta_2^{(n)})\ldots(z-\zeta_{k-1}^{(n)})\quad(z-\zeta_{k+1}^{(n)})\ldots(z-\zeta_{n+1}^{(n)})\ (\equiv \omega_n(z)/(z-\zeta_k^{(n)}))$$

considered as a function of z takes its maximum absolute value $|\omega_n'(\zeta_k^{(n)})|$ on E for $z = \zeta_k^{(n)}$. Thus we must have

$$\left|\frac{\omega_n(z)}{(z-\zeta_k^{(n)})\ \omega_n'(\zeta_k^{(n)})}\right| \leqq 1, \qquad z \in E. \tag{4.5.21}$$

For the interpolatory polynomial $H_n(z)$ constructed above we therefore have the estimate

$$|H_n(z)| \leqq \Sigma_{k=1}^{n+1} |F(\zeta_k^{(n)})|, \qquad z \in E. \tag{4.5.22}$$

We now have

Theorem 4.5.3. *If* $\{\zeta_k^{(n)}\}$, *given by* (4.5.19), *are the Van der Monde points for the set* E, *then*

$$\lim_{n\to\infty} |\omega_n(z)|^{1/(n+1)} = |\Omega(z)| \tag{4.5.23}$$

uniformly on compact subsets of E' *and*

$$\lim_{n\to\infty} M_n^{1/(n+1)} = \rho \tag{4.5.24}$$

where $\omega_n(z) = \Pi_{k=1}^{n+1} (z-\zeta_k^{(n)})$ *and* $M_n = \sup_{z\in E} |\omega_n(z)|$.

Proof. Let f be arbitrary analytic function in E; then there exists an $R > \rho$ such that f is analytic in E_R. By Theorem 4.2.7, for every $n \geq 0$ there exists a polynomial $Q_n(z)$ of degree not exceeding n such that

$$|f(z) - Q_n(z)| < M\left(\frac{\rho}{R}\right)^n , \qquad z \in E \tag{4.5.25}$$

where M is a constant independent of z and n. We now introduce the function $F(z) = f(z) - Q_n(z)$ and construct for it an interpolatory polynomial $H_n(z)$ of degree n having interpolatory points at $\{\zeta_1^{(n)}, \ldots, \zeta_{n+1}^{(n)}\}$. We then have, from (4.5.22) and (4.5.25),

$$|H_n(z)| \leqq \Sigma_{k=1}^{n} |F(\zeta_k^{(n)})| \leqq (n+1)M\left(\frac{\rho}{R}\right)^n , \qquad z \in E. \tag{4.5.26}$$

Consider $L_n^*(z) = H_n(z) + Q_n(z)$; $L_n^*(z)$ is a polynomial of degree not exceeding n. At the Van der Monde points $\{\zeta_k^{(n)}\}$, we have

$$L_n^*(\zeta_k^{(n)}) = Q_n(\zeta_k^{(n)}) + (f(\zeta_k^{(n)}) - Q(\zeta_k^{(n)})) = f(\zeta_k^{(n)}), \qquad k = 1,\ldots,n+1,$$

showing that $L_n^*(z)$ is an interpolatory polynomial at the Van der Monde points for f(z). Further, by (4.5.25) and (4.5.26), for $z \in E$, we have

$$\begin{aligned} |f(z)-L_n^*(z)| = |f(z)-Q_n(z)-H_n(z)| &\leqq |f(z)-Q_n(z)|+|H_n(z)| \\ &\leqq M\left(\frac{\rho}{R}\right)^n + (n+1)\, M\left(\frac{\rho}{R}\right)^n \\ &= (n+2)\, M\left(\frac{\rho}{R}\right)^n \\ &\to 0 \text{ as } n \to \infty. \end{aligned}$$

This shows that the sequence of interpolating polynomials $L_n^*(z)$ at the Van der Monde points converges uniformly on E to f(z). By Theorem 4.5.2 we then have that (4.5.23) and (4.5.24) hold for the points (4.5.19). □

Our next lemma gives the relation between the errors when one considers the best approximation of a regular function f on E in the sup norm by polynomials and its approximation in the same norm by interpolatory polynomials at the Van der Monde points of E.

Lemma 4.5.2. *We have*

$$\Delta_n^{(\infty)}(f) \leqq \sigma_n(f) \leqq (n+2)\, \Delta_n^{(\infty)}(f) \tag{4.5.27}$$

where

$$\Delta_n^{(\infty)}(f) = \inf_{g \in P_n} \| f-g \|_\infty = \inf_{g \in P_n} \sup_{z \in E} |f(z) - g(z)|,$$

P_n *being the set of all polynomials of degree not exceeding* n *and*

$$\sigma_n(f) = \| f-L_n^* \|_\infty = \sup_{z \in E} |f(z) - L_n^*(z)|,$$

$L_n^*(z)$ *being interpolatory polynomial for* f *having Van der Monde points as interpolatory points.*

Proof. The left-hand inequality of (4.5.27) being obvious, we prove the right-hand inequality. Let $p_n(z)$ be the best approximating polynomial of degree $\leqq$ n in sup norm, i.e.,

$$\Delta_n^{(\infty)}(f) = \| f-p_n \|_\infty$$

and let $H_n(z)$ be the polynomial that interpolates to $f-p_n$ at the Van der Monde points of E. Then $L_n^*(z) = H_n(z) + p_n(z)$ is the nth degree polynomial that interpolates to f at Van der Monde points; for

$$L_n^*(\zeta_k^{(n)}) = H_n(\zeta_k^{(n)}) + p_n(\zeta_k^{(n)}) = f(\zeta_k^{(n)}), \qquad k = 1,2,\ldots,n+1.$$

Therefore

$$\sigma_n(f) = \| f-L_n^* \|_\infty \leqq \| f-p_n \|_\infty + \| p_n \|_\infty$$

$$\leqq \Delta_n^{(\infty)}(f) + \Sigma_{k=1}^n |f(\zeta_k^{(n)}) - p_n(\zeta_k^{(n)})| \qquad \text{(by (4.5.22))}$$

$$\leqq \Delta_n^{(\infty)}(f) + (n+1)\, \Delta_n^{(\infty)}(f) = (n+2)\, \Delta_n^{(\infty)}(f)$$

which gives (4.5.27). □

In view of inequality (4.5.27) we see that, as far as the growth measurement of an analytic function in terms of $\sigma_n(f)$ is concerned, the rate of decrease of $\sigma_n(f)$ is of the same order of magnitude as $\Delta_n^{(\infty)}(f)$. Thus the corresponding theorems on $\Delta_n^{(\infty)}(f)$ lead immediately to the following growth estimates for an $f \in H(E;R)$ in terms of $\sigma_n(f)$.

Theorem 4.5.4. *A regular function* f *on* E *is in* $H(E;R)$ *if and only if*

$$\limsup_{n \to \infty} \left[\sigma_n(f)\right]^{1/n} = \frac{\rho}{R} . \tag{4.5.28}$$

Theorem 4.5.5. *If* $f \in H(E;R)$ *is of* E-*order* α_o $(0 \le \alpha_o \le \infty)$, *then*

$$\frac{\alpha_o}{\alpha_o+1} = \limsup_{n \to \infty} \left\{ \frac{\log^+\log^+(\sigma_n(f)(R/\rho)^n)}{\log n} \right\} \tag{4.5.29}$$

where the left-hand side in (4.5.29) *is to be interpreted as* 1 *for* $\alpha_o = \infty$.

Theorem 4.5.6. *Let* $f \in H(E;R)$ *be of* E-*order* α_o $(0 < \alpha_o < \infty)$ *and of* E-*type* τ $(0 \le \tau \le \infty)$. *Set*

$$J = \limsup_{n \to \infty} \frac{\left[\log^+(\sigma_n(f)(R/\rho)^n)\right]^{\alpha_o+1}}{n^{\alpha_o}} . \tag{4.5.30}$$

If $0 < J < \infty$, *then* f *is of* E-*order* α_o *and* E-*type* τ *if and only if*

$$\left(\frac{R}{\rho}\right)^{\alpha_o} \frac{(\alpha_o+1)^{\alpha_o+1}}{\alpha_o^{\alpha_o}} \tau = J. \tag{4.5.31}$$

If $J = 0$ *or* ∞, *the function* f *is respectively of* E-*growth* $(\alpha_o, 0)$ *or of* E-*growth not less than* (α_o, ∞) *and conversely.*

Theorem 4.5.7. *Let* $f \in H(E;R)$ *be of lower* E-*order* β $(0 \le \beta \le \infty)$. *Then for any increasing sequence* $\{n_k\}$ *of natural numbers,*

$$1+\beta \geqq \liminf_{k \to \infty} \frac{\log n_{k-1}}{\log n_k - \log^+\log^+(\sigma_{n_k}(f)(R/\rho)^{n_k})} . \tag{4.5.32}$$

Further, if f *has* E-*order* α_o $(0 < \alpha_o < \infty)$ *and lower* E-*type* ν $(0 \le \nu \le \infty)$, *then*

$$\frac{(\alpha_o+1)^{\alpha_o+1}}{\alpha_o^{\alpha_o}}\left(\frac{R}{\rho}\right)^{\alpha_o}\nu$$

$$\geqq \liminf_{k\to\infty}\{n_{k-1}(\log^+(\sigma_{n_k}(f)(R/\rho)^{n_k})^{1/n_k})^{\alpha_o+1}\}. \qquad (4.5.33)$$

Theorem 4.5.8. *Let* $f \in H(E;R)$ *be of* E-*order* α_o $(0 < \alpha_o \leq \infty)$ *and lower* E-*order* β $(0 \leq \beta \leq \infty)$ *such that* $g(z) = \sum_{n=0}^{\infty} (\sigma_n(f)\left(\frac{R}{\rho}\right)^n)z^n$ *is an admissible function, then*

$$1+\beta = \sup_{\{n_k\}}\left\{\liminf_{k\to\infty}\frac{\log n_{k-1}}{\log n_k - \log^+\log^+(\sigma_{n_k}(f)(R/\rho)^{n_k})}\right\} \qquad (4.5.34)$$

where supremum is taken over all increasing sequences of natural numbers.

Theorem 4.5.9. *Let* $f \in H(E;R)$ *be of* E-*order* α_o *and lower* E-*type* ν *such that* $\sigma_n(f)/\sigma_{n+1}(f)$ *forms a nondecreasing sequence for* $n > n_o$, *then*

$$\frac{(\alpha_o+1)^{\alpha_o+1}}{\alpha_o^{\alpha_o}}\left(\frac{R}{\rho}\right)^{\alpha_o}\nu \leqq \liminf_{n\to\infty}\frac{[\log^+(\sigma_n(f)(R/\rho)^n)]^{\alpha_o+1}}{n^{\alpha_o}}. \qquad (4.5.35)$$

The problem of interpolation leads to interesting results when the set of interpolating points is finite, consisting, say, of $\alpha_1,\alpha_2,\ldots,\alpha_m$, and the function $f(z)$ to be interpolated is regular inside a lemniscate determined by these points. In this case not only can the function be expanded in a series of polynomials converging uniformly on compact subsets inside the lemniscate but it also leads to a new method of growth measurement of such functions.

Let

$$q(z) = (z-\alpha_1)(z-\alpha_2)\ \ldots\ (z-\alpha_m), \qquad (4.5.36)$$

where α_k $(k = 1,\ldots,m)$ are fixed but not necessarily distinct points, be a polynomial of degree m. A locus of the form

$$\Gamma_\mu = \{z : |q(z)| = \mu > 0\} \qquad (4.5.37)$$

is called a *lemniscate*. It is easy to see that when μ is sufficiently small and there are precisely ℓ distinct points α_k, the lemniscate Γ_μ consists of ℓ ovals, each oval surrounding precisely one point α_k. When μ increases, these ovals increase in size, so that when μ is sufficiently large the locus (4.5.37) consists of precisely one Jordan curve. In fact, any lemniscate that includes in its interior the smallest convex set containing the zeros of $q(z)$ (and hence of $q'(z)$) must consist of a single Jordan analytic curve. We shall assume throughout that $\alpha > 0$ is so large that Γ_α consists of a single Jordan analytic curve.

Suppose that f is analytic inside and on a lemniscate Γ_μ $(\mu > \alpha)$, given by (4.5.37). The inside of the lemniscate Γ_μ we shall denote by Γ_μ^o. Let $S_n(z)$ be the polynomial of degree mn - 1 which interpolates to f(z) at the points α_k $(k = 1,\ldots,m)$, each counted of multiplicity n. The polynomial $S_n(z)-S_{n-1}(z)$ is then a polynomial of degree mn-1, and has, at each α_k, a zero of order (n-1). Thus $S_n(z)-S_{n-1}(z)$ is a product of $[q(z)]^{n-1}$ with a polynomial $p_n(z)$ of degree at most m-1, i.e., $S_n(z)$ has the form

$$S_n(z) = \Sigma_{k=1}^n p_k(z) \, [q(z)]^{k-1}, \tag{4.5.38}$$

where each $p_k(z)$ is a polynomial of degree not exceeding m-1. By (4.5.4), we must then have

$$f(z)-S_n(z) = \frac{1}{2\pi i} \int_{\Gamma_\mu} \frac{[q(z)]^n \, f(t) \, dt}{[q(t)]^n \, (t-z)}, \qquad z \in \Gamma_\mu^o. \tag{4.5.39}$$

Thus, if $z \in \Gamma_{\mu'}^o \cup \Gamma_{\mu'}$ $(\alpha < \mu' < \mu)$, then

$$|f(z)-S_n(z)| \leqq M \frac{\mu'^n}{\mu^n} \tag{4.5.40}$$

where M is a constant independent of n and z. The above relation shows that the sequence $\{S_n(z)\}$ converges to f(z) uniformly on compact subsets of Γ_μ^o and so, for $z \in \Gamma_\mu^o$, f has an expansion of the form

$$f(z) = p_1(z)+p_2(z) \, [q(z)]+\ldots+ p_n(z) \, [q(z)]^{n-1}+\ldots \tag{4.5.41}$$

where each $p_n(z)$ is a uniquely determined polynomial of degree not exceeding m-1 and q(z) is given by (4.5.36).

We now consider the class F_R of functions regular in Γ_R^o that have a singularity on Γ_R $(R > \alpha)$. As observed above, each such function can be expressed in a polynomial series of the form (4.5.41) that converges uniformly on compact subsets of Γ_R^o. For $\alpha < r < R$ and $f \in F_R$, set

$$M(\Gamma_r) \equiv M(\Gamma_r, f) = \max_{z \in \Gamma_r} |f(z)|; \tag{4.5.42}$$

it follows, from the remarks made above, that $M(\Gamma_r)$ is an increasing function of r. We now introduce the concepts of L-order, L-type etc. for functions in F_R with the help of $M(\Gamma_r)$. Thus $f \in F_R$ is said to be of *L-order* γ $(0 \leq \gamma \leq \infty)$ if

$$\gamma = \limsup_{r \to R} \frac{\log^+ \log^+ M(\Gamma_r)}{\log (Rr/(R-r))^{1/m}}; \tag{4.5.43}$$

$f \in \Gamma_R$ is said to be of *lower L-order* δ $(0 \leq \delta \leq \infty)$ if

$$\delta = \liminf_{r \to R} \frac{\log^+ \log^+ M(\Gamma_r)}{\log (Rr/(R-r))^{1/m}}. \tag{4.5.44}$$

If $0 < \gamma < \infty$, $f \in F_R$ is said to be of *L-type* T_o and *lower L-type* ν_o if

$$\begin{matrix} T_o \\ \nu_o \end{matrix} = \lim_{r \to R} \begin{matrix} \sup \\ \inf \end{matrix} \frac{\log M(\Gamma_r)}{[Rr/(R-r)]^{\gamma/m}}. \tag{4.5.45}$$

Before determining the above growth constants in terms of the polynomial coefficients of the series (4.5.41) we shall prove two lemmas which extend Cauchy's inequality and the Cauchy-Hadamard formula to the case of polynomial series.

Lemma 4.5.3. *The polynomial coefficients* $p_n(z)$ *in* (4.5.41) *satisfy, for* $\alpha < r < R$,

$$\| p_n \|_\alpha \leq A \frac{M(\Gamma_r)}{r^n} \| \Gamma_r \|, \qquad n = 1,2,... \tag{4.5.46}$$

where $\| p_n \|_\alpha = \max_{z \in \Gamma_\alpha} |p_n(z)|$, $\| \Gamma_r \|$ *is the length of* Γ_r, $M(\Gamma_r)$ *is defined by* (4.5.42) *and* A *is a constant independent of* n *and* r.

Proof. By (4.5.39) we have, for $\alpha < r < R$,

$$f(z) - \Sigma_{k=1}^{n} p_k(z) \, [q(z)]^{k-1} = \frac{1}{2\pi i} \int_{\Gamma_r} \frac{[q(z)]^n \, f(t)dt}{[q(t)]^n \, (t-z)}, \quad z \in \Gamma_r^o .$$

This leads to

$$p_n(z) = \frac{1}{2\pi i} \int_{\Gamma_r} \frac{[q(t)-q(z)] \, f(t)dt}{[q(t)]^n \, (t-z)}, \qquad z \in \Gamma_r^o.$$

Since $q(t)-q(z)$ is a polynomial in t of degree m which vanishes at $t = z$, we may write $q(t)-q(z) = (t-z)Q(t,z)$ where $Q(t,z)$ is a polynomial in each of the variables t and z of degree $m-1$. Thus we have, for $z \in \Gamma_\alpha$, $\alpha < r < R$,

$$|p_n(z)| \leq \frac{1}{2\pi} \cdot \frac{M(\Gamma_r)}{r^n} \, \| \Gamma_r \| \max_{(t,z)\in\Gamma_r \times \Gamma_\alpha} |Q(t.z)|$$

which gives

$$\| p_n \|_\alpha \leq A \frac{M(\Gamma_r)}{r^n} \| \Gamma_r \|$$

where $A = \frac{1}{2\pi} \max_{(t,z)\in\Gamma_R \times \Gamma_R} |Q(t,z)|$. □

Lemma 4.5.4. *A function* f, *defined by* (4.5.41), *belongs to* F_R $(R > \alpha)$ *if and only if*

$$\limsup_{n \to \infty} \| p_n \|_\alpha^{1/n} = \frac{1}{R} \tag{4.5.47}$$

where $\| p_n \|_\alpha$ *is as defined in Lemma* 4.5.3.

Proof. Let $f \in F_R$. (4.5.46) then gives

$$\limsup_{n \to \infty} \| p_n \|_\alpha^{1/n} \leq \frac{1}{r} \text{ for every } r \text{ satisfying } \alpha < r < R,$$

so that

$$\limsup_{n \to \infty} \| p_n \|_\alpha^{1/n} \leq \frac{1}{R} . \tag{4.5.48}$$

Let, if possible,

$$\limsup_{n\to\infty} \| p_n \|_\alpha^{1/n} = \frac{1}{R_o} < \frac{1}{R} \tag{4.5.49}$$

and let R_1 be such that $R < R_1 < R_o$. (4.5.48) then **gives**

$$\| p_n \|_\alpha < \frac{1}{R_1^n} \qquad \text{for} \qquad n \geq n_o = n_o(R_1). \tag{4.5.50}$$

Consider the function $\Psi(z) = [p_n(z)]^m / \{q(z)\}^{m-1}$; Ψ is analytic in the exterior of Γ_α, even at infinity, if suitably defined. Thus, by the maximum modulus principle, it follows that

$$\max_{z\in\Gamma_r} \left| \frac{[p_n(z)]^m}{\{q(z)\}^{m-1}} \right| \leqq \max_{z\in\Gamma_\alpha} \left| \frac{[p_n(z)]^m}{\{q(z)\}^{m-1}} \right|, \qquad r > \alpha.$$

So

$$\max_{z\in\Gamma_r} |p_n(z)| \leqq \left(\frac{r}{\alpha}\right)^{\frac{m-1}{m}} \| p_n \|_\alpha \qquad \text{for} \quad r > \alpha. \tag{4.5.51}$$

(4.5.50) and (4.5.51) then give

$$\max_{z\in\Gamma_r} |p_n(z)| \leqq \left(\frac{r}{\alpha}\right)^{\frac{m-1}{m}} \left(\frac{1}{R_1^n}\right), \qquad r > \alpha. \tag{4.5.52}$$

(4.5.52) guarantees that (4.5.41) converges uniformly on compact subsets of $\Gamma^o_{R_1}$. However this contradicts the fact that $f \in F_R$. Hence (4.5.47) must hold. The converse part of the lemma is obvious. □

We now obtain formulae connecting the growth of $f \in F_R$ with the coefficients of its Jacobi series expansion (4.5.41). First we have

Theorem 4.5.10. *Let* $f \in F_R$ *be given by* (4.5.41) *and be of* L-*order* γ $(0 \leq \gamma \leq \infty)$ *in* Γ^o_R. *Then*

$$\gamma = m \limsup_{n\to\infty} \left\{ \frac{\log^+ \log^+ (\| p_n \|_\alpha R^n)}{\log n - \log^+ \log^+ (\| p_n \|_\alpha R^n)} \right\}. \tag{4.5.53}$$

Proof. We consider the function g defined as

$$g(z) = \Sigma_{n=1}^{\infty} (\| p_n \|_\alpha R^n) z^{n-1}. \tag{4.5.54}$$

By Lemma 4.5.4, g is analytic in $U = \{z: |z| < 1\}$ and has a singularity on $|z| = 1$. Further, since

$$|f(z)| \leqq \Sigma_{n=1}^{\infty} \| p_n \|_r r^{n-1} \quad \text{for} \quad z \in \Gamma_r,$$

it follows, by (4.5.51), that

$$M(\Gamma_r) \leqq \left(\frac{r}{\alpha}\right)^{\frac{m-1}{m}} \Sigma_{n=1}^{\infty} \| p_n \|_\alpha r^{n-1} < \left(\frac{R}{\alpha}\right)^{\frac{m-1}{m}} R^{-1} \Sigma_{n=1}^{\infty} (\| p_n \|_\alpha R^n) \left(\frac{r}{R}\right)^{n-1},$$

so that

$$M(\Gamma_r) \leqq \alpha^{-1} \left(\frac{\alpha}{R}\right)^{1/m} M(\tfrac{r}{R} ; g), \tag{4.5.55}$$

where $M(\frac{r}{R} ; g) = \max_{|z|=\frac{r}{R}} |g(z)|$.

Hence, using (4.5.55) and (4.5.43), we get

$$\gamma \leqq m \limsup_{r \to R} \frac{\log^+ \log^+ M(r/R;g)}{\log(Rr/(R-r))}$$

$$= m \limsup_{t \to 1} \frac{\log^+ \log^+ M(t;q)}{-\log (1-t)}$$

$$= m \limsup_{n \to \infty} \frac{\log^+ \log^+ (\| p_n \|_\alpha R^n)}{\log n - \log^+ \log^+ (\| p_n \|_\alpha R^n)} \tag{4.5.56}$$

where we have used (2.2.8) for g defined by (4.5.54).

For the reverse inequality it is sufficient to consider the case when

$$\beta^* = \limsup_{n \to \infty} \frac{\log^+ \log^+ (\| p_n \|_\alpha R^n)}{\log n - \log^+ \log^+ (\| p_n \|_\alpha R^n)} > 0. \tag{4.5.57}$$

Let $\varepsilon > 0$ be any number such that $0 < \varepsilon < \beta^*$, then (4.5.57) gives an increasing sequence $\{n_k\}$ such that

$$\log^+(\| p_{n_k} \|_\alpha R^{n_k}) \geq n_k^{(\beta^*-\varepsilon)/(1+\beta^*-\varepsilon)} \quad \text{for} \quad k = 1,2,\ldots . \tag{4.5.58}$$

Using (4.5.46) and (4.5.58) we get, for $R > r > \alpha$ and $k = 1,2,\ldots,$

$$n_k^{(\beta^*-\varepsilon)/(1+\beta^*-\varepsilon)} \leq \log (A\| \Gamma_r \|) + \log M(\Gamma_r) - n_k \log \frac{r}{R} . \tag{4.5.59}$$

Now let us choose a sequence $r_k \to R$ as

$$r_k = R\left[1-n_k^{-1/\beta^*}\right], \qquad k = 1,2,\ldots . \tag{4.5.60}$$

(4.5.59) and (4.5.60) then give

$$\limsup_{k \to \infty} \frac{\log^+ \log^+ M(\Gamma_{r_k})}{\log (R\, r_k/(R-r_k))} \geq \beta^*$$

and so (4.5.43) gives

$$\gamma \geq m\beta^* . \tag{4.5.61}$$

Combining (4.5.56) and (4.5.61), we get (4.5.53). □

The following theorem for L-type can also be proved in a similar manner by using (4.5.46), (4.5.55) and the corresponding result for the function g.

Theorem 4.5.11. *Let* $f \in F_R$ *be given by* (4.5.41) *and be of* L-*order* γ $(0 < \gamma < \infty)$ *and* L-*type* T_o $(0 \leq T_o \leq \infty)$. *Set*

$$J_o = \limsup_{n \to \infty} \frac{\left[\log^+(\| p_n \|_\alpha R^n)\right]^{(\gamma/m)+1}}{n^{\gamma/m}} . \tag{4.5.62}$$

If $0 < J_o < \infty$, *then* f *is of* L-*order* γ *and* L-*type* T_o *if and only if*

$$R^{\gamma/m} \frac{((\gamma/m)+1)^{(\gamma/m)+1}}{(\gamma/m)^{\gamma/m}} T_o = J_o . \tag{4.5.63}$$

If $J_o = 0$ *or* ∞, *the function* f *is respectively of* L-*growth* $(\gamma,0)$ *or of* L-*growth not less than* (γ,∞) *and conversely*.

Now we have the following theorem for lower q-order.

Theorem 4.5.12. *Let* $f \in F_R$ *be given by* (4.5.41) *and be of* L*-order* γ $(0 < \gamma \leq \infty)$, *then its lower* L*-order* δ $(0 \leq \delta \leq \infty)$ *satisfies*

$$1 + \frac{\delta}{m} \geq \sup_{\{n_k\}} \left[\liminf_{k \to \infty} \frac{\log n_{k-1}}{\log n_k - \log^+ \log^+ (\| p_{n_k} \|_\alpha R^{n_k})}\right] \tag{4.5.64}$$

where supremum is taken over all increasing sequences $\{n_k\}$ *of positive integers. Further, if* g, *defined by* (4.5.54), *is an admissible function then equality holds in* (4.5.64).

Proof. For any increasing sequence $\{n_k\}$ of positive integers, set

$$\beta' \equiv \beta'(\{n_k\}) = \limsup_{k \to \infty} \left\{ \frac{\log n_k - \log^+ \log^+ (\| p_{n_k} \|_\alpha R^{n_k})}{\log n_{k-1}} \right\}. \tag{4.5.65}$$

Let $0 < \varepsilon < 1 - \beta'$, then (4.5.65) gives

$$\log^+ (\| p_{n_k} \|_\alpha R^{n_k}) > n_k n_{k-1}^{-(\beta'+\varepsilon)} \qquad \text{for } k \geq k_o. \tag{4.5.66}$$

Choose a sequence $r_k \to R$ defined by

$$\log (R/r_k) = \frac{1}{e} n_{k-1}^{-(\beta'+\varepsilon)} \qquad \text{for } k = 2,3,\ldots, \tag{4.5.67}$$

then (4.5.46) and (4.5.66) give

$$\log M(\Gamma_r) \geq \frac{1}{e} n_{k-1}^{-(\beta'+\varepsilon)} - \log(A \| \Gamma_r \|) + n_k \log \frac{r_k}{R} \quad \text{for } k \geq k_o. \tag{4.5.68}$$

Using (4.5.67) and (4.5.68), we get, for k sufficiently large and for $r_k \leq r \leq r_{k+1}$,

$$\frac{\log^+ \log^+ M(\Gamma_r)}{\log(Rr/(R-r))} \geq \left(1 - \frac{1}{\beta'+\varepsilon}\right) \frac{1 + \log \log (R/r)}{\log (Rr/(R-r))} + o(1).$$

On proceeding to limits, this gives, in view of (4.5.44),

$$1 + \frac{\delta}{m} \geq \frac{1}{\beta'}$$

which leads to (4.5.64). If g, defined by (4.5.54), is an admissible function then, using (4.5.55), (4.5.44) and the result (2.3.15) for the function g, we get equality in (4.5.64). □

The corresponding theorem about lower L-type also follows in a similar manner. We leave its formulation and proof as an exercise for the reader.

We have seen above that if the set of interpolatory points is finite, the lemniscates determined by these points play a significant role in studying the growth of an analytic function. Still another instructive way of effectively determining interpolatory polynomials and studying growth is through Newton's interpolation series. Let $\alpha_1, \alpha_2, \ldots, \alpha_n, \ldots$ be a sequence of, not necessarily distinct, points and let f be a function that is analytic at each α_i, $i = 1,2,\ldots$. Let $s_n(z)$ be the unique polynomial of degree not exceeding n such that $s_n(z)$ interpolates to f(z) in the points $\alpha_1, \alpha_2, \ldots, \alpha_{n+1}$. Then $s_{n+1} - s_n(z)$, being a polynomial of degree not greater that n+1, must be a constant multiple of $(z-\alpha_1)\ldots(z-\alpha_n)(z-\alpha_{n+1})$ since the polynomials $s_n(z)$ and $s_{n+1}(z)$ coincide in the points $\alpha_1, \ldots, \alpha_{n+1}$. Thus $s_n(z)$ is the sum of the first (n+1) terms of a series of the form

$$f(z) = a_o + a_1(z-\alpha_1) + a_2(z-\alpha_1)(z-\alpha_2) + \ldots + a_n(z-\alpha_1)\ldots(z-\alpha_n) + \ldots \quad (4.5.69)$$

The expansion (4.5.69) is called the Newton's interpolation series expansion of f(z) with respect to the sequence of interpolation points $\{\alpha_n\}$ and $s_n(z)$ are called Newton's interpolation polynomials. These interpolatory polynomials have the great advantage in numerical computation that, for the determination of the next higher degree interpolatory polynomial, only a single coefficient needs to be determined.

The coefficients a_n in (4.5.69) can be computed by means of the usual divided difference methods. Thus $a_o = f(\alpha_1)$, $a_1 = (f(\alpha_2) - f(\alpha_1))/(\alpha_2 - \alpha_1)$ and so on. Since the nth partial sum of (4.5.69) interpolates to f(z) in the finite sequence $\alpha_1, \ldots, \alpha_{n-1}$, it follows that the Newton series expansion of f(z) always converges to f(z) at the points α_i.

To seek convergence of Newton's series (4.5.69) to f(z) at other points, one may readily obtain the Cauchy integral representation for the coefficients a_n, partial sums $s_n(z)$ and remainder $f(z) - s_n(z)$. Thus, if f is analytic within and on a simple closed contour C whose interior contains all the points α_i, then use of Hermite's interpolation formula yields the following:

$$s_n(z) = \frac{1}{2\pi i} \int_C \frac{\omega_{n+1}(t) - \omega_{n+1}(z)}{\omega_{n+1}(t)(t-z)} f(t)dt, \; z \in \text{Int}(C) \tag{4.5.70}$$

$$a_n = \frac{1}{2\pi i} \int_C \frac{f(t)}{\omega_{n+1}(t)} dt \tag{4.5.71}$$

$$f(z) - s_n(z) = \frac{1}{2\pi i} \int_C \frac{\omega_{n+1}(z)f(t)}{\omega_{n+1}(t)(t-z)} dt, \quad z \in \text{Int}(C), \tag{4.5.72}$$

where $\omega_{n+1}(z) = \Pi_{i=1}^{n+1} (z-\alpha_i)$.

If $w = \Omega(z)$ is the univalent analytic function that maps the exterior of C onto $|w| > \rho$ and if

$$\lim_{n\to\infty} |\omega_{n+1}(z)|^{1/(n+1)} = |\Omega(z)|$$

then, by Theorem 4.5.2, the Newton series (4.5.69) converges uniformly to f(z) inside and on C. One could then consider the characterization of the growth of functions analytic inside and on C in terms of Newton's series coefficients a_n and error of approximation $\|f-s_n\|$. However, we shall not go into the details of these but leave them as an exercise for the reader.

EXERCISES 4.5

1. Show that, in general, the sequence of polynomials $\{p_n(z)\}$, where $p_n(z)$ is a polynomial of degree not exceeding n that interpolates at the points $z_k^{(n)}$, $k = 1,\ldots,n+1$, to an analytic function f(z) in a domain D, need not converge to f(z) in D.

   Let points $x_k^{(n)}$, $k = 1,2,\ldots,n+1$, $n = 0,1,\ldots$, lie on an interval $I = [a,b]$ and $N^{(n)}(c,d)$ denote the number of points $x_k^{(n)}$ lying in the interval $[c,d] \subset [a,b]$. The points $\{x_k^{(n)}\}$ are said to be *uniformly distributed* over I if $\lim_{n\to\infty} \frac{N^{(n)}(c,d)}{N^{(n)}(a,b)} = \frac{d-c}{b-a}$ for every subinterval $[c,d] \subset I$.

2. Show that the points $x_k^{(n)}$, $k = 1,\ldots,n+1$, $n = 0,1,\ldots$, are uniformly distributed over the interval $[a,b]$ if and only if, for any function f continuous on $[a,b]$,

$$\lim_{n\to\infty} \frac{1}{n+1} \Sigma_{k=1}^{n+1} f(x_k^{(n)}) = \frac{1}{b-a} \int_a^b f(x)dx.$$

3. Let D be a simply connected domain bounded by a closed Jordan curve C and let points $z_k^{(n)}$, $k = 1,\ldots,n+1$, $n = 0,1,\ldots$, lie on C. Let $w = \Omega(z)$ be the analytic univalent function mapping exterior of C onto $|w| > \rho$ such that $\Omega(\infty) = \infty$, $\Omega'(\infty) = 1$. Show that the points $z_k^{(n)}$ are uniformly distributed on C if and only if (4.5.8) holds.

For a compact set F and points $z_1,\ldots,z_n \in F$ construct the polynomial $p(z) = \Pi_{k=1}^n (z-z_k)$ and set $\lambda_n = \inf_{\{z_k\}} \sup_{z\in F} |p(z)|$. The polynomial $p^*(z) = \Pi_{k=1}^n (z-z_{k,n}^*)$, $z_{k,n}^* \in F$ for which $\lambda_n = \sup_{z\in F} |p^*(z)|$ is called the *Chebyshev polynomial* for F and the points $\{z_{k,n}^*\}$ are called the *Chebyshev points* for F.

4. Let E be as in Theorem 4.5.2 and $\{z_{k,n}^*\}$ be Chebyshev interpolation points for E. Then prove that (4.5.8) holds. Consequently, a polynomial sequence $\{H_n(z)\}$ interpolating a regular function f(z) in E at Chebyshev interpolation points converges uniformly to f(z) on E.

5. For $f \in F_R$, given by the Jacobi series (4.5.41), introduce the concepts of (α,β)-L-order, generalized order etc. and characterize the various growth parameters thus introduced in terms of $(\|p_n\|_\alpha R^n)$.

(Juneja and Kapoor [2])

6. Let f be analytic in a domain D bounded by a simple closed contour C and let points $\alpha_1,\alpha_2,\ldots$ be in D. If $f \in H(D \cup C;R)$, characterize its various growth parameters in terms of Newton series coefficients of f.

(Juneja [5])

7. Let f be as in Exercise 6. Characterize various growth parameters of f in terms of $\| f-s_n\|$, where $s_n(z)$ denotes the nth partial sum of Newton's series of f.

(Juneja [6])

REFERENCES

Bergman [1],
Curtiss [1],
Goodman [1],
Juneja [3], [4], [5], [6],
Juneja and Kapoor [1], [2],
Juneja and Rizvi [1],
Kovari and Pommerenke [1],
Markushevich [1],
Nehari [1],
Rice [1],
Rizvi [1], [2], [3],
Rizvi and Juneja [1], [2],
Smirnov and Lebedev [1],
Suffridge [1],
Ullman [1],
Walsh [1],
Winiarski [1].

5 Analytic functions of several complex variables

5.1. INTRODUCTION

Let $G^n(\tilde{z}^o,\bar{R}) = \{\tilde{z} = (z_1,\ldots,z_n) : |z_i - z_i^o| < R_i,\ i=1,\ldots,n\}$ be a polydisc in $\mathbb{C}^n$ with centre $\tilde{z}^o = (z_1^o,\ldots,z_n^o)$ and polyradius $\bar{R} = (R_1,\ldots,R_n)$, where z_i^o are complex numbers and R_i are fixed real numbers for $i = 1,\ldots,n$. In this chapter we are concerned with the growth aspects of the functions $f:\mathbb{C}^n \to \mathbb{C}$ which are analytic in $G^n(\tilde{z}^o,\tilde{R})$. For the sake of simplicity, we assume throughout that $\tilde{z}^o = \tilde{0}$, $\bar{R} = \bar{1} \equiv (1,\ldots,1)$ and denote $G \equiv G^n(\tilde{0},\bar{1})$.

A function f, analytic in G, is uniquely expressed as an absolutely convergent power series (cf. Appendix A.16)

$$f(\tilde{z}) = \sum_{||\bar{k}||=0}^{\infty} b_{\bar{k}}\, \tilde{z}^{\bar{k}}, \quad \tilde{z} \in G, \tag{5.1.1}$$

where, $\bar{k} = (k_1,\ldots,k_n)$ is n-tuple of nonnegative integers, $\tilde{z}^{\bar{k}} = z_1^{k_1}\ldots z_n^{k_n}$, and $||\bar{k}|| = k_1+\ldots+k_n$. With $\bar{r} = (r_1,\ldots,r_n)$, $0 \le r_i < 1$, $i = 1,\ldots,n$, let $M(\bar{r}) \equiv M(\bar{r},f)$, defined as

$$M(\bar{r},f) = \max_{\substack{|z_i| \le r_i \\ i=1,\ldots,n}} |f(\tilde{z})| = \max_{\substack{|z_i| = r_i \\ i=1,\ldots,n}} |f(\tilde{z})|, \tag{5.1.2}$$

be the maximum of modulus of f on the closed polydisc $\bar{G}(\bar{r}) \equiv \bar{G}^n(\tilde{0},\bar{r})$. Since $|f(\tilde{z})| \le M(\bar{r})$ for all $\tilde{z} \in \bar{G}(\bar{r})$, the analogue of Cauchy inequality in one variable case for $f(\tilde{z})$ is given (cf. Appendix A.17) by

$$|b_{\bar{k}}| \le \frac{M(\bar{r})}{\bar{r}^{\bar{k}}} \tag{5.1.3}$$

where $\bar{r}^{\bar{k}} = r_1^{k_1} \ldots r_n^{k_n}$ and $0 \le r_i < 1$, $i = 1,\ldots,n$.

The growth of a function f, analytic in G, as determined by its maximum modulus function $M(\bar{r})$, can be studied in several different ways. Thus, to measure the growth of f with respect to all the variables simultaneously, we introduce the concepts of G-order and G-type in Section 5.2. For a comprehensive view of growth of f, the concepts of system of associated orders, system of associated types and their hypersurfaces are introduced in

Section 5.3. The growth of analytic functions f in G with respect to each of the variables separately is studied by defining partial orders of f in Section 5.4. The geometry of hypersurfaces of systems of associated orders and of associated types is investigated in Section 5.3. The relation between hypersurface of associated orders and G-order and that between system of associated orders and partial orders and between G-order and partial orders are found in Sections 5.3 and 5.4. Sections 5.2 and 5.3 also contain the complete characterization of the growth parameters in terms of the coefficients $b_{\overline{k}}$ in the power series expansion (5.1.1) of the function.

5.2. G-ORDER, G-TYPE AND COEFFICIENT CHARACTERIZATIONS

Let f be analytic in G and $M(\overline{r})$ be its maximum modulus function defined by (5.1.2). Set

$$M_G(t,f) = \max_{\overline{r} \in tG} M(\overline{r},f), \quad 0 < t < 1,$$

where

$$G \equiv |G| = \{(r_1,\dots,r_n) \in \mathbb{R}^n : 0 \leq r_i < 1,\ i=1,\dots,n\}.$$

We define the *G-order* ρ_G of f as

$$\rho_G = \limsup_{t \to 1} \frac{\log^+ \log^+ M_G(t,f)}{-\log (1-t)} . \tag{5.2.1}$$

If $0 < \rho_G < \infty$, the *G-type* T_G of f is defined as

$$T_G = \limsup_{t \to 1} \frac{\log^+ M_G(t,f)}{(1-t)^{-\rho_G}} . \tag{5.2.2}$$

For example, the function $\exp\{(1/(1-z_1))(1/(1-z_2))\}$, analytic in $G^2(\tilde{0},\overline{1})$, has G-order 2 and G-type 1.

The following theorem gives characterizations of G-order of a function f analytic in the polydisc G in terms of the coefficients $b_{\overline{k}}$ in its power series expansion (5.1.1).

Theorem 5.2.1. *Let* $f(z) = \sum_{\|\overline{k}\| = 0}^{\infty} b_{\overline{k}} \tilde{z}^{\overline{k}}$ *be analytic in the polydisc* G *and have* G*-order* ρ_G, $0 \leq \rho_G \leq \infty$. *Then*

$$\frac{\rho_G}{\rho_G+1} = \limsup_{\|\overline{k}\| \to \infty} \left\{ \frac{\log^+ \log^+ |b_{\overline{k}}|}{\log \|\overline{k}\|} \right\} \tag{5.2.3}$$

where $\|\overline{k}\| = k_1 + \ldots + k_n$ *and the left-hand side in* (5.2.2) *is interpreted as* 1 *if* $\rho_G = \infty$.

Proof. We consider the function $\phi(\tilde{z}, \omega)$ of (n+1) complex variables $z_1, \ldots, z_n, \omega$ defined by

$$\phi(\tilde{z}, \omega) = f(\omega \tilde{z}) = \sum_{\|\overline{k}\| = 0}^{\infty} b_{\overline{k}} \, \omega^{\|\overline{k}\|} \, \tilde{z}^{\overline{k}}$$

where $|\omega| < 1$. Set

$$P_m(\tilde{z}) = \sum_{\|\overline{k}\| = m} b_{\overline{k}} \tilde{z}^{\overline{k}} .$$

Then

$$\phi(\tilde{z}, \omega) = \sum_{m=0}^{\infty} P_m(\tilde{z}) \, \omega^m$$

is an analytic function of ω in U. Applying Cauchy inequalities for the coefficients of a power series in one variable, we get, for $0 < t < 1$ and $\tilde{z} \in G$,

$$|P_m(\tilde{z})| \leq \frac{\max_{|\omega|=t} |\phi(\tilde{z}, \omega)|}{t^m} . \tag{5.2.4}$$

Now, for $\overline{r} = (r_1, \ldots, r_n) \in G$,

$$\begin{aligned} \max_{|\omega|=t} |\phi(\tilde{z}, \omega)| &\leq \max_{|\omega|=t} \max_{\substack{|z_i|=r_i \\ 1 \leq i \leq n}} |f(\omega \tilde{z})| \\ &= \max_{\substack{|\zeta_i|=tr_i \\ 1 \leq i \leq n}} |f(\tilde{\zeta})| \\ &\leq \max_{\overline{r}^* \in tG} M(\overline{r}^*, f) = M_G(t, f) \end{aligned}$$

where $\tilde{\zeta} = (\zeta_1,\ldots,\zeta_n)$. Therefore (5.2.4) gives

$$|P_m(\tilde{z})| \le \frac{M_G(t,f)}{t^m}. \tag{5.2.5}$$

Since (5.2.5) holds for every $\tilde{z} \in G$, we get

$$\begin{aligned} M_G(1,P_m) &= \max_{\bar{r}\in G} M(\bar{r},P_m) \\ &= \max_{\bar{r}\in G} \max_{\substack{|z_i|=r_i \\ i=1,\ldots,n}} |P_m(\tilde{z})| \\ &= \frac{M_G(t,f)}{t^m}. \end{aligned}$$

Thus, for all t, $0 < t < 1$, and every positive integer m,

$$M_G(1,P_m) \le \frac{M_G(t,f)}{t^m}. \tag{5.2.6}$$

Now first let $\rho_G < \infty$. Then it follows that, for any $\varepsilon > 0$, there exists a constant $A \equiv A(\varepsilon)$ such that, for all t satisfying $0 < t < 1$,

$$\log^+ M_G(t,f) < (1-t)^{-\rho_G-\varepsilon} + \log A.$$

Therefore (5.2.6) gives, for $0 < t < 1$ and every positive integer m,

$$M_G(1,P_m) \le A \exp\,((1-t)^{-\rho_G-\varepsilon})t^{-m}. \tag{5.2.7}$$

Minimizing the right-hand side of inequality (5.2.7) gives

$$M_G(1,P_m) \le A \exp\,\{(1+\rho_G+\varepsilon)\left(\frac{m}{\rho_G+\varepsilon}\right)^{(\rho_G+\varepsilon)/(\rho_G+\varepsilon+1)}\}. \tag{5.2.8}$$

We now find an upper bound on the coefficients of the polynomial $P_m(z)$. For $||\bar{k}|| = m$ and any $\bar{r} \in G$,

$$|b_{\bar{k}}| \le \frac{M(\bar{r},P_m)}{\bar{r}^{\bar{k}}}.$$

Minimizing the right-hand side of this inequality for all $\bar{r} \in G$, it follows that, for any $\bar{k}$ with $||\bar{k}|| = m$,

$$|b_{\bar{k}}| \leq M_G(1,P_m).$$

Combining this inequality with (5.2.8), we have

$$|b_{\bar{k}}| \leq A \exp \left\{(1+\rho_G+\varepsilon)\left(\frac{\|\bar{k}\|}{\rho_G+\varepsilon}\right)^{\frac{\rho_G+\varepsilon}{\rho_G+\varepsilon+1}}\right\}$$

so that

$$\limsup_{\|\bar{k}\|\to\infty} \frac{\log^+\log^+|b_{\bar{k}}|}{\log \|\bar{k}\|} \leq \frac{\rho_G}{\rho_G+1}. \qquad (5.2.9)$$

If $\rho_G = \infty$, we proceed with an arbitrary large number in place of $\rho_G+\varepsilon$ and get 1 on the right-hand side of (5.2.9).

For the reverse inequality, we let

$$\limsup_{\|\bar{k}\|\to\infty} \frac{\log^+\log^+|b_{\bar{k}}|}{\log \|\bar{k}\|} = \mu \qquad (5.2.10)$$

so that we have to prove $\mu \geq \rho_G/(\rho_G+1)$. If $\mu = 1$, this is obvious. Let $\mu < 1$. Then it follows from (5.2.10) that for any, $0 < \varepsilon < 1-\mu$, there exists a nonnegative integer $m(\varepsilon)$ such that, for $\|\bar{k}\| \geq m(\varepsilon)$,

$$\frac{\log^+\log^+|b_{\bar{k}}|}{\log \|\bar{k}\|} \leq \mu + \varepsilon < 1.$$

Consequently,

$$|b_{\bar{k}}| \leq \exp\left(\|\bar{k}\|^{\mu+\varepsilon}\right).$$

This inequality implies that, for $0 < t < 1$,

$$M_G(t,f) \leq \max_{\bar{r}\in G} \sum_{\|\bar{k}\|=0}^{\infty} |b_{\bar{k}}| \, \bar{r}^{\bar{k}} \, t^{\|\bar{k}\|}$$

$$\leq \sum_{\|\bar{k}\|=0}^{\infty} |b_{\bar{k}}| \, t^{\|\bar{k}\|}$$

$$\leq \sum_{\|\bar{k}\|\leq m(\varepsilon)} t^{\|\bar{k}\|} |b_{\bar{k}}| + \sum_{\|\bar{k}\|>m(\varepsilon)} t^{\|\bar{k}\|} \exp(\|\bar{k}\|^{\mu+\varepsilon})$$

$$< c_1 t^{m(\varepsilon)} + c_2 + \Sigma_{m=0}^{\infty} t^m (1+m)^n \exp(m^{\mu+\varepsilon}) \tag{5.2.11}$$

where c_1 and c_2 are constants. Now the function

$$F(z) = \Sigma_{m=0}^{\infty} (1+m)^n \exp(m^{\mu+\varepsilon}) z^m$$

is analytic in U. By Theorem 2.2.1, the order of F is $(\mu+\varepsilon)/(1-\mu-\varepsilon)$. Therefore, by the definition of order in one variable case, for any $\varepsilon' > 0$ and t sufficiently close to 1,

$$M(t,F) < \exp \{(1-t)^{-(\frac{\mu+\varepsilon}{1-\mu-\varepsilon} + \varepsilon')}\}.$$

But, by (5.2.11),

$$M_G(t,f) < c_1 t^{m(\varepsilon)} + c_2 + M(t,F),$$

so that we have, for t sufficiently close to 1,

$$M_G(t,f) < c_1 t^{m(\varepsilon)} + c_2 + \exp \{(1-t)^{-(\frac{\mu+\varepsilon}{1-\mu-\varepsilon} + \varepsilon')}\}.$$

Since ε and ε' are arbitrary, the above inequality gives that

$$\rho_G \leq \frac{\mu}{1-\mu}$$

so that

$$\frac{\rho_G}{1+\rho_G} \leq \mu.$$

The last inequality together with (5.2.9) gives (5.2.3). □

Our next result is the characterization of G-type of a function f, analytic in G, in terms of the coefficients in its power series (5.1.1).

Theorem 5.2.2. *Let* $f(\tilde{z}) = \sum_{\|\bar{k}\|=0}^{\infty} b_{\bar{k}} \tilde{z}^{\bar{k}}$ *be analytic in the polydisc* G, *have* G-*order* ρ_G $(0 < \rho_G < \infty)$ *and* G-*type* T_G $(0 \leq T_G \leq \infty)$. *Then*

$$\frac{(\rho_G+1)^{\rho_G+1}}{\rho_G^{\rho_G}} T_G = \limsup_{\|\bar{k}\|\to\infty} \left\{ \frac{(\log^+|b_{\bar{k}}|)^{\rho_G+1}}{\|\bar{k}\|^{\rho_G}} \right\}. \tag{5.2.12}$$

Proof. The techniques employed in the proof of this theorem are essentially the same as those in Theorem 5.2.1. In place of inequalities (5.2.7) and (5.2.8), we now have the inequalities

$$M_G(1,P_m) \leq A \exp\{(T_G+\varepsilon)\ (1-t)^{-\rho_G}\}\ t^{-m}$$

and

$$M_G(1,P_m) \leq A \exp\{(T_G+\varepsilon)^{1/(\rho_G+\varepsilon)}\ \frac{(\rho_G+1)}{\rho_G}\ m^{\rho_G/(\rho_G+1)}\}.$$

Consequently, with $\|\bar{k}\| = m$,

$$|b_{\bar{k}}| \leq M_G(1,P_m)$$

$$\leq A \exp\{(T_G+\varepsilon)^{1/(\rho_G+\varepsilon)}\ \frac{(\rho_G+1)}{\rho_G}\ \|\bar{k}\|^{\rho_G/(\rho_G+1)}\}$$

which yields

$$\frac{(\rho_G+1)^{\rho_G+1}}{\rho_G^{\rho_G}} T_G \leq \limsup_{\|\bar{k}\|\to\infty} \left\{ \frac{(\log^+|b_{\bar{k}}|)^{\rho_G+1}}{\|\bar{k}\|^{\rho_G}} \right\}.$$

Similarly, to prove the reverse inequality, we estimate $M_G(t,f)$ in terms of the number δ defined by the right-hand side of (5.2.12) to get an inequality analogous to (5.2.11). By considering

$$F^*(z) = \Sigma_{m=0}^{\infty} \exp\{(\delta+\varepsilon)^{1/(\rho_G+1)}\ m^{\rho_G/(\rho_G+1)}\}\ z^m$$

instead of F(z) in the proof of Theorem 5.2.1 and applying Theorem 2.2.2 instead of Theorem 2.2.1, we get the desired inequality. □

EXERCISES 5.2

1. Let, for $\alpha_i > 0$, $i = 1,\ldots,n$

$$F_1(\tilde{z}) = \exp\,(\Sigma_{i=1}^{n}\ (1-z_i)^{-\alpha_i})$$

$$F_2(\tilde{z}) = \exp\,(\Pi_{i=1}^{n}\ (1-z_i)^{-1}).$$

Prove that the G-orders of the functions F_1 and F_2 are respectively $\max\,(\alpha_1,\ldots,\alpha_n)$ and $1/n$.

2. Find the G-order of the function

$$F_3(\tilde{z}) = \exp\,(\Pi_{i=1}^{n}\ (1-z_i)^{-\alpha_i})$$

where $\alpha_i > 0$, $i = 1,\ldots,n$.

3. Prove that the function

$$F_4(\tilde{z}) = \sum_{\|\bar{k}\|=0}^{\infty} \exp\,(\beta(k_1+\ldots+k_n)^{\alpha})\tilde{z}^{\bar{k}},$$

where $0 < \alpha < 1$ and $0 < \beta < \infty$, has G-order $\alpha/(1-\alpha)$ and G-type $(1-\alpha)\alpha^{\alpha/(1-\alpha)}\beta^{1/(1-\alpha)}$.

4. If f_1 and f_2 are analytic in G, find relations of G-orders of the functions f_1+f_2 and f_1f_2 with those of f_1 and f_2.

For a function g, analytic in $G^n(0,\bar{R})$, set

$$M(\bar{r},g) = \max_{\substack{|\zeta_i|=r_i \\ i=1,\ldots,n}} |g(\tilde{z})|$$

and

$$M_{G_{\bar{R}}}(t,g) = \max_{\bar{r}\in tG_{\bar{R}}} M(\bar{r},g)$$

where $0 \leq r_i < R$ for $i=1,\ldots,n$ and $0 < t < 1$. The $G_{\bar{R}}$*-order* $\rho_{G,\bar{R}}$ and $G_{\bar{R}}$*-type* $T_{G,\bar{R}}$ of g are defined as

$$\rho_{G,\bar{R}} = \limsup_{t\to 1} \frac{\log^+\log^+ M_{G_{\bar{R}}}(t,g)}{-\log\,(1-t)}$$

and, for $0 < \rho_{G,\overline{R}} < \infty$,

$$T_{G,\overline{R}} = \limsup_{t \to 1} \frac{\log^+ M_{G_{\overline{R}}}(t,g)}{(1-t)^{-\rho_{G,\overline{R}}}} .$$

5. Prove that $G_{\overline{R}}$-order $\rho_{G,\overline{R}}$ and $G_{\overline{R}}$-type $T_{G,\overline{R}}$ of a function

$g(\tilde{z}) = \sum_{\|\overline{k}\|=0}^{\infty} b_{\overline{k}} \tilde{z}^{\overline{k}}$, analytic in $G^n(0,\overline{R})$, are equal to G-order and G-type of the function $f(\tilde{z}) = g(\overline{R}\tilde{z})$, where $\overline{R}\tilde{z} = (R_1 z_1, \ldots, R_n z_n)$. Hence, using Theorems 5.2.1 and 5.2.2, prove that

$$\rho_{G,\overline{R}} = \limsup_{\|\overline{k}\| \to \infty} \left\{ \frac{\log^+ \log^+ (|b_{\overline{k}}| \overline{R}^{\overline{k}})}{\log \|\overline{k}\|} \right\}$$

and

$$\frac{(\rho_{G,\overline{R}}+1)^{\rho_{G,\overline{R}}+1}}{(\rho_{G,\overline{R}})^{\rho_{G,\overline{R}}}} T_{G,\overline{R}} = \limsup_{\|\overline{k}\| \to \infty} \left\{ \frac{(\log^+ |b_{\overline{k}}| \overline{R}^{\overline{k}})^{\rho_{G,\overline{R}}+1}}{\|\overline{k}\|^{\rho_{G,\overline{R}}}} \right\} .$$

A function $\rho_G(r)$ is said to be a *proximate order of a function* f, analytic in G and having nonzero finite G-order ρ_G, if $\rho_G(r)$ satisfies the conditions (1.6.1)-(1.6.3) and T_G^*, defined as

$$T_G^* = \limsup_{t \to 1} \frac{\log^+ M_G(t,f)}{(1-t)^{-\rho_G(t)}} ,$$

is nonzero finite. The growth parameter T_G^* is called *G-type of f with respect to the proximate order* $\rho_G(r)$.

6. Let $f(\tilde{z}) = \sum_{\|\overline{k}\|=0}^{\infty} b_{\overline{k}} \tilde{z}^{\overline{k}}$ be analytic in G. By analogy with Theorem 2.6.1, find a characterization of G-type T_G^* of f, with respect to a proximate order $\rho_G(r)$, in terms of the coefficients $b_{\overline{k}}$.

(Juneja and Kapoor [7])

For a function f, analytic in G, define the *lower G-order* λ_G and *lower G-type* t_G of f as

$$\lambda_G = \liminf_{t \to 1} \frac{\log^+ \log^+ M_G(t,f)}{-\log(1-t)}$$

and, for $0 < \rho_G < \infty$,

$$t_G = \liminf_{t \to 1} \frac{\log^+ M_G(t,f)}{(1-t)^{-\rho_G}} .$$

*7. If $f(\tilde{z}) = \sum_{\|\bar{k}\|=0}^{\infty} b_{\bar{k}} \tilde{z}^{\bar{k}}$ is analytic in G and has lower G-order λ_G and lower G-type t_G, find the coefficient characterizations of λ_G and t_G analogous to those found in Theorems 2.3.1 to 2.3.10 for one variable case.

For a function f, analytic in G, define

$$\rho_G(q) = \limsup_{t \to 1} \frac{\log^{[q]} M_G(t,f)}{-\log(1-t)}$$

and, for $0 < \rho_G(q) < \infty$,

$$T_G(q) = \limsup_{t \to 1} \frac{\log^{[q-1]} M_G(t,f)}{(1-t)^{-\rho_G(q)}}$$

where $q = 2,3,\ldots$ is such that $\rho_G(q-1) = \infty$ and $\rho_G(q) < \infty$.

8. Let $f(\tilde{z}) = \sum_{\|\bar{k}\|=0}^{\infty} b_{\bar{k}} \tilde{z}^{\bar{k}}$ be analytic in G. Determine $\rho_G(q)$ and $T_G(q)$ in terms of the coefficients $b_{\bar{k}}$.

(Juneja and Kapoor [5])

A function f, analytic in G, is said to have logarithmic G-order ρ_G^o and logarithmic G-type T_G^o if

$$\rho_G^o = \limsup_{t \to 1} \frac{\log^+ \log^+ M_G(t,f)}{\log \log 1/(1-t)}$$

and, for $1 < \rho_G^o < \infty$,

$$T_G^o = \limsup_{t \to 1} \frac{\log^+ M_G(t,f)}{(\log 1/(1-t))^{\rho_G^o}} .$$

9. Let $f(\tilde{z}) = \sum_{\|\bar{k}\|=0}^{\infty} b_{\bar{k}} \tilde{z}^{\bar{k}}$ be analytic in G. Find the characterization of the logarithmic G-order ρ_G^o and logarithmic G-type T_G^o of f in terms of the coefficients $b_{\bar{k}}$.

(Juneja and Kapoor [6])

5.3. SYSTEMS OF ASSOCIATED ORDERS, ASSOCIATED TYPES AND COEFFICIENT CHARACTERIZATIONS.

Let

$$\mathbb{R}_+^n = \{(r_1,\ldots,r_n) \in \mathbb{R}^n : \ r_i \geq 0, \ \ i = 1,\ldots,n\}$$

$$I^n = \{(r_1,\ldots,r_n) \in \mathbb{R}^n : \ 0 \leq r_i < 1, \ \ i = 1,\ldots,n\}$$

and, for $\bar{r} = (r_1,\ldots,r_n) \in \mathbb{R}_+^n$, denote $|\bar{r}| = \max(r_1,\ldots,r_n)$. For a function f, analytic in G, let $B_{\bar{\rho}} = B_{\bar{\rho}}(f)$ be the set of all points $\bar{a} \in \mathbb{R}_+^n$ such that, for $|\bar{r}| \to 1$,

$$\log M(\bar{r},f) < (1-r_1)^{-a_1} + (1-r_2)^{-a_2} + \ldots + (1-r_n)^{-a_n} \qquad (5.3.1)$$

where $M(\bar{r},f)$ is the maximum modulus function of f defined by (5.1.2). It is obvious that if $\bar{a}' \in B_{\bar{\rho}}$, the set $B_{\bar{\rho}}$ contains the entire hyperoctant $\{\bar{a} \in \mathbb{R}^n : a_i > a_i', \ i = 1,\ldots,n\}$. The sets in $\mathbb{R}^n$ possessing this property are called as *octant-like*.

It is also clear that if $\bar{a}' \notin B_\rho$, then any point $\bar{a} \in \mathbb{R}_+^n$ with $a_i \leq a_i'$ is also not in $B_{\bar{\rho}}$. Thus the boundary points of the set $B_{\bar{\rho}}(f)$ form a certain hypersurface $S_{\bar{\rho}} = S_{\bar{\rho}}(f)$. This hypersurface divides the hyperoctant $\mathbb{R}_+^n$ into two parts, in one of which inequality (5.3.1) holds while in the other it is false. The hypersurface $S_{\bar{\rho}}$ characterizes the growth of the function $M(\bar{r},f)$ and is called the hypersurface of associated orders of f. Any system of numbers $\rho_1,\ldots,\rho_n$ such that $(\rho_1,\ldots,\rho_n) \in S_{\bar{\rho}}$ is called a system of associated orders of f. If $\rho_i > 0$, $i = 1,\ldots,n$, then $\rho_1,\ldots,\rho_n$ form a system of associated orders of f if and only if

$$\limsup_{|\bar{r}|\to 1} \frac{\log^+\log^+ M(\bar{r},f)}{\log\{(1-r_1)^{-\rho_1}+\ldots+(1-r_n)^{-\rho_n}\}} = 1.$$

For example, the point (2,2) is on the hypersurface of associated orders of the function $\exp\left(\left(\frac{1}{1-z_1}\right)\left(\frac{1}{1-z_2}\right)\right)$.

Similarly, if $\rho_1,\ldots,\rho_n$, $0 < \rho_i < \infty$, $i = 1,\ldots,n$, is a system of associated orders of f, we consider the set $B_{\bar{T}} \equiv B_{\bar{T}}(f,\bar{\rho})$ consisting of all $\bar{b} \in \mathbb{R}^n_+$ such that, for $|\bar{r}| \to 1^-$,

$$\log M(\bar{r},f) < b_1(1-r_1)^{-\rho_1}+\ldots+b_n(1-r_n)^{-\rho_n}. \tag{5.3.2}$$

Like the set $B_{\bar{\rho}}$, the set $B_{\bar{T}}$ is also octant-like and its boundary $S_{\bar{T}} \equiv S_{\bar{T}}(f,\bar{\rho})$ is a hypersurface. We call $S_{\bar{T}}$ the hypersurface of associated types for associated orders $\rho_1,\ldots,\rho_n$. If $(T_1,\ldots,T_n) \in S_{\bar{T}}$, the numbers $T_1,\ldots,T_n$ are called a system of associated types with respect to the associated orders $(\rho_1,\ldots,\rho_n)$. As in the case of associated orders, it is clear that the positive numbers $T_1,\ldots,T_n$ form a system of associated types with respect to associated orders $\rho_1,\ldots,\rho_n$ if and only if

$$\limsup_{|\bar{r}|\to 1} \frac{\log^+ M(\bar{r},f)}{T_1(1-r_1)^{-\rho_1}+\ldots+T_n(1-r_n)^{-\rho_n}} = 1.$$

A simple relationship exists between the hypersurface of associated orders and the G-order of a function analytic in G. In fact, G-order of f is equal to the value of the parameter t for which the hypersurface $S_{\bar{\rho}}$ of f intersects the ray $\{\bar{a}:\bar{a} \in \mathbb{R}^n_+,\ a_i = t,\ i = 1,\ldots,n,\ 0 \le t < \infty\}$. To see this, let $(t,t,\ldots,t) \in S_{\bar{\rho}}(f)$. Then, for any $\varepsilon > 0$ and $|\bar{r}|$ sufficiently close to 1,

$$\log^+ M(\bar{r},f) < (1-r_1)^{-(t+\varepsilon)}+\ldots+(1-r_n)^{-(t+\varepsilon)}$$

and so, for $\bar{r}^* = (r,r,\ldots,r)$ such that r is sufficiently close to 1,

$$\log^+ M(\bar{r}^*,f) < n(1-r)^{-(t+\varepsilon)} < (1-r)^{-(t+2\varepsilon)}. \tag{5.3.3}$$

On the other hand, there exists a sequence $\bar{r}_n^* = (r_n,\ldots,r_n)$ such that $|\bar{r}_n^*| \to 1$ as $n \to \infty$ and

$$\log^+ M(\bar{r}_n^*,f) > (1-r_n)^{-(t-\varepsilon)}, \tag{5.3.4}$$

for otherwise we have, for all $\bar{r}^*$ with $|\bar{r}^*|$ sufficiently close to 1,

$$\log^+ M(\bar{r}^*,f) \leq (1-r)^{-(t-\varepsilon)} \tag{5.3.5}$$

which implies that, for $\bar{r} = (r_1,\ldots,r_n)$ with $|\bar{r}|$ sufficiently close to 1,

$$\begin{aligned}\log^+ M(\bar{r},f) &\leq \log M(\bar{r}^{**},f)\\ &\leq \max\{(1-r_1)^{-(t-\varepsilon)},\ldots,(1-r_n)^{-(t-\varepsilon)}\}\\ &< (1-r_1)^{-(t-\varepsilon)}+\ldots+(1-r_n)^{-(t-\varepsilon)}\end{aligned}$$

where $\bar{r}^{**} = (|\bar{r}|,\ldots,|\bar{r}|)$. But the above inequality contradicts the fact that $(t,\ldots,t) \in S_{\bar{\rho}}(f)$. Since $M(\bar{r}^*,f) = M_G(r,f)$, it follows from (5.3.3) and (5.3.4) that $\rho_G = t$. □

Our following theorem gives a characterization of the system of associated orders of a function f, analytic in G, in terms of the coefficients in its power series (5.1.1).

Theorem 5.3.1. *The positive numbers* $\rho_1,\ldots,\rho_n$ *form a system of associated orders of a function* $f(\tilde{z}) = \sum_{\|\bar{k}\|=0}^{\infty} b_{\bar{k}} \tilde{z}^{\bar{k}}$, *analytic in the polydisc* G, *if and only if*

$$\limsup_{\|\bar{k}\|\to\infty} \left\{\frac{\log^+\log^+ |b_{\bar{k}}|}{\log\{\Sigma_{i=1}^n k_i^{\rho_i/(\rho_i+1)}\}}\right\} = 1. \tag{5.3.6}$$

We need the following lemma:

Lemma 5.3.1. *Let* $f(\tilde{z}) = \sum_{\|\bar{k}\|=0}^{\infty} b_{\bar{k}} \tilde{z}^{\bar{k}}$ *be analytic in the polydisc* G. *Then a point* $(c_1,\ldots,c_n)$ *is in the interior* $B_{\bar{\rho}}^o$ *of the set* $B_{\bar{\rho}}$, *defined by* (5.3.1), *if and only if*

$$\limsup_{\|\bar{k}\| \to \infty} \left\{ \frac{\log^+ \log^+ |b_{\bar{k}}|}{\log \{\Sigma_{i=1}^n k_i^{c_i/(c_i+1)}\}} \right\} < 1. \tag{5.3.7}$$

Proof. Let $(c_1, \ldots, c_n) \in B^o_{\bar{\rho}}$. Then, there exist an $\varepsilon > 0$ and a constant $A(\varepsilon)$ such that, for all $\bar{r} = (r_1, \ldots, r_n) \in I^n_{+}$,

$$M(\vec{r}, f) \leq A(\varepsilon) \exp \{\Sigma_{i=1}^n (1-r_i)^{-(c_i-\varepsilon)}\}$$

so that, on using Cauchy's inequality (cf. (5.1.3)), it follows that, for any $\bar{r} \in I^n_{+}$,

$$|b_{\bar{k}}| \leq \frac{A(\varepsilon) \exp \{\Sigma_{i=1}^n (1-r_i)^{-(c_i-\varepsilon)}\}}{r_1^{k_1} \ldots r_n^{k_n}}. \tag{5.3.8}$$

Now, for $0 < \alpha,\ \beta < \infty$,

$$\min_{0<t<1} \frac{\exp(1-t)^{-\alpha}}{t^\beta} = \frac{\exp\{(\beta/\alpha)^{\alpha/(\alpha+1)}\}}{\{1 - (\alpha/\beta)^{1/(\alpha+1)}\}^\beta},$$

therefore inequality (5.3.8) gives that

$$|b_{\bar{k}}| \leq A(\varepsilon) \Pi_{i=1}^n \frac{\exp \{(k_i/(c_i-\varepsilon))^{(c_i-\varepsilon)/(c_i-\varepsilon+1)}\}}{\{1-((c_i-\varepsilon)/k_i)^{1/(c_i-\varepsilon+1)}\}^{k_i}}.$$

It is easily seen from the above estimate of $|b_{\bar{k}}|$ that, for $|\bar{k}|$ sufficiently large, we have

$$\log|b_{\bar{k}}| \leq \Sigma_{i=1}^n \{(\frac{k_i}{c_i-\varepsilon})^{(c_i-\varepsilon)/(c_i-\varepsilon+1)}$$

$$- k_i \log \{1-(\frac{c_i-\varepsilon}{k_i})^{1/(c_i-\varepsilon+1)}\}\} + o(1)$$

$$\simeq \Sigma_{i=1}^{n} \{(\frac{k_i}{c_i-\varepsilon})^{(c_i-\varepsilon)/(c_i-\varepsilon+1)}$$

$$+ (c_i-\varepsilon)^{1/(c_i-\varepsilon+1)} k_i^{(c_i-\varepsilon)/(c_i-\varepsilon+1)}\} + o(1)$$

$$= \{\Sigma_{i=1}^{n} (1+c_i-\varepsilon)(\frac{k_i}{c_i-\varepsilon})^{(c_i-\varepsilon)/(c_i-\varepsilon+1)}\} + o(1).$$

The above inequality clearly gives that

$$\limsup_{\|\bar{k}\| \to \infty} \frac{\log^+\log^+|b_{\bar{k}}|}{\log \{\Sigma_{i=1}^{n} (1+c_i-\varepsilon)(k_i/(c_i-\varepsilon))^{(c_i-\varepsilon)/(c_i-\varepsilon+1)}\}} \le 1$$

and so we have

$$\limsup_{\|\bar{k}\| \to \infty} \frac{\log^+\log^+|b_{\bar{k}}|}{\log \{\Sigma_{i=1}^{n} (1+c_i)(k_i / c_i)^{c_i/(c_i+1)}\}} < 1. \tag{5.3.9}$$

Since, as $\|\bar{k}\| \to \infty$,

$$\log \{\Sigma_{i=1}^{n} (1+c_i)(k_i / c_i)^{c_i/(c_i+1)}\} \sim \log \{\Sigma_{i=1}^{n} k_i^{c_i/(c_i+1)}\}, \tag{5.3.10}$$

inequality (5.3.9) implies (5.3.7) and this proves the necessity part of the lemma.

To prove the sufficiency part, assume that the positive numbers $c_1,\ldots,c_n$ satisfy relation (5.3.7) so that

$$\tau \equiv \limsup_{\|\bar{k}\| \to \infty} \frac{\log^+\log^+|b_{\bar{k}}|}{\log \{\Sigma_{i=1}^{n} k_i^{c_i/(c_i+1)}\}} < 1.$$

Then, for any $\varepsilon > 0$ such that $\tau(1+\varepsilon) < 1$, there exists a number $m(\varepsilon)$ such that, for $\|\bar{k}\| > m(\varepsilon)$,

$$|b_{\bar{k}}| \le \exp \{\Sigma_{i=1}^{n} k_i^{c_i/(c_i+1)}\}^{\tau(1+\varepsilon)}$$

$$\le \exp \{A\, \Sigma_{i=1}^{n} k_i^{c_i\tau(1+\varepsilon)/(c_i+1)}\}$$

where A is a positive constant. This estimate of the coefficients $b_{\bar{k}}$ implies that

$$M(\bar{r},f) \leq \sum_{\|\bar{k}\| \leq m(\varepsilon)} b_{\bar{k}} \bar{r}^{\bar{k}}$$

$$+ \sum_{\|\bar{k}\|=0}^{\infty} \exp\{A \Sigma_{i=1}^{n} k_i^{c_i\tau(1+\varepsilon)/(c_i+1)}\} r_1^{k_1} \cdots r_n^{k_n}$$

$$\leq \sum_{\|\bar{k}\| \leq m(\varepsilon)} b_{\bar{k}} \bar{r}^{\bar{k}} + \Pi_{i=1}^{n} \Sigma_{k_i=0}^{\infty} \exp\{Ak_i^{c_i\tau(1+\varepsilon)/(c_i+1)}\} r_i^{k_i} \quad (5.3.11)$$

Since, for each i=1,...,n the function

$$f_i(z_i) = \Sigma_{k_i=0}^{\infty} \exp\{Ak_i^{c_i\tau(1+\varepsilon)/(c_i+1)}\} z_i^{k_i}$$

is analytic in $|z_i| < 1$ and, by Theorem 2.2.1, has order less than $c_i\tau(1+\varepsilon)$, it follows that, for $|z_i| = r_i$ sufficiently close to 1,

$$M(r_i,f_i) = \Sigma_{k_i=0}^{\infty} \exp\{Ak_i^{c_i\tau(1+\varepsilon)/(c_i+1)}\} r_i^{k_i}$$

$$\leq B_i \exp\{(1-r_i)^{-c_i\tau(1+2\varepsilon)}\}$$

where B_i are positive constants. Thus it follows from (5.3.11) that

$$M(\bar{r},f) \leq \sum_{\|\bar{k}\| \leq m(\varepsilon)} |b_{\bar{k}}| \bar{r}^{\bar{k}} + D \exp\{\Sigma_{i=1}^{n} (1-r_i)^{-c_i\tau(1+2\varepsilon)}\}$$

where D is a positive constant. The above inequality implies that the point $\bar{c}^*$ with coordinates $c_i^* = c_i(1+3\varepsilon)\tau$ belongs to the set $B_{\bar{\rho}}$ for any $\varepsilon > 0$. Now, choosing ε so small that $(1+3\varepsilon)\tau < 1$, we see that the point $\bar{c} = (c_1,\ldots,c_n) \in B^{o}_{\bar{\rho}}$, in view of the property of $B_{\bar{\rho}}$ that, if $\bar{c}^* \in B_{\bar{\rho}}$, then the entire hyperoctant $\{c^{**} = (c_1^{**},\ldots,c_n^{**}): c_i^{**} > c_i^*, i = 1,\ldots,n\}$ is also in $B_{\bar{\rho}}$. □

Proof of Theorem 5.3.1. Let positive numbers $\rho_1,\ldots,\rho_n$ satisfy (5.3.6). By Lemma 5.3.1, $(\rho_1,\ldots,\rho_n) \notin B^o_{\bar\rho}$. However, it is obvious that this point is the limit of points $(\rho_1 t,\ldots,\rho_n t) \in B^o_{\bar\rho}$ as $t \to 1^+$. Thus $(\rho_1,\ldots,\rho_n) \in S_{\bar\rho}$.

Conversely, let $(\rho_1,\ldots,\rho_n) \in S_{\bar\rho}$ and $\rho_i > 0$, $i = 1,\ldots,n$. Then, using Lemma 5.3.1 once again, it follows that

$$\limsup_{\|\bar k\|\to\infty} \frac{\log^+\log^+|b_{\bar k}|}{\log\{\Sigma_{i=1}^n k_i^{\rho_i/(\rho_i+1)}\}} \geq 1. \tag{5.3.12}$$

Since the closure of the set $B_{\bar\rho}$ is octantlike, for $t > 1$, the points $(\rho_1 t,\ldots,\rho_n t)$ are in $B^o_{\bar\rho}$. Therefore, for all $t > 1$,

$$\limsup_{\|\bar k\|\to\infty} \frac{\log^+\log^+|b_{\bar k}|}{\log\{\Sigma_{i=1}^n k_i^{\rho_i t/(\rho_i t+1)}\}} < 1.$$

The above inequality gives that, for all $t > 1$,

$$\limsup_{\|\bar k\|\to\infty} \frac{\log^+\log^+|b_{\bar k}|}{\log\{\Sigma_{i=1}^n k_i^{\rho_i/(\rho_i+1)}\}} < t$$

and so

$$\limsup_{\|\bar k\|\to\infty} \frac{\log^+\log^+|b_{\bar k}|}{\log\{\Sigma_{i=1}^n k_i^{\rho_i/(\rho_i+1)}\}} \leq 1.$$

This inequality together with (5.3.12) implies (5.3.6). □

Proceeding on the lines of the proof of Theorem 5.3.1, for a function f analytic in *G*, we can also characterize the system of associated types with respect to associated orders in terms of the coefficients $b_{\bar k}$ in the power series (5.1.1) of f. Thus we have

Theorem 5.3.2. *The positive numbers* $T_1,\ldots,T_n$ *form a system of associated types with respect to associated orders* $\rho_1,\ldots,\rho_n$ *of a function*

$f(\tilde{z}) = \sum_{\|\overline{k}\|=0}^{\infty} b_{\overline{k}} \tilde{z}^{\overline{k}}$, *analytic in the polydisc G, if and only if*

$$\limsup_{\|\overline{k}\|\to\infty} \frac{\log^{+} |b_{\overline{k}}|}{\Sigma_{i=1}^{n} (1+\rho_i)b_i^{1/(1+\rho_i)} (k_i / \rho_i)^{\rho_i/(1+\rho_i)}} = 1. \tag{5.3.13}$$

The proof of Theorem 5.3.2 differs from that of Theorem 5.3.1 essentially in details and is left as an exercise for the reader.

The hypersurfaces of associated orders and associated types have interesting geometrical properties for a class of analytic functions. For the study of this aspect of hypersurfaces, we first have some definitions.

A set $E \subset \mathbb{R}_+^n$ is called a *complete domain* in $\mathbb{R}_+^n$ if, together with $\overline{r}^o = (r_1^o,\dots,r_n^o) \in E$, it contains all points $\overline{r} = (r_1,\dots,r_n)$ with $0 \leq r_i \leq r_i^o$ and it contains no point of the closure of its complement in $\mathbb{R}_+^n$. A set $E \subset \mathbb{R}_+^n$ is said to be *convex* if, for every pair of points $(a_1,\dots,a_n)$, $(b_1,\dots,b_n)$ in E, the entire segment $(\lambda a_1+\mu b_1,\dots, (\lambda a_n+ \mu b_n)$ is also in E, where $0 \leq \lambda, \mu \leq 1$ and $\lambda+\mu = 1$. A function $\phi(\overline{r}) = \phi(r_1,\dots,r_n)$ defined on I_+^n is said to be *pluriconvex* in $\log(1/(1-r_1))\dots,\log(1/(1-r_n))$ if $\phi(\overline{r})$ is a convex function of the variables $\log(1/(1-r_1)),\dots,\log(1/(1-r_n))$. Then for every $\overline{t} = (t_1,\dots,t_n)$ and $\overline{s} = (s_1,\dots,s_n)$ in I_+^n and for all λ,μ with $0 \leq \lambda, \mu \leq 1$ and $\lambda+\mu = 1$,

$$\phi(1-(1-r_1)^{\lambda}(1-s_1)^{\mu},\dots,1-(1-t_n)^{\lambda}(1-s_n)^{\mu}) \leq \lambda\phi(t_1,\dots,t_n) + \mu\phi(s_1,\dots,s_n). \tag{5.3.14}$$

We consider the class C(G) consisting of all functions f, analytic in G, for which the function $\log M(\overline{r},f)$ is a pluriconvex function in $\log(1/(1-r_1)),\dots,\log(1/(1-r_n))$, where $M(\overline{r},f)$ is defined by (5.1.2). For example, the function $\exp((1/(1-z_1))(1/(1-z_2)))$ is easily seen to be in C(G). Our next result gives a geometrical property of the hypersurface of associated orders for a function f in $C(G^2(0,1))$.

Theorem 5.3.3. *Let* $f \in C(G)$ *and* $B_{\overline{\rho}}^o$ *denote the interior of the set* $B_{\overline{\rho}} \equiv B_{\overline{\rho}}(f)$, *defined by* (5.3.1). *Then the image* $B_{\overline{\rho}}^{-1}$ *of* $B_{\overline{\rho}}^o$ *under the transformation* $a' = 1/a$ *is a complete convex domain.*

Proof. $B^{-1}_{\bar{\rho}}$ is obviously a complete domain. To prove that it is convex, let $\bar{a} = (a_1,\ldots,a_n)$ and $\bar{b} = (b_1,\ldots,b_n)$ be in $B^o_{\bar{\rho}}$. Set, for $i=1,\ldots,n$,

$$1-t_i=(1-r_i)^{a_i/(\lambda a_i + \mu b_i)}, \quad 1-s_i=(1-r_i)^{b_i/(\lambda a_i+ \mu b_i)}.$$

Then $r_i = 1-(1-t_i)^\lambda(1-s_i)^\mu$, $i = 1,\ldots,n$. Therefore, since $f \in C(G)$, we have that for $|\bar{r}|$ sufficiently close to 1,

$$\begin{aligned}\log M(\bar{r},f) &\equiv \log M(r_1,\ldots,r_n)\\ &= \log M(1-(1-t_1)^\lambda(1-s_1)^\mu,\ldots,1-(1-t_n)^\lambda(1-s_n)^\mu)\\ &\le \lambda \log M(t_1,\ldots,t_n) + \mu \log M(s_1,\ldots,s_n)\\ &\le \lambda \Sigma^n_{i=1} (1-t_i)^{-b_i} + \mu \Sigma^n_{i=1} (1-s_i)^{-a_i}\\ &= \Sigma^n_{i=1} \left(\frac{1}{1-r_i}\right)^{1/(\frac{\lambda}{b_i} + \frac{\mu}{a_i})}.\end{aligned}$$

Thus, for any λ and μ with $\lambda,\mu \ge 0$ and $\lambda+\mu = 1$, the point $((\lambda/b_1) + (\mu/a_1)),\ldots,((\lambda/b_n) + (\mu/a_n))$ is in $B^{-1}_{\bar{\rho}}$ and so $B^{-1}_{\bar{\rho}}$ is convex. □

The geometrical properties of hypersurfaces of associated types can also be found by analogy and are given by the following theorem.

Theorem 5.3.4. *Let* $f \in C(G)$ *and* $B^o_{\bar{T}}$ *denote the interior of the set* $B_{\bar{T}} \equiv B_{\bar{T}}(f,\bar{\rho})$, *defined by* (5.3.2). *Then the image* $\log B_{\bar{T}}$ *of* $B^o_{\bar{T}}$ *under the transformation* $b'_i = \log b_i$ *is octant-like and convex.*

Proof. It is clear that $\log B_{\bar{T}}$ is octant-like. To prove convexity, let the points $\bar{b}$ and $\bar{b}'$ be arbitrarily chosen in $B^o_{\bar{T}}$. Set, for $i = 1,\ldots,n$,

$$(1-t_i) = (1-r_i)\left(\frac{b'_i}{b_i}\right)^{\mu/\rho_i}; \quad (1-s_i) = (1-r_i)\left(\frac{b_i}{b'_i}\right)^{\lambda/\rho_i}$$

where $0 \le \lambda$, $\mu \le 1$ and $\lambda+\mu = 1$. Then $r_i = 1-(1-t_i)^\lambda(1-s_i)^\mu$, $i = 1,\ldots,n$.

Using the definition of the set $B_{\bar{T}}$ and the fact that $f \in C(G)$, we get that, for $|\bar{r}|$ sufficiently close to 1,

$$\log M(\overline{r},f) \equiv \log M(r_1,\ldots,r_n)$$

$$= \log M(1-(1-t_1)^{\lambda}(1-s_1)^{\mu},\ldots, 1-(1-t_n)^{\lambda}(1-s_n)^{\mu})$$

$$\leq \lambda \log M(t_1,\ldots,t_n) + \mu \log M(s_1,\ldots,s_n)$$

$$\leq \lambda \Sigma_{i=1}^{n} b_i(1-t_i)^{-\rho_i} + \mu \Sigma_{i=1}^{n} b_i' (1-s_i)^{-\rho_i}$$

$$= \Sigma_{i=1}^{n} b_i^{\lambda} b_i'^{\mu}(1-r_i)^{-\rho_i} .$$

But this estimate of $\log M(\overline{r},f)$ gives that $(b_1'^{\mu} b_1^{\lambda},\ldots,b_n'^{\mu} b_n^{\lambda})$ $B_{\overline{T}}$, and so the points $(\mu \log b_1' + \lambda \log b_1,\ldots,\mu \log b_n' + \lambda \log b_n) \in \log B_{\overline{T}}$. Since $\overline{b}$, $\overline{b}'$ are arbitrary points of $B_{\overline{T}}^{o}$, we get that $\log B_{\overline{T}}$ is convex. □

EXERCISES 5.3

1. Show that the hypersurface of associated orders of the function F_1 in Exercise 5.2.1 is the boundary of the hyperoctant

   $\{\overline{a} = (a_1,\ldots,a_n): a_i > \alpha_i, i = 1,\ldots,n\}$.

2. Show that the hypersurface of associated orders of the function F_2 in Exercise 5.2.1 is $\{\overline{a} = (a_1,\ldots,a_n): \overline{a} \in \mathbb{R}_+^n, \Sigma_{i=1}^{n}(1/a_i) = 1\}$. Hence find the sypersurface of associated orders of the function F_3 in Exercise 5.2.2.

3. Find the systems of associated orders and associated types for the function F_4 in Exercise 5.2.3.

4. If f_1 and f_2 are analytic in G, find relations of the systems of associated orders of f_1+f_2 and f_1f_2 with those of f_1 and f_2.

*5. Does the following converse of Theorem 5.3.3 hold? If B is a domain in $\mathbb{R}_+^n$ and if the image B^{-1} of B under the mapping $a_i' = 1/a_i$, $i = 1,\ldots,n$ is a complete convex domain, then the boundary S of the domain B is the hypersurface of associated orders for some function $f \in C(G)$.

6. By analogy with Exercise 5.2.5 for $G_{\overline{R}}$-order and $G_{\overline{R}}$-type, define systems of $\overline{R}$-associated orders $(\rho_{1,\overline{R}},\ldots,\rho_{n,\overline{R}})$ and systems of $\overline{R}$-associated types $(T_{1,\overline{R}},\ldots,T_{n,\overline{R}})$ for a function g, analytic in $G^n(0, \overline{R})$.

(a) How are systems of $\overline{R}$-associated orders and $\overline{R}$-associated types related respectively with the systems of associated orders and associated types of the function $f(\tilde{z}) = g(\overline{R}\,\tilde{z})$?

(b) Find the characterizations of the system of $\overline{R}$-associated orders $(\rho_{1,\overline{R}},\ldots,\rho_{n,\overline{R}})$ and $\overline{R}$-associated types $(T_{1,\overline{R}},\ldots,T_{n,\overline{R}})$ of the coefficients in the power series of g.

Let f be analytic in G and $B_{\overline{\lambda}}(f)$ denote the set of all points $\overline{a} \in R^n_+$ such that, as $|\overline{r}| \to 1$,

$$\log^+ M(\overline{r}) > \Sigma_{i=1}^{n} (1-r_i)^{-a_i}.$$

Then the boundary points of the set $B_{\overline{\lambda}}(f)$ form a certain hypersurface $S_{\overline{\lambda}}(f)$. We call $S_{\overline{\lambda}}(f)$ the *hypersurface of associated lower orders* of f. Any system of numbers $(\lambda_1,\ldots,\lambda_n) \in S_{\overline{\lambda}}(f)$ is called a *system of associated lower orders* of f.

Similarly, for a function f, analytic in G, any system of numbers $(t_1,\ldots,t_n)$ is said to be a system of associated lower types of f with respect to a system $(\rho_1,\ldots,\rho_n)$ of associated orders of f, if $(t_1,\ldots,t_n)$ belongs to the hypersurface $S_{\overline{t}}(f)$ consisting of the boundary points of the set $B_{\overline{t}}(f)$ of all points $\overline{b} \in R^n_+$ such that, for $|\overline{r}| \to 1$,

$$\log^+ M(\overline{r}) > \Sigma_{i=1}^{n} b_i (1-r_i)^{-\rho_i}.$$

7. Prove that the positive numbers $\lambda_1,\ldots,\lambda_n$ form a system of associated lower orders of a function f, analytic in G, if and only if

$$\liminf_{|\overline{r}|\to 1} \frac{\log^+\log^+ M(\overline{r})}{\log(\Sigma_{i=1}^{n} (1-r_i)^{-\lambda_i})} = 1.$$

*8. Let $f(\tilde{z}) = \sum_{\|\overline{k}\|=0}^{\infty} b_{\overline{k}} \tilde{z}^{\tilde{k}}$ be analytic in G. Find the characterization of a system $(\lambda_1,\ldots,\lambda_n)$, $\lambda_i > 0$, $i = 1,\ldots,n$, of associated lower orders of f, in terms of the coefficients $b_{\overline{k}}$.

9. Find a characterization, analogous to that in Exercise 5.3.7, of associated lower types of functions analytic in G.

*10. Let $f(\tilde{z}) = \sum_{\|\bar{k}\|=0}^{\infty} b_{\bar{k}} z^{\tilde{\bar{k}}}$ be analytic in G. Find the characterizations of systems of associated lower types of f in terms of the coefficients $b_{\bar{k}}$.

For a function f, analytic in G, let $B_{\bar{\rho}}(q',f)$ denote the set of all points $\bar{a} \in R_+^n$ such that, for some $q' = 2,3,\ldots$ and $|\bar{r}| \to 1$,

$$\log^+ M(\bar{r}) < \Sigma_{i=1}^n \exp^{[q'-2]} (1-r_i)^{-a_i} .$$

Let $q = \inf q'$. Then the boundary $S_{\bar{\rho}}(q,f)$ of the set $B_{\bar{\rho}}(q,f)$ is called the *hypersurface of associated* q-*orders* of f. Any system of numbers $(\rho_1(q),\ldots,\rho_n(q)) \in S_{\bar{\rho}}(q)$ is called a *system of associated* q-*orders of* f.

11. If f is analytic in G, prove that the positive numbers $(\rho_1(q),\ldots,\rho_n(q))$, $q = 2,3,\ldots$, form a system of associated q-orders if and only if

$$\limsup_{|\bar{r}|\to 1} \left\{ \frac{\log^+ \log^+ M(\bar{r})}{\log \Sigma_{i=1}^n \exp^{[q-2]} (1-r_i)^{-\rho_i(q)}} \right\} = 1 .$$

12. Let $f(\tilde{z}) = \sum_{\|\bar{k}\|=0}^{\infty} b_{\bar{k}} z^{\tilde{\bar{k}}}$ be analytic in G. Find the characterization of a system $(\rho_1(q),\ldots,\rho_n(q))$, $\rho_i(q) > 0$, $i = 1,\ldots,n$, of associated q-orders of f, in terms of the coefficients $b_{\bar{k}}$.

*13. Does a result analogous to Theorem 5.3.3 concerning the hypersurfaces of associated q-orders of a function $f \in C(G)$ hold?

If f is analytic in G, define the hypersurface $S_{\bar{\rho}}^{0}(f)$ as the boundary of the set of all points $\bar{a} \in R_+^n$ for which, as $|\bar{r}| \to 1$,

$$\log^+ M(\bar{r}) < \Sigma_{i=1}^n \log(1-r_i)^{-a_i}.$$

We call $S_{\bar{\rho}^o}(f)$ the *hypersurface of associated logarithmic orders* of f. Any system of numbers $(\rho_1^o,\dots,\rho_n^o) \in S_{\bar{\rho}^o}(f)$ is called a *system of associated logarithmic orders of* f.

14. Let f be analytic in G. Prove that the positive numbers $(\rho_1^o,\dots,\rho_n^o)$ form a system of associated logarithmic orders of f, if and only if,

$$\limsup_{|\bar{r}| \to 1} \left\{ \frac{\log^+\log^+ M(\bar{r})}{\log(\sum_{i=1}^n \log(1-r_i)^{-\rho_i^o})} \right\} = 1.$$

15. Let $f(\tilde{z}) = \sum_{\|\bar{k}\|=0}^{\infty} b_{\bar{k}} \tilde{z}^{\bar{k}}$ be analytic in G. Find the characterization of a system $(\rho_1^o,\dots,\rho_n^o)$ of associated logarithmic orders of f, in terms of the coefficients $b_{\bar{k}}$.

*16. Does a result analogous to Theorem 5.3.3 concerning the hypersurfaces of logarithmic orders of a function $f \in C(G)$ hold?

5.4. GROWTH MEASUREMENT WITH RESPECT TO ONE OF THE VARIABLES

The concepts of G-order, G-type and systems of associated orders and associated types, defined in the previous sections, characterize the growth of a function f, analytic in G, with respect to all the variables simultaneously. However, in many cases, some characterization of the growth of f with respect to one of the variables is needed. In the present section we study this aspect of the growth of analytic functions of several variables.

Suppose $f \in C(G)$, i.e., f is analytic in G and $\log M(\bar{r},f)$ is pluriconvex (cf. Section 5.3) in $\log(1/(1-r_1)),\dots,\log(1/(1-r_n))$. Then, f is said to have the *partial order* ρ_i in the variable z_i if, for $|z_j| = r_j$ remaining fixed, $j \neq i$,

$$\hat{\rho}_i = \limsup_{r_i \to 1} \frac{\log^+\log^+ M(\bar{r},f)}{-\log(1-r_i)} \tag{5.4.1}$$

where $|z_i| = r_i$ and $\bar{r} = (r_1,\dots,r_n)$.

We first show that $\hat{\rho}_i$ is well defined by (5.4.1) in the sense that the value of $\hat{\rho}_i$ given by the right-hand side of (5.4.1) does not depend on the value of other fixed variables.

Theorem 5.4.1. *Let* f ∈ C(G). *Then the partial order* $\hat{\rho}_i$ *of* f *in the variable* z_i, *as given by* (5.4.1), *is independent of the values of other fixed variables* z_j, $j \neq i$.

Proof. Without loss of generality, we assume that i = n. Set,

$$\rho(r_1,\ldots,r_{n-1}) = \limsup_{r_n \to 1} \frac{\log^+\log^+ M(r_1,\ldots,r_{n-1},r_n)}{-\log(1-r_n)}.$$

Since the function $M(\bar{r},f) \equiv M(r_1,\ldots,r_n)$ is monotonically increasing in each of the variables, the function $\rho(r_1,\ldots,r_{n-1})$ is also monotonically increasing in each of the variables. Therefore, for every $\varepsilon > 0$, there exists $A_\varepsilon(r_1,\ldots,r_{n-1}) < \infty$ such that, for $0 \leq r_i < 1$, $i = 1,\ldots,n$,

$$\log^+ M(\bar{r},f) \leq A_\varepsilon(r_1,\ldots,r_{n-1}) + (1-r_n)^{-(\rho(r_1,\ldots,r_{n-1})+\varepsilon)} . \qquad (5.4.2)$$

Since f ∈ C(G), $\phi(\bar{r}) = \log^+ M(r_1,\ldots,r_n)$ satisfies inequality (5.3.14). In this inequality we set, for $0 \leq r_i < r_i' < 1$,

$$t_i = r_i,\ s_i = 1 - \left(\frac{1-r_i'}{(1-r_i)^\lambda}\right)^{1/\mu},\ i = 1,\ldots,n-1$$

$$t_n = 1 - (1-r_n')^{1/\lambda},\ s_n = 0$$

so that, on using (5.4.2), we have

$$\begin{aligned}\log^+ M(r_1',\ldots,r_{n-1}',r_n') &= \log^+ M(1-(1-s_1)^\mu(1-t_1)^\lambda,\ldots,1-(1-s_n)^\mu(1-t_n)^\lambda) \\ &\leq \lambda \log^+ M(r_1,\ldots,r_{n-1},1-(1-r_n')^{1/\lambda}) \\ &\quad + \mu \log^+ M\left(1-\left(\frac{1-r_1'}{(1-r_1)^\lambda}\right)^{1/\mu},\ldots, 1-\left(\frac{1-r_{n-1}'}{(1-r_{n-1})^\lambda}\right)^{1/\mu},0\right)\end{aligned}$$

$$\leq A_\varepsilon(r_1,\ldots,r_{n-1})+(1-r'_n)^{-(\rho(r_1,\ldots,r_{n-1})+\varepsilon)} +$$

$$+ \log^+ M\left(1-\left(\frac{1-r'_1}{(1-r_1)^\lambda}\right)^{1/\mu},\ldots,1-\left(\frac{1-r'_{n-1}}{(1-r_{n-1})^\lambda}\right)^{1/\mu},0\right).$$

Therefore, for any λ satisfying $0 < \lambda < 1$, $\varepsilon > 0$ and $0 \leq r_i, r'_i < 1$, we have

$$\rho(r'_1,\ldots,r'_{n-1}) \leq \frac{1}{\lambda}\,(\rho(r_1,\ldots,r_{n-1}) + \varepsilon).$$

Since $\varepsilon > 0$ and λ satisfying $0 < \lambda < 1$ are arbitrary, the above inequality gives

$$\rho(r'_1,\ldots,r'_{n-1}) \leq \rho(r_1,\ldots,r_{n-1})$$

and since the numbers r_i, r'_i in $0 \leq r_i < r'_i < 1$ are arbitrary, we get that

$$\rho(r'_1,\ldots,r'_{n-1}) = \rho(r_1,\ldots,r_{n-1}). \qquad \square$$

The function $\exp((1/(1-z_1))^2(1/(1-z_2)))$, analytic in $G^2(0,1)$, is easily seen to have the partial order 2 in the variable z_1 and the partial order 1 in the variable z_2.

There are simple relationships that exist between G-order and the systems of associated orders with the partial order in one of the variables, for a function $f \in C(G)$.

Theorem 5.4.2. *Let* $f \in C(G)$ *have* G-*order* ρ_G *and let* $\hat{\rho}_i$ *denote the partial order of* f *with respect to the variable* z_i, $i = 1,\ldots,n$. *Then,*

$$\rho_G \leq \Sigma_{i=1}^n \hat{\rho}_i. \tag{5.4.3}$$

Proof. Since $f \in C(G)$, $\log M(r_1,\ldots,r_n) \equiv \log M(\overline{r},f)$ is a convex function of the variables $\log(1/(1-r_1)),\ldots,\log(1/(1-r_n))$. It follows therefore that, for $\lambda_i \geq 0$, $\lambda_1+\ldots+\lambda_n = 1$ and $0 \leq t_{ij} < 1$, $1 \leq i,j \leq n$, the function $\log M(r_1,\ldots,r_n)$ satisfies the inequality

$$\log^+ M(1-(1-t_{11})^{\lambda_1}\ldots(1-t_{1n})^{\lambda_n},\ldots,1-(1-t_{n1})^{\lambda_1}\ldots(1-t_{nn})^{\lambda_n})$$

$$\leq \Sigma_{i=1}^n \lambda_i \log^+ M(t_{i1},\ldots,t_{in}). \tag{5.4.4}$$

Set, for $0 < r < 1$,

$$t_{ii} = 1-(1-r)^{1/\lambda_i} \quad \text{and} \quad t_{ij} = 0 \quad \text{for} \quad j \neq i.$$

Then inequality (5.4.4) gives that

$$\log^+ M(r,\ldots,r) \leq \Sigma_{i=1}^n \lambda_i \log^+ M(0,\ldots,0,1-(1-r)^{1/\lambda_i},0,0).$$

Now, by the definition of the order $\hat{\rho}_i$ with respect to one of the variables, for every $\varepsilon > 0$ and every r in $0 < r < 1$ there exists a constant $A(\varepsilon)$, such that

$$\log^+ M(r,\ldots,r) \leq A(\varepsilon) + \Sigma_{i=1}^n (1-r)^{-(\hat{\rho}_i+\varepsilon)/\lambda_i}.$$

Since $\lambda_i \geq 0$ for $i = 1,\ldots,n$ and $\lambda_1+\ldots+\lambda_n = 1$, we can choose, for $i=1,\ldots,n$,

$$\lambda_i = \frac{\hat{\rho}_i + \varepsilon}{\hat{\rho}_1 +\ldots+\hat{\rho}_n+n\varepsilon}$$

so that the above estimate of $\log^+ M(r,\ldots,r)$ becomes

$$\log^+ M(r,\ldots,r) \leq A(\varepsilon) + n(1-r)^{-(\Sigma_{i=1}^n \hat{\rho}_i+n\varepsilon)}.$$

Thus we have

$$\rho_G = \limsup_{r \to 1} \frac{\log^+ \log^+ M(r,\ldots,r)}{-\log(1-r)} \leq \Sigma_{i=1}^n \hat{\rho}_i . \qquad \square$$

Theorem 5.4.3. *Let* $S_{\hat{\rho}}(f)$ *denote the hypersurface of associated orders of a function* f *in* $C(G)$ *and let* $\hat{\rho}_i$ *denote the partial order of* f *with respect to the variable* z_i, $i = 1,\ldots,n$. *Then, for* $i = 1,\ldots,n$,

$$\hat{\rho}_i = \inf \rho_i \tag{5.4.5}$$

where infimum is taken over all $(\rho_1,\ldots,\rho_n) \in S_{\bar{\rho}}(f)$.

Proof. Without loss of generality, we may assume that $i = n$. By the definitions of G-order ρ_G and the order $\hat{\rho}_n$ in the variable z_n, for any $\varepsilon > 0$ and r and r_n sufficiently close to 1, we have

$$\log^+ M(r,\ldots,r) < (1-r)^{-(\rho_G+\varepsilon)}$$

and

$$\log^+ M(0,\ldots,0,r_n) < (1-r_n)^{-(\hat{\rho}_n+\varepsilon)} .$$

Since $f \in C(G)$, we use inequality (5.3.14) with $r = \max(r_1,\ldots,r_n)$, $t_i = 1-(1-r)^{1/\lambda}$, $s_i = 0$ for $i = 1,\ldots,n-1$ and $t_n = 0$, $s_n = 1-(1-r_n)^{1/\mu}$ together with the monotonicity of the function $M(r_1,\ldots,r_n) \equiv M(\bar{r},f)$, to get

$$\begin{aligned}\log^+ M(r_1,\ldots,r_n) &\le \log^+ M(r,\ldots,r,r_n)\\ &\le \lambda \log^+ M(1-(1-r)^{1/\lambda}, 1-(1-r)^{1/\lambda},\ldots,1-(1-r)^{1/\lambda}, 0)\\ &\quad + \mu \log^+ M(0,\ldots,0,1-(1-r_n)^{1/\mu})\\ &< \lambda(1-r)^{-(\rho_G+\varepsilon)/\lambda} + \mu(1-r_n)^{-(\hat{\rho}_n+\varepsilon)/\lambda}\\ &< \Sigma_{i=1}^{n-1} (1-r_i)^{-(\rho_G+\varepsilon)/\lambda} + (1-r_n)^{-(\hat{\rho}_n+\varepsilon)/\lambda} . \end{aligned} \tag{5.4.6}$$

Therefore, by the definition of the set $B_{\bar{\rho}}(f)$, we have that

$$\left(\frac{\rho_G+\varepsilon}{\lambda}, \frac{\rho_G+\varepsilon}{\lambda}, \ldots, \frac{\rho_G+\varepsilon}{\lambda}, \frac{\bar{\rho}_n+\varepsilon}{\lambda}\right)$$

is in $B_{\bar{\rho}}(f)$. But this implies that there exists a point $(\rho_1,\ldots,\rho_n)$ on the hypersurface $S_{\bar{\rho}}(f)$ such that

$$\rho_n \le \frac{\hat{\rho}_n+\varepsilon}{\lambda} .$$

Since $\varepsilon > 0$ and λ in $0 < \lambda < 1$ are arbitrary, we have

$$\inf_{(\rho_1,\ldots,\rho_n) \in S_{\bar{\rho}}(f)} \{\rho_n\} \le \hat{\rho}_n .$$

The reverse inequality

$$\hat{\rho}_n \le \inf_{(\rho_1,\ldots,\rho_n) \in S_{\bar{\rho}}(f)} \{\rho_n\}$$

is obviously true and hence (5.4.5) follows. □

EXERCISES 5.4

1. Find the partial orders of the function F_1 in Exercise 5.2.1 and the function F_3 in Exercise 5.2.2.

   Let f, in $C(G)$, have partial orders $\hat{\rho}_i$, $i = 1,\ldots,n$. The *partial type* $\hat{T}_i$, $i = 1,\ldots,n$, of f with resepct to the partial order $\hat{\rho}_i$ is defined as

$$\hat{T}_i = \limsup_{r_i \to 1} \frac{\log^+ M(\bar{r})}{(1-r_i)^{-\hat{\rho}_i}} .$$

2. Let $\phi(\omega)$ be analytic in the whole complex plane and its order $\rho_\infty = \limsup_{r \to \infty} (\log \log M(r)/\log r)$ satisfy $1/2 < \rho_\infty < 1$. Define

$$F(\tilde{z}) = F(z_1,z_2) = \phi\left(\left(\frac{1}{1-z_1}\right)\left(\frac{1}{1-z_2}\right)\right).$$

   Prove that

   (a) the partial orders $\hat{\rho}_i$, $i = 1,2$, of F are both equal to 1;

   (b) the partial types $\hat{T}_i$, $i = 1,2$, of F do not exceed m/n, where $\rho_\infty < m/n < 1$ and m and n are positive integers;

   (c) G-type of F is $2\rho_\infty > 1$.

3. Show that analogues of Theorems 5.4.2 and 5.4.3 do not hold for partial types, a system of associated types or G-type of a function $f \in C(G)$.

4. Let f_1 and f_2 be in $C(G)$. Find the relations of partial orders and partial types of the functions f_1+f_2 and $f_1 f_2$ with those of the functions f_1 and f_2, whenever f_1+f_2 and $f_1 f_2$ are in $C(G)$.

*5. Let $f(\tilde{z}) = \sum_{\|\bar{k}\|=0}^{\infty} b_{\bar{k}} z^{\tilde{\bar{k}}}$ be in $C(G)$ and have partial orders $\hat{\rho}_i$ and partial types $\hat{T}_i$, $i = 1,\ldots,n$. Find the characterizations of $\hat{\rho}_i$ and $\hat{T}_i$ in terms of the coefficients $b_{\bar{k}}$.

   The *partial q-order* $\hat{\rho}_i(q)$, $i = 1,\ldots,n$, of a function f, in $C(G)$, is defined as

$$\hat{\rho}_i(q) = \limsup_{r_i \to 1} \frac{\log^{[q]} M(\bar{r})}{-\log(1-r_i)}$$

where, $q = 2,3,\ldots$ is such that $\hat{\rho}_i(q) < \infty$ and $\hat{\rho}_i(q-1) = \infty$. If $0 < \hat{\rho}_i(q) < \infty$, the *partial q-type* $\hat{T}_i(q)$ of f with respect to the partial order $\hat{\rho}_i(q)$ is defined as

$$\hat{T}_i(q) = \limsup_{r_i \to 1} \frac{\log^{[q-1]} M(\bar{r})}{(1-r_i)^{-\hat{\rho}_i(q)}} .$$

6. Find the relations analogous to those obtained in Theorems 5.4.2 and 5.4.3 for partial q-orders, a system of associated q-orders (cf. Exercises 5.3) and the order $\rho_G(q)$ (cf. Exercises 5.2) of a function $f \in C(G)$.

*7. Let $f(\tilde{z}) = \sum_{\|\bar{k}\|=0}^{\infty} b_{\bar{k}} \tilde{z}^{\bar{k}}$ be in $C(G)$ and have partial q-order $\hat{\rho}_i(q)$ and partial q-type $\hat{T}_i(q)$, $i = 1,\ldots,n$. Find the characterization of $\hat{\rho}_i(q)$ and $\hat{T}_i(q)$ in terms of the coefficients $b_{\bar{k}}$.

The partial logarithmic order $\hat{\rho}_i^o$, $i = 1,\ldots,n$, of a function f, in $C(G)$, is defined as

$$\hat{\rho}_i^o = \limsup_{r_i \to 1} \frac{\log \log M(\bar{r})}{\log \log(1/(1-r_i))} .$$

If $1 < \hat{\rho}_i^o < \infty$, the partial logarithmic type $\hat{T}_i^o$ of f with respect to partial logarithmic order $\hat{\rho}_i^o$ is defined as

$$T_i^o = \limsup_{r_i \to 1} \frac{\log M(\bar{r})}{(\log(1/(1-r_i)))^{\hat{\rho}_i}} .$$

8. Find the relations analogous to those obtained in Theorems 5.4.2 and 5.4.3 for partial logarithmic orders, a system of associated logarithmic orders (cf. Exercises 5.3) and the order ρ_G^o (cf. Exercises 5.2) of a function $f \in C(G)$.

*9. Let $f(\tilde{z}) = \sum_{\|\overline{k}\|=0}^{\infty} b_{\overline{k}} \tilde{z}^{\overline{k}}$ be in $C(G)$ and have partial logarithmic order $\hat{\rho}_i^o$ and partial logarithmic type $\hat{T}_i^o$. Find the characterizations of $\hat{\rho}_i^o$ and $\hat{T}_i^o$ in terms of the coefficients $b_{\overline{k}}$.

<u>REFERENCES</u>: Juneja and Kapoor [3], [4], [5], [6], [7].

Appendix: Some basic results

A.1 Weierstrass Approximation Theorem

Let $f: [a,b] \to \mathbb{C}$ be a continuous function. Then there exists a sequence $\{p_n\}$ of polynomials such that $p_n \to f$ uniformly in $[a,b]$.

A.2 Helley's Selection Principle

Let $\{\Psi_\alpha(x)\}$, $0 < \alpha < 1$, be a family of functions of uniformly bounded variation on $[a,b]$, i.e., there exists a constant A independent of α such that, for every partition $a=t_o < t_1 < \ldots < t_{n-1} < t_n = b$

$\Sigma_{k=1}^{n} |\Psi_\alpha(t_k) - \Psi_\alpha(t_{k-1})| \leq A.$

Then there exists a sequence $\{r_k\}$ tending to 1 and a function $\Psi(x)$ of bounded variation on $[a,b]$ such that $\Psi_{r_k}(x) \to \Psi(x)$ uniformly on $[a,b]$.

A.3 Fatou's Lemma

Let $\{f_n\}$ be a sequence of nonnegative Lebesgue measurable functions defined on a Lebesgue measurable set $E \subset \mathbb{R}$. If, for $x \in E$, $f(x) = \liminf_{n \to \infty} f_n(x)$, then

$$\int_E f \, d\mu \leq \liminf_{n \to \infty} \int_E f_n d\mu$$

where μ is the Lebesgue measure.

A.4 Weierstrass Theorem for Analytic Functions

Let $\{f_n\}$ be a sequence of functions analytic in a domain D and $f_n(z) \to f(z)$ in D, the convergence being uniform on every compact subset of D. Then, f is analytic in D. Moreover, f_n' converges to f' on every compact subset of D.

In addition, if $f_n(z) \neq 0$ in D for $n \geq n_o$, then either $f \equiv 0$ or $f(z) \neq 0$ in D *(Hurwitz Theorem)*.

A.5 Maximum Modulus Principle (for function of one complex variable)

Let f be analytic in a bounded domain $D \subset \mathbb{C}$ and continuous on the closure $\overline{D}$ of D. Then

$$\max_{z \in \overline{D}} |f(z)| = \max_{z \in \partial D} |f(z)|$$

where, ∂D is the boundary of D.

A.6 Minimum Modulus Principle

Let f be analytic in a bounded domain $D \subset \mathbb{C}$ and continuous on the closure $\overline{D}$ of D. Let $f(z) \neq 0$ for all $z \in D$. Then

$$\min_{z \in \overline{D}} |f(z)| = \min_{z \in \partial D} |f(z)| .$$

A.7 Montel's Theorem

A family $\{f_\alpha\}$ of functions analytic in a domain $D \subset \mathbb{C}$ is a normal family if and only if, for each compact set $K \subset D$, there exists a constant M independent of α such that

$$|f_\alpha(z)| \leq M$$

for every α and all $z \in K$.

A.8 Poisson Integral Formula

Let $U(re^{i\theta})$ be harmonic in $\mathcal{U}$ and continuous on $\overline{\mathcal{U}}$. Then, for $0 \leq r < 1$,

$$U(re^{i\theta}) = \frac{1}{2\pi} \int_0^{2\pi} \frac{1-r^2}{1+r^2-2r\cos(\theta-t)} U(e^{it})dt.$$

Conversely, if $f(e^{it})$ is integrable on $[0,2\pi]$, then the function $U(re^{i\theta})$ defined by

$$U(re^{i\theta}) = \frac{1}{2\pi} \int_0^{2\pi} \frac{1-r^2}{1+r^2-2r\cos(\theta-t)} f(e^{it})dt$$

is harmonic in $\mathcal{U}$ and $U(re^{i\theta}) = f(e^{i\theta})$ a.e. in $[0, 2\pi]$.

A.9 Jensen's Formula

Let f be analytic in $|z| \leq r$ and suppose that $a_1,\ldots,a_n$ are the zeros of f in $|z| < r$ repeated according to their multiplicity. If $f(0) \neq 0$, then

$$\frac{1}{2\pi} \int_0^{2\pi} \log |f(re^{i\theta})| d\theta = \log |f(0)| + \Sigma_{k=1}^{n} \log \left(\frac{r}{|a_k|}\right) .$$

A.10 Poisson-Jensen Formula

Let f be analytic in $|z| \leq R$, except possibly having poles at $b_1,\ldots,b_m$ in $|z| < R$ repeated according to the multiplicity. Let f have zeros at $a_1,\ldots,a_n$ in $|z| < R$ repeated according to the multiplicity. Then, for $z = re^{i\theta}$, $0 \leq r < R$,

$$\log|f(z)| = \frac{1}{2\pi}\int_o^{2\pi} \frac{R^2-r^2}{R^2+r^2-2Rr\cos(\theta-t)} \log|f(Re^{it})|dt +$$

$$+ \Sigma_{i=1}^{m} \log\left|\frac{R^2-\bar{b}_i z}{R(z-b_i)}\right| - \Sigma_{j=1}^{n} \log\left|\frac{R^2-\bar{a}_j z}{R(z-a_j)}\right| .$$

A real valued continuous function v on a domain $D \subset \mathbb{C}$ is said to be subharmonic in D if, for every $\{|z-a| \leq r\} \subset D$,

$$v(a) \leq \frac{1}{2\pi}\int_o^{2\pi} v(a+re^{i\theta})d\theta.$$

A.11 Maximum Principle for Subharmonic Functions

Let $v:D \to R$ be continuous in the domain D. Then v is subharmonic in D if and only if, for every bounded domain D_1, with its closure $\bar{D}_1 \subset D$, and for every continuous function $U_1 : \bar{D}_1 \to R$ which is harmonic in D_1 and satisfies $v(z) \leq U_1(z)$ on ∂D_1, we have $v(z) \leq U_1(z)$ for all z in D_1.

A.12 Riemann Mapping Theorem

Let D be a simply connected domain with at least two boundary points. Let z_o be a fixed point in D, then there exists a unique function f, analytic and one-to-one in D, such that $f(D) = \mathcal{U}$, $f(z_o) = 0$, $f'(z_o) > 0$.

A.13 Distortion Theorem

Let f be analytic and univalent (i.e. one-to-one) in $\mathcal{U}$. Suppose $f(0) = 0$ and $f'(0) = 1$. Then, for $z = re^{i\theta} \in \mathcal{U}$,

$$\frac{1-r}{1+r} \leq \left|\frac{zf'(z)}{f(z)}\right| \leq \frac{1+r}{1-r} .$$

A.14 Kernel of Sequence of Domains

Let $\{D_n\}$ be a sequence of simply connected domains in $\mathbb{C}$ such that a fixed disc $\{|z-z_o|<r\} \subset D_n$ for all n = 1,2,... . The *kernel* D_{z_o} of the

sequence $\{D_n\}$ is defined as the largest domain containing z_o such that, if F is any compact set contained in D_{z_o}, then $F \subset D_n$ for all $n > n_o(F)$.

The sequence of simply connected domains $\{D_n\}$ with kernel D_{z_o} with respect to the point z_o is said to converge to D_{z_o} if every subsequence of $\{D_n\}$ also has the same kernel D_{z_o} with respect to the point z_o.

A.15 <u>Caratheodory's Mapping Theorem</u>

Let $\{D_n\}$ be a sequence of simply connected domains such that, for all $n = 1,2,\ldots$, $\{|z-z_o|<R_o\} \subset D_n \subset \{|z| < R\}$ and let D_{z_o} be the kernel of the sequence $\{D_n\}$ with respect to the point z_o. Let $w = f_n(z)$ be the function mapping D_n conformally onto $U = \{w: |w| < 1\}$ such that $f_n(z_o) = 0$, $f_n'(z_o) > 0$. Furthermore, let $w = f(z)$ map D_{z_o} conformally onto U such that $f(z_o) = 0$, $f'(z_o) > 0$. Then the sequence of domains $\{D_n\}$ converges to D_{z_o} if and only if the sequence of functions $\{f_n(z)\}$ converges uniformly on compact subsets of D_{z_o} to the function $f(z)$.

A.16 <u>Maximum Modulus Principle for Analytic Functions of Several Complex Variables</u>

Let f be analytic in the polydisc $G^n(\overline{0},\overline{r})$. If there exists a point $\tilde{z}^o \in G^n(\overline{0},\overline{r})$ such that $|f(\tilde{z})| \leq |f(\tilde{z}^o)|$ for all $\tilde{z} \in G^n(\overline{0},\overline{r})$, then $f(\tilde{z}) \equiv f(\tilde{z}^o)$ for all $\tilde{z} \in G^n(\overline{0},\overline{r})$.

A.17 <u>Power Series Expansion for Analytic Functions in Polydisc</u>

Let f be analytic in the polydisc $G^n(\overline{0},\overline{R})$. Then f can be uniquely expanded as an absolutely convergent power series

$$f(\tilde{z}) = \sum_{\|\overline{k}\|=0}^{\infty} b_{\overline{k}} \tilde{z}^{\overline{k}}, \quad \tilde{z} \in G^n(\overline{0},\overline{R})$$

where

$$b_{\overline{k}} = \frac{1}{\overline{k}!} \left. \frac{\partial^{\|\overline{k}\|} f}{\partial z_1^{k_1} \cdots \partial z_n^{k_n}} \right|_{\tilde{z}=\overline{0}} .$$

A.18 Cauchy's Inequality for Analytic Functions of Several Complex Variables

Let $f(\tilde{z}) = \sum_{\|\bar{k}\|=0}^{\infty} b_{\bar{k}} \tilde{z}^{\bar{k}}$ be analytic in the polydisc $G^n(\bar{0},\bar{R})$ and $|f(\tilde{z})| \leq M$ for all $\tilde{z} \in G^n(\bar{0},\bar{R})$. Then,

$$|b_{\bar{k}}| \leq \frac{M}{\bar{r}^{\bar{k}}} = \frac{M}{r_1^{k_1} \cdots r_n^{k_n}} .$$

REFERENCES:

Conway [1],
Rudin [1],
Grauert and Fritzsche [1].

References

1. B. J. Aborn and H. Shankar, [1] Generalized growth for functions analytic in a finite disc. *Pure and Appl. Math. Sci.* 12 (1980), 83-94.

2. S. K. Bajpai and J. Tanne, [1] Analogues of entire function inequalities for an analytic function. *Canadian J. Math.* 27(2), (1965), 286-293.

3. S. K. Bajpai, J. Tanne, and D. Whittier, [1] A decomposition theorem for an analytic function. *J. Math. Anal. Appl.* 48 (1974), 736-742.

4. S. Bergman, [1] The Kernel Function and Conformal Mapping. *Amer. Math. Soc., N.Y.*, 1950.

5. F. Beuermann, [1] Wachstumsordnung, Koeffizientenwachstum und Nullstellendichte bei Potenzreihen mit endlichem Konvergenzkreis. *Math. Z.* 33(1931), 98-108.

6. R. A. Bogda and H. Shankar, [1] The maximum term of a power series. *Math. Japon.* 19 (1974), 195-203.

   [2] Convolutions of growth numbers of analytic functions. *Rocky Mountain J. Math.* 10 (1980), 475-483.

7. J. Clunie and W. K. Hayman, [1] The maximum term of a power series II. *J. D'Analyse Math.* 14 (1965), 15-65.

8. J. B. Conway, [1] *Functions of One Complex Variable.* Springer Verlag, N. Y., 1973.

9. J. H. Curtiss, [1] Faber polynomials and the Faber series. *Amer. Math. Monthly.* 78 (1971), 577-596.

10. A. Dinghas, [1] Uber einige Monotoniesatze in der Theorie der schlichten Funktionen. *Avhdl Norske Vid. Akad. Oslo.* I (1959), No. 1, pp. 18.

11. P. L. Duren, [1] *Theory of H^p Spaces.* Academic Press, N.Y., 1970.

12. B. Epstein, [1] *Orthogonal Families of Analytic Functions.* Macmillan, N.Y., 1965.

13. A. W. Goodman, [1] A note on the zeros of Faber polynomials. *Proc. Amer. Math. Soc.* 49 (1975), 407-410.

14. K. Gopal, [1] A study in the growth properties and mean values of analytic functions. Thesis, Meerut University, Meerut, 1979.

15. H. Grauert and K. Fritzsche, [1] *Several Complex Variables.* Springer Verlag, Berlin, 1976.

16. W. K. Hayman, [1] The minimum modulus of large integral functions. *Proc. London. Math. Soc.* (3), 2 (1952), 469-512.

[2] A generalization of Stirling's formula. *J. Reine Angewandte Math.* 196 (1956), 67-95.

17. M. Heins, [1] The minimum modulus of bounded analytic functions. *Duke Math. J.* 14 (1947), 179-215.

18. K. Hoffman, [1] *Banach Spaces of Analytic Functions.* Prentice Hall, Inc., N. J., 1962.

19. O. P. Juneja, [1] A note on analytic functions with prescribed asymptotic growth (to appear).

[2] A unified approach to the study of coefficient problems for analytic functions (to appear).

[3] On the growth of analytic functions having Faber series expansion (to appear).

[4] On Fourier coefficients of analytic functions (to appear).

[5] Newton series expansions of analytic functions (to appear).

[6] Interpolation and approximation of analytic functions (to appear).

20. O. P. Juneja and G. P. Kapoor, [1] On approximation of analytic functions in L^p-norm (to appear).

[2] On Jacobi series expansions of analytic functions (to appear).

[3] The growth measurement and coefficients of functions analytic in a polydisc (to appear).

[4] Systems of associated growth parameters and their hypersurfaces for analytic functions in a polydisc (to appear).

[5] On the growth measurement and coefficients of functions analytic in a polydisc having fast rates of growth (to appear).

[6] On the measurement of growth of functions analytic in a polydisc having slow rates of growth (to appear).

[7] On proximate order of functions analytic in a polydisc (to appear).

21. O. P. Juneja and S. R. H. Rizvi, [1] Interpolation and approximation of entire functions. *Proc. Royal Irish. Acad., Sec. A.*, 79 (1979), 37-47.

22. G. P. Kapoor, [1] A note on the proximate order of functions analytic in the unit disc. *Istanbul Univ. Fen. Fak. Mec., Ser. A.*, 36 (1971), 35-40.

[2] A study in the growth properties and coefficients of analytic functions. Indian Institute of Technology Kanpur, Thesis, 1972.

[3] On the lower order of functions analytic in the unit disc. *Math. Japon.*, 17 (1), (1972), 49-54.

[4] On the proximate order and maximum term of analytic functions in the unit disc. *Rev. Roumaine Math. Pures Appl.* 18 (1973), 1207-1215.

[5] On extreme rates of growth of functions analytic in a unit disc. *Riv. Mat. Univ. Parma* (4), 2 (1976), 101-106.

23. G. P. Kapoor and K. Gopal, [1] On the coefficients of functions analytic in the unit disc having fast rates of growth. *Annali di Mat. Pura ed. Applicata* (IV), CXXI (1979), 337-349.

[2] Decomposition theorems for analytic functions having slow rates of growth. *J. Math. Anal. Appl.* 74 (1980), 446-455.

[3] On the maximum term and rank of analytic functions and their derivatives. *Yokohama Math. J.* 29 (1981), 37-46.

24. G. P. Kapoor and O. P. Juneja, [1] On the lower order of functions analytic in the unit disc II. *Indian J. Pure Appl. Math.* 7 (3), (1976), 241-246.

25. G. P. Kapoor and A. Nautiyal, [1] On the coefficients of functions analytic in the unit disc having slow rates of growth. *Annali di Mat. Pura ed. Applicata* (IV), CXXXI (1982), 281-290.

[2] Polynomial coefficients and generalized orders of an entire function. *Math.Japon.* 29(6), (1984) (to appear).

26. T. Kovari, [1] On the maximum modulus and maximum term of functions analytic in the unit disc. *J. London Math. Soc.* 41 (1966), 129-137.

27. T. Kovari and Ch. Pommerenke, [1] On Faber polynomials and Faber expansions. *Math. Z.* 99 (1967), 193-206.

28. C. N. Linden, [1] The minimum modulus of functions regular and of finite order in the unit circle. *Quart. J. Math., Oxford Ser. 2,* 7 (1956), 196-216.

[2] Minimum modulus theorems for functions regular in the unit circle. *Quart. J. Math., Oxford Ser. 2,* 12 (1961), 1-16.

[3] The representation of regular functions *J. London. Math. Soc.,* 39 (1964), 19-30.

[4] The distribution of zeros of regular functions. *Proc. London. Math. Soc.* (3), 15 (1965), 301-322.

[5] Functions analytic in a disc having prescribed growth properties. *J. London Math. Soc.* (2), 2, (1970), 262-272.

[6] The minimum modulus of functions of slow growth in the unit disc. *Mathematical Essays dedicated to A. J. Macintyre,* Ohio Univ. Press, Athens, Ohio, 1970, 237-246.

[7] Integral means and zero distributions of Blaschke product. *Canadian J. Math.* 24 (1972), 755-760.

29. G. R. MacLane, [1] *Asymptotic Values of Holomorphic Functions.* Rice University Studies, Houston, 1963.

30. A. I. Markushevich, [1] *Theory of Functions of a complex Variable,* Vol. III. Prentice Hall, N.Y., 1967.

31. A. Nautiyal, [1] Growth and approximation of analytic functions and solutions of certain partial differential equations. Indian Institute of Technology Kanpur, Thesis, 1981.

32. Z. Nehari, [1] *Conformal Mapping.* McGraw Hill, N.Y., 1952.

33. R. Nevanlinna, [1] *Analytic Functions.* Springer-Verlag, 1970.

34. P. J. Nicholls and L. R. Sons, [1] Minimum modulus and zeros of functions in the unit disc. *Proc. London Math. Soc.* (3), 31 (1975), 99-113.

35. G. Polya and G. Szego, [1] *Problems and Theorems in Analysis;* Vol. I, Springer-Verlag, 1972; Vol. II, Springer-Verlag, 1976.

36. P. Porceilli, [1] *Linear Spaces of Analytic Functions.* Rand McNally, Chicago, 1966.

37. J. R. Rice, [1] The degree of convergence for entire functions. *Duke Math.* 38 (1971), 429-440.

38. S. R. H. Rizvi, [1] Some aspects of expansions, approximation and interpolation of analytic functions. Thesis, Indian Institute of Technology, Kanpur, 1978.

[2] On the degree of approximation of analytic functions. *Ganita,* 30 (1979), 12-19.

[3] On Jacobi series expansions of analytic functions.

39. S. R. H. Rizvi and O. P. Juneja, [1] On the Fourier expansions of entire functions. *Bull. Inst. Math. Acad. Sin.* 7 (1979), 187-199.

[2] A contribution to best approximation in the L_2-norm. *Publ. Inst. Math.* 28 (1980), 167-177.

40. P. C. Rosenbloom, [1] *Probability and Entire Functions. Studies in Mathematical Analysis.* G. Szego et al., Stanford University, 1963.

41. W. Rudin, [1] *Real and Complex Analysis.* McGraw Hill Book Company, N.Y., 1966.

42. M. N. Seremeta, [1] Connection between the growth of a function analytic in a disc and moduli of the coefficients of its Taylor series (Ukrainian). *Visnik L'viv. Derz Univ. Ser. Meh. Mat.,* vyp. 2, (1965), 101-110.

43. J. H. Shapiro and A. L. Shields, [1] Unusual topological properties of the Nevanlinna class. *Amer. J. Math.* 97 (4), (1975), 915-936.

44. D. F. Shea, [1] Functions analytic in a finite disc having asymptotically prescribed characteristic. *Pacific J. Math.* 17 (1966), 549-560.

45. V. I. Smirnov and N. A. Lebedev, [1] *Functions of a Complex Variable : Constructive Theory*. Iliffe, London, 1968.

46. L. R. Sons, [1] Regularity of growth and gaps. *J. Math. Anal. Appl.* 24 (1968), 298-306.

[2] Corrigendum to 'Regularity of growth and gaps'. *J. Math. Anal. Appl.* 58 (1977), 232.

[3] Value distribution of canonical products in the unit disc. *Math. Japon.* 22 (1977), 27-38.

[4] Zeros of functions with slow growth in the unit disc. *Math. Japon.* 24 (1979), 271-282.

47. G. S. Srivastava and O. P. Juneja, [1] The maximum term of a power series. *J. Math. Anal. Appl.* 81 (1981), 1-7.

48. T. J. Suffridge, [1] On the kernel function for the intersection of two simply connected domains. *Proc. Amer. Math. Soc.* 22 (1969), 37-41.

49. M. Tsuji, [1] Canonical product for a meromorphic function in the unit circle. *J. Math. Soc. Japan.* 8 (1956), 7-21.

[2] *Potential Theory in Modern Function Theory*. Maruzen, Tokyo, 1959.

50. J. L. Ullman, [1] The location of the zeros of the derivatives of Faber polynomials. *Proc. Amer. Math. Soc.* 34 (1972), 422-424.

51. G. Valiron, [1] *Fonctions Analytiques*. Paris Presses Universitaires de France, 1954.

52. J. L. Walsh, [1] Interpolation and approximation by rational functions in the complex domain. *Amer. Math. Soc.*, Providence, RI, 1965.

53. A. Wiman, [1] Uber den Zusammenhang zwischen dem Maximalbetrage einer analytischen Funktion und dem grossten Betrage bei gegebenen Argumente der Funktion. *Acta Math.* 37 (1914), 305-326.

54. T. Winiarski, [1] Approximation and interpolation of entire functions. *Annales Polon. Math.* 23 (1970), 259-273.

55. S. Yamashita, [1] Criteria for functions to be of Hardy class H^p. *Proc. Amer. Math. Soc.* 75 (1979), 69-72.

[2] The meromorphic Hardy class in the Nevanlinna class. *J. Math. Anal. Appl.* 30 (2), (1981), 298-304.

Subject index